Micromachines for Dielectrophoresis, Volume II

Micromachines for Dielectrophoresis, Volume II

Editor

Rodrigo Martinez-Duarte

MDPI • Basel • Beijing • Wuhan • Barcelona • Belgrade • Manchester • Tokyo • Cluj • Tianjin

Editor
Rodrigo Martinez-Duarte
Mechanical Engineering
Clemson University
Clemson
United States

Editorial Office
MDPI
St. Alban-Anlage 66
4052 Basel, Switzerland

This is a reprint of articles from the Special Issue published online in the open access journal *Micromachines* (ISSN 2072-666X) (available at: www.mdpi.com/journal/micromachines/special_issues/Dielectrophoresis_Volume_II).

For citation purposes, cite each article independently as indicated on the article page online and as indicated below:

LastName, A.A.; LastName, B.B.; LastName, C.C. Article Title. *Journal Name* **Year**, *Volume Number*, Page Range.

ISBN 978-3-0365-7301-4 (Hbk)
ISBN 978-3-0365-7300-7 (PDF)

Contents

About the Editor

Rodrigo Martinez-Duarte

Rodrigo Martinez-Duarte is an Associate Professor in the Department of Mechanical Engineering at Clemson University in the US and Head of the Multiscale Manufacturing Laboratory. His group's expertise lies at the interface between micro/nanofabrication, carbonaceous materials, electrokinetics, and microfluidics.

Rodrigo is known as the pioneer of carbon-electrode dielectrophoresis (carbonDEP), a technique for bioparticle manipulation using carbon electrodes and microfluidic devices, with applications in diagnostics and therapeutics. He is also internationally known for pushing for the use of renewable materials and non-traditional techniques such as origami and robocasting to manufacture shaped geometries that serve as precursors to architected carbon and carbide structures. His group is innovating ways to use microbial factories as nanoweavers of biofibers. A recurrent theme in his Multiscale Manufacturing Laboratory is assessing the effect of system parameters on intended performance, as applied to healthcare diagnostics and manufacturing. For example, how does processing impact the properties of carbonaceous materials and structures at multiple length scales, particularly towards tailoring their performance?

At Clemson University, he teaches manufacturing processes and their application, as well as fundamentals of micro/nanofabrication, a course he introduced to the university curricula. His pedagogical approach focuses on teamwork; critical thinking; and project-based learning that emphasizes sustainability.

Besides the US, Rodrigo has lived and worked in Switzerland, Spain, India, Mexico, and South Korea and has a track record of service and leadership to different constituencies. He is the recipient of awards for public impact, teaching, and service.

Preface to "Micromachines for Dielectrophoresis, Volume II"

Dielectrophoresis (DEP) continues to rise as an effective technique for the label-free identification and manipulation of targeted particles in multiple applications. This second volume of the "Micromachines for Dielectrophoresis" series includes 13 novel contributions to the field in aspects ranging from fundamentals and theoretical modeling to applications. Thank you to all the authors, reviewers, and academic editors that have made this contribution possible.

Rodrigo Martinez-Duarte
Editor

micromachines

Editorial

Editorial for the Special Issue on Micromachines for Dielectrophoresis, Volume II

Rodrigo Martinez-Duarte

Multiscale Manufacturing Laboratory, Department of Mechanical Engineering, Clemson University, Clemson, SC 29634, USA; rodrigm@g.clemson.edu

Dielectrophoresis (DEP) remains an effective technique for the label-free identification and manipulation of targeted particles ranging from sizes from nano to micrometers and from inert particles to biomolecules and cells. This Special Issue includes 13 novel contributions to the field in aspects ranging from fundamentals and theoretical modeling to applications.

Several contributions advance the DEP fundamentals in a significant manner. In Protein Dielectrophoresis: A Tale of Two Clausius–Mossottis—Or Something Else? [1], Ron Pethig postulates an empirical theory to predict the DEP response of a protein from the magnitude of the dielectric β-dispersion produced by its relaxing permanent dipole moment using the equivalent of a molecular version of the macroscopic Clausius–Mossotti factor. His contribution responds to the observation that the standard DEP theory fails to describe the DEP data obtained for different proteins in the past several years. In the first of two related contributions, Gimsa in Active, Reactive, and Apparent Power in Dielectrophoresis: Force Corrections from the Capacitive Charging Work on Suspensions Described by Maxwell–Wagner's Mixing Equation [2], introduces a new expression for the DEP force derived from the electrical work in a charge-cycle model that allows the field-free transition of a single object between the centers of two adjacent cubic volumes in an inhomogeneous field. The comparison of this expression with the classical solution provides a new perspective on the notorious problem of electrostatic modeling of AC electrokinetic effects in lossy media and provides insight into the relationships between active, reactive, and apparent power in DEP force generation. In a second contribution, Gimsa and Radai in Dielectrophoresis from the System's Point of View: A Tale of Inhomogeneous Object Polarization, Mirror Charges, High Repelling and Snap-to-Surface Forces and Complex Trajectories Featuring Bifurcation Points and Watersheds [3] build upon the first contribution to describe the polarizability of an entire DEP system as a function of the position of the object with a numerical "conductance field" and argue that such an approach can explain experimental findings, such as the accumulation of viruses and proteins, where the dipole approach cannot account for sufficiently high holding forces to defeat the Brownian motion. Another important contribution is that from Zaman et al. titled Modeling Brownian Microparticle Trajectories in Lab-on-a-Chip Devices with Time Varying Dielectrophoretic or Optical Forces [4], where they present a generalized computational physics model to simulate the trajectories of particles under the influence of both DEP and optical forces. Of note, their model considers time varying applied forces, Brownian motion, fluid flow, collision mechanics, and hindered diffusion caused by hydrodynamic interactions and is shown to be capable of simulating the time evolution of the positions and velocities of an arbitrarily large number of particles simultaneously. For the benefit of the community, this model is made freely available through a GitHub repository. Miloh and Nagler present Travelling-Wave Dipolophoresis: Levitation and Electrorotation of Janus Nanoparticles, which is a theoretical study of the hydrodynamic and electrokinetic response of both metallic spherical polarized colloids as well as metallodielectic Janus particles. In this work, they consider cases of linear and helical time-harmonic travelling-wave

Citation: Martinez-Duarte, R. Editorial for the Special Issue on Micromachines for Dielectrophoresis, Volume II. *Micromachines* **2023**, *14*, 769. https://doi.org/10.3390/mi14040769

Received: 25 March 2023
Accepted: 25 March 2023
Published: 30 March 2023

excitations and provide explicit expressions for the resulting DEP and charge-induced [5] electrophoretic forces and moments exerted on freely suspended particles. Last, Williams et al. present their work on the observation of simultaneous DEP and electrostatic forces in the implementation of an electric curtain. In Particle-Induced Electrostatic Repulsion within an Electric Curtain Operating below the Paschen Limit [6], they describe how an electric curtain is implemented and polarized to produce an electric field below the Paschen limit. Experiments conducted with micrometer-sized soda lime glass beads show how individual particles themselves can trigger electrostatic repulsion in an otherwise dielectric system.

Important advances on different applications are also reported. In Dielectrophoretic Micro-Organization of Chondrocytes to Regenerate Mechanically Anisotropic Cartilaginous Tissue [7], Takeuchi and Miyata leverage negative DEP to pattern Bovine chondrocytes into line-array cell clusters in an agarose gel. Their results show how the embedded chondrocytes remained viable and reconstructed cartilaginous tissue along the patterned cell array and how the cell-containing agarose gel demonstrated mechanical anisotropy. In One-Dimensional Flow of Bacteria on an Electrode Rail by Dielectrophoresis: Toward Single-Cell-Based Analysis [8], Yamaguchi and Yamamoto describe the numerical and experimental evaluation of a device utilizing a DEP force to array bacteria in a single line to allow their facile numeration, a feat desired in biotechnology and medicine to count bacteria with single-cell resolution. In A Study of Dielectrophoresis-Based Liquid Metal Droplet Control Microfluidic Device [9] by Tian et al., an array of liquid gallium-based alloy electrodes driven at ±1000 V is demonstrated to control a liquid metal droplet, physically isolated from the electrodes, at velocities up to 1 mm/s and depending on the droplet diameter. Towards the use of DEP in practical applications, Ettehad and Wenger in Characterization and Separation of Live and Dead Yeast Cells Using CMOS-Based DEP Microfluidics [10] present a DEP system enabled by CMOS to separate viable from non-viable cells. This is significant given the wide application of CMOS fabrication processes in everyday electronics. Moving into the manipulation of nanometric objects using DEP, Dimaki and co-authors in Sub–100 nm Nanoparticle Upconcentration in Flow by Dielectrophoretic Forces [11] present a novel microfluidics system for the concentration of sub–100 nm nanoparticles in a flow using electrical forces generated by a DC or AC electrical field. The authors show how using different electrode configurations can lead to the concentration of particles, as low as 47 nm, by a factor of up to 11. In Dielectrophoresis-Based Positioning of Carbon Nanotubes for Wafer-Scale Fabrication of Carbon Nanotube Devices [12], Kimbrough et al. report the use of DEP for the wafer-scale fabrication of carbon nanotube field-effect transistors, or FET. They note a relatively high rate of FET functionality, up to 87%, in a DEP-based process that is amenable to integration with processes used in semiconductor manufacturing. Last, in Design of Driving Waveform Based on a Damping Oscillation for Optimizing Red Saturation in Three-Color Electrophoretic Displays [13], Yi et al. present a study on the waveform to drive three-color electrophoretic displays (EPDs), particularly their red saturation. To this end, they present an optimized waveform composed of multiple stages, including erasing, particle activation, the purification of red particles, and the display of red color. The experimental results show how the maximum red saturation reaches up to 27.57% improvement compared with more traditional driving waveforms.

I encourage the reader to explore these excellent and meaningful contributions to the field. I also wish to express thanks to all authors for contributing to this second installment of /Micromachines/ for Dielectrophoresis; the reviewers, whose insightful feedback helped improve the impact of these contributions; and the Academic Editors Antonio Ramos, Xiangchun Xuan, Nam-Trung Nguyen, and Aiqun Liu for their important feedback.

Conflicts of Interest: The author declares no conflict of interest.

References

1. Pethig, R. Protein Dielectrophoresis: A Tale of Two Clausius–Mossottis or Something Else? *Micromachines* **2022**, *13*, 261. [CrossRef] [PubMed]
2. Gimsa, J. Active, Reactive, and Apparent Power in Dielectrophoresis: Force Corrections from the Capacitive Charging Work on Suspensions Described by Maxwell-Wagner's Mixing Equation. *Micromachines* **2021**, *12*, 738. [CrossRef] [PubMed]
3. Gimsa, J.; Radai, M.M. Dielectrophoresis from the System's Point of View: A Tale of Inhomogeneous Object Polarization, Mirror Charges, High Repelling and Snap-to-Surface Forces and Complex Trajectories Featuring Bifurcation Points and Watersheds. *Micromachines* **2022**, *13*, 1002. [CrossRef] [PubMed]
4. Zaman, M.A.; Wu, M.; Padhy, P.; Jensen, M.A.; Hesselink, L.; Davis, R.W. Modeling Brownian Microparticle Trajectories in Lab-on-a-Chip Devices with Time Varying Dielectrophoretic or Optical Forces. *Micromachines* **2021**, *12*, 1265. [CrossRef] [PubMed]
5. Miloh, T.; Nagler, J. Travelling-Wave Dipolophoresis: Levitation and Electrorotation of Janus Nanoparticles. *Micromachines* **2021**, *12*, 114. [CrossRef] [PubMed]
6. Williams, S.J.; Schneider, J.D.; King, B.C.; Green, N.G. Particle-Induced Electrostatic Repulsion within an Electric Curtain Operating below the Paschen Limit. *Micromachines* **2022**, *13*, 288. [CrossRef] [PubMed]
7. Takeuchi, Y.; Miyata, S. Dielectrophoretic Micro-Organization of Chondrocytes to Regenerate Mechanically Anisotropic Cartilaginous Tissue. *Micromachines* **2021**, *12*, 1098. [CrossRef] [PubMed]
8. Yamaguchi, Y.; Yamamoto, T. Micromachines One-Dimensional Flow of Bacteria on an Electrode Rail by Dielectrophoresis: Toward Single-Cell-Based Analysis. *Micromachines* **2021**, *12*, 123. [CrossRef] [PubMed]
9. Tian, L.; Ye, Z.; Gui, L. A Study of Dielectrophoresis-Based Liquid Metal Droplet Control Microfluidic Device. *Micromachines* **2021**, *12*, 340. [CrossRef] [PubMed]
10. Ettehad, H.M.; Wenger, C. Characterization and Separation of Live and Dead Yeast Cells Using Cmos-Based Dep Microfluidics. *Micromachines* **2021**, *12*, 270. [CrossRef] [PubMed]
11. Dimaki, M.; Olsen, M.H.; Rozlosnik, N.; Svendsen, W.E. Sub–100 Nm Nanoparticle Upconcentration in Flow by Dielectrophoretic Forces. *Micromachines* **2022**, *13*, 866. [CrossRef] [PubMed]
12. Kimbrough, J.; Williams, L.; Yuan, Q.; Xiao, Z. Dielectrophoresis-Based Positioning of Carbon Nanotubes for Wafer-Scale Fabrication of Carbon Nanotube Devices. *Micromachines* **2021**, *12*, 12. [CrossRef] [PubMed]
13. Yi, Z.; Zeng, W.; Ma, S.; Feng, H.; Zeng, W.; Shen, S.; Shui, L.; Zhou, G.; Zhang, C. Design of Driving Waveform Based on a Damping Oscillation for Optimizing Red Saturation in Three-Color Electrophoretic Displays. *Micromachines* **2021**, *12*, 162. [CrossRef] [PubMed]

micromachines

Article

Protein Dielectrophoresis: A Tale of Two Clausius-Mossottis—Or Something Else?

Ronald Pethig

Institute for Integrated Micro and Nano Systems, School of Engineering & Electronics, The University of Edinburgh, The King's Buildings, Edinburgh EH9 3JF, UK; ron.pethig@ed.ac.uk

Abstract: Standard DEP theory, based on the Clausius–Mossotti (CM) factor derived from solving the boundary-value problem of macroscopic electrostatics, fails to describe the dielectrophoresis (DEP) data obtained for 22 different globular proteins over the past three decades. The calculated DEP force appears far too small to overcome the dispersive forces associated with Brownian motion. An empirical theory, employing the equivalent of a molecular version of the macroscopic CM-factor, predicts a protein's DEP response from the magnitude of the dielectric β-dispersion produced by its relaxing permanent dipole moment. A new theory, supported by molecular dynamics simulations, replaces the macroscopic boundary-value problem with calculation of the cross-correlation between the protein and water dipoles of its hydration shell. The empirical and formal theory predicts a positive DEP response for protein molecules up to MHz frequencies, a result consistently reported by electrode-based (eDEP) experiments. However, insulator-based (iDEP) experiments have reported negative DEP responses. This could result from crystallization or aggregation of the proteins (for which standard DEP theory predicts negative DEP) or the dominating influences of electrothermal and other electrokinetic (some non-linear) forces now being considered in iDEP theory.

Keywords: Clausius–Mossotti function; dielectric beta-dispersion; dielectrophoresis; electrokinetics; Lorentz cavity; Maxwell cavity; molecular dynamics simulations; proteins

Citation: Pethig, R. Protein Dielectrophoresis: A Tale of Two Clausius-Mossottis—Or Something Else?. *Micromachines* **2022**, *13*, 261. https://doi.org/10.3390/mi13020261

Academic Editor: Rodrigo Martinez-Duarte

Received: 19 January 2022
Accepted: 3 February 2022
Published: 6 February 2022

1. Introduction

The subject of protein dielectrophoresis (DEP) is at an important stage where a maturing theory, supported by molecular dynamics (MD) simulations of solvated proteins [1] has clarified aspects that have largely remained unresolved since the pioneering studies reported in 1994 by Washizu et al. [2]. In particular, the applied electric fields and field gradients are in many cases far too weak to generate DEP forces capable of overcoming the thermal (Brownian) force acting on a protein molecule [3–5]. There have also been inconsistencies regarding the polarity (positive or negative) of the observed DEP response for the protein most commonly studied (bovine serum albumin) [4]. These aspects have implications for the general field of molecular DEP. Are there generic lessons to be learnt from the evolving theory for proteins that can be applied to the DEP of small DNA fragments or ribosomal RNA, for example? Or perhaps each class of biomolecule will present its own particular challenge? Based on an empirical theory, it was proposed that the DEP response (including the DEP cross-over) of a globular protein can be predicted from the magnitude and frequency profile of its dielectric β-dispersion, which reflects the protein's squared dipole moment and its relaxation time [4]. Support for this proposal is given by MD simulations for lysozyme and ubiquitin, which show that the β-dispersion also encompasses cross-correlations of the protein dipole with its hydration shell [1]. The corresponding protein–water Kirkwood correlation factor is found to be close to unity. This implies little correlation, in line with an earlier assumption for myoglobin [6], but is impossible to prove using conventional dielectric spectroscopy.

Figure 1 serves to identify the two 'Clausius–Mossottis', $[CM]_{macro}$ and $[CM]_{molecular}$, of this article's title and to indicate that the new theory holds the key for them both [1]. This figure also prompts clarification of the assertion that $[CM]_{macro}$ should not be employed to analyze the DEP responses of proteins and other biomolecular particles that possess a permanent dipole moment [3]. On account of the Janus nature of the DEP susceptibility factor derived by Heyden and Matyushov for proteins [1], this question is relevant and can be answered as follows.

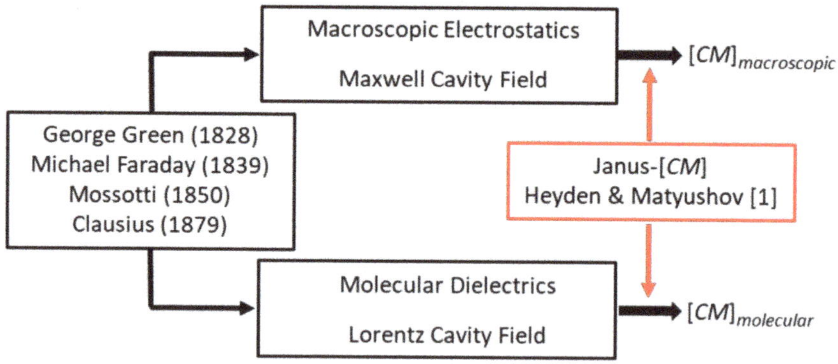

Figure 1. Standard DEP theory employs macroscopic electrostatics to calculate an induced dipole moment and the Clausius–Mossotti factor, $[CM]_{macroscopic}$. The DEP of particles possessing a permanent dipole moment is better formulated within the context of the Clausius–Mossotti law of molecular dielectrics ($[CM]_{molecular}$). The new theory [1] holds the key to transitioning between the two [*CM*]s, whose origins trace back to Green [7], Faraday [8], Mossotti [9] and Clausius [10].

The word 'dielectrophoresis' (DEP) implies a particle being carried (i.e., moved) by a dielectric force. As taught in early texts, this is referred to as the action of a ponderomotive force (able to move an object of mass) and is defined in terms of the gradient of the particle's potential energy U when placed in an electric field gradient [11,12]:

$$\mathrm{F}_P = -\frac{\partial(U - U_0)}{\partial r} = -\nabla U_P \tag{1}$$

The force F_p thus drives the particle towards places of field strength where the particle's free energy (chemical potential) is reduced. Although this relationship can be used to analyse the DEP of a particle possessing both an induced and a permanent dipole moment, standard DEP theory defines this in terms of the force acting on its moment m in an electric field gradient $\nabla \mathrm{E}_m$ [13,14]:

$$\mathrm{F}_{DEP} = m \cdot \frac{\partial \mathrm{E}_m}{\partial r} = (m \cdot \nabla)\mathrm{E}_m \tag{2}$$

Equations (1) and (2) are equivalent because the potential energy of a *rigid* dipole (i.e., unable to rotate) is given by the following relationship [15]:

$$U = -(m \cdot \mathrm{E}_m) \tag{3}$$

This equivalence provided by Equation (3) is restricted to where the dipole's polar angle of orientation θ with respect to the applied field remains fixed. A change of θ results in a change of U and produces a restoring torque $\mathrm{T} = -\frac{\partial U}{\partial \theta}$ that acts on the moment. The work done (i.e., energy gained) is the product of force applied over a distance, so that for a rigid dipole Equations (1) and (2) are connected through the relationship $U_{DEP} = -\int \mathrm{F}_{DEP}$. This predicts that if the electric polarizability of the particle *exceeds* that of the fluid medium it has displaced, it will minimize its potential energy by seeking a region of *high* field strength near an electrode. Work must be done on the particle to move it away, against the field, from an electrode (i.e., the field does negative work). By convention this is termed *positive* DEP. Negative DEP corresponds to where a particle of lower polarizability than its surrounding medium moves to a local field minimum and away from the electrodes. In practical applications, such as the selective separation of a target particle from a mixture, it is important to know how the DEP force depends on the frequency of the applied field. In standard DEP theory for macroscopic particles (e.g., cells, bacteria or polymer beads) this information is carried by the macroscopic Clausius–Mossotti factor $[CM]_{macro}$. This theory has worked well for DEP studies that have progressed from the micron-scale of mammalian cells, microalgae and bacteria, down to the submicron-scale of virions and small vesicles [16]. These particles do not carry a permanent dipole moment. Replacing the moment m of Equation (2) with the known value of the permanent dipole moment of a protein does not solve the problem as to how the DEP force can overcome the energy kT of thermal 'noise' and Brownian motion (Boltzmann constant k and absolute temperature T) [4]. The aspect absent from Equations (2) and (3) is the fact that globular proteins are usually free to 'tumble' in a medium such as water. It is now clear that an essential aspect of DEP theory, for particles possessing a permanent dipole moment that can rotate, is to replace the rigid dipole moment m of Equation (2) with its time-averaged, field-oriented moment proportional to $(m^2/3kT)E_m$ [1].

Pohl in his research articles and book [13] does not adopt the $[CM]$ factor. Herman Schwan, a pioneer of electrical bioimpedance, is probably responsible for its use in electrorotation theory, where he refers to it as '*effectively a macroscopic application of the Clausius–Mossotti factor*' [17]. Later, Jones [14] adopts the '*Clausius–Mossotti function*' in the theory for DEP. However, this practice is not to be confused with application of the Clausius–Mossotti law [10]—identified in Figure 1 as $[CM]_{molecular}$. This law relates the relative permittivity of a dielectric material to the number density of polarizable elements in its molecular structure—of which tumbling molecular dipoles are particularly important elements [18,19]. The genesis of $[CM]_{molecular}$ differs significantly from that of $[CM]_{macro}$.

The theoretical foundations of $[CM]_{macro}$ and $[CM]_{molecular}$, are Green's theoretical formulation of electrical potential [7], Faraday's discovery of specific inductive capacity (i.e., relative permittivity ε_r) [8], and Mossotti's hypothesis that the electric fluid residing in a "corpuscle" within a dielectric is displaced under the action of a local field to form an electric doublet (i.e., induced dipole) [9]. Green derived the electric force exerted by a charged spheroid (electrical density ρ), on a volume element at a distant point r, in terms of a potential function $V = \iiint \left(\frac{\rho}{r}\right) dv$. Outside the spheroid $\nabla^2 V = 0$ (Laplace's equation) and inside it $\nabla^2 V = -\rho/\varepsilon_0$ (Poisson's equation), given here in SI units with ε_0 the permittivity of vacuum. Clausius used these concepts to derive his eponymous factor $[CM]_{molecular}$ [10]. Mossotti's hypothesis influenced Maxwell's concept of the displacement current density, $D = \varepsilon_0 \varepsilon_r E_m$ [20]. The derivation of $[CM]_{macro}$ is obtained by solving Laplace's equation [3], using boundary rules for Green's potential function (it must be continuous) and for Maxwell's displacement current density (its normal component must be continuous) at the interface between different dielectric materials. An important innovation of the new theory is to identify that, for protein DEP theory, these boundary rules of macroscopic electrostatics should be replaced by calculation of cross-correlations of the protein dipole with water dipoles in its hydration shell [1]. The new theory also takes into account the polarization of the protein's hydration shell induced by the protein's permanent dipole moment. These should be considered as generic aspects for molecular DEP, and highlight the importance of dielectric spectroscopy and MD simulations going forward.

Finally, aspects of 'Something Else' included in this article's title address experimental issues such as control of protein and suspending medium stability, together with determinations of the magnitudes of the applied field and field gradient. Experimental guidelines for protein DEP were set by Washizu et al. [2], such as an upper limit for protein concentration (0.1 μg/mL) and solvent conductivity (1 mS/m). Researchers engaged in eDEP, where conductive electrodes generate the required field gradient, have tended to follow these guidelines. In such cases, there are consistent reports of a positive DEP response observed for the 22 different proteins so far studied [4], in line with the new theory plus MD simulations [1] and as predicted by the empirical theory [4]. Experimenters who employ insulating structures to create the field gradients (iDEP) have sometimes stretched these experimental guidelines to explore new opportunities for exploiting protein DEP. For example, protein concentrations of 1 gm/mL and solvent conductivities of 100 mS/m have been used. In such cases, negative rather than positive protein DEP has been reported (e.g. [21–23]). As indicated in Figure 1, and discussed in Section 2, the theory underpinning $[CM]_{macro}$ employs the Maxwell cavity field, with the limitation of a *minimum* cavity size *below* which the rules of classical electrostatics theory may not hold. For the Lorentz cavity field employed in the derivation of $[CM]_{molecular}$, there is a *maximum* cavity size *above* which cavity field theory breaks down. This interesting situation [24] could provide an experimental approach to determining and controlling a threshold for the particle size above which protein DEP is governed by $[CM]_{macro}$ rather than $[CM]_{molecular}$. For example, in an assessment of future protein-based drug carriers, standard DEP theory correctly predicts the reported separation by negative DEP of *micron*-sized designer protein particles according to their size and shape [25]. Whilst, for BSA assumed to be in its monomolecular form, *positive* iDEP is observed at a concentration of 7 nM [26,27] and *negative* iDEP at a concentration of 0.15 mM [21]. As discussed in Section 5, these conflicting results were obtained using direct current (DC) iDEP experiments, where non-linear electrophoretic and electroosmosis effects can overcome DEP. Finally, the phenomenon of pearl chaining, where particles aggregate under the influence of an applied field, was identified and experimentally avoided as a possible confounding issue for molecular DEP by Washizu et al. [2]. The possibility that field-driven protein aggregation can occur at a threshold of protein concentration, which may provide new opportunities of exploitation, also merits further consideration under the category of 'Something Else'.

2. Limitations of Macroscopic Electrostatics with Respect to Protein DEP

2.1. The Maxwell Cavity Susceptibility and $[CM]_{macro}$

$[CM]_{macro}$ is derived from calculation of the field inside a particle (relative permittivity ε_p) residing within a dielectric medium (relative permittivity ε_m) in which a uniform field E_m has been established. Charges are created at the dielectric discontinuity represented by the particle–medium interface, to produce an interface dipole moment, M^{int}. The polarization of the particle, P_p, expressed as the induced dipole moment per unit volume, is given by:

$$P_p = \frac{M^{int}}{V_p} = (\varepsilon_m - \varepsilon_p)\varepsilon_0 E_i = \chi_p \varepsilon_o E_i \tag{4}$$

where χ_p is termed the electric susceptibility of a volume of a material. As described in Section 3, dielectric theory often employs the susceptibility rather than the permittivity of a material. In standard DEP theory, the induced interface dipole moment, M^{int}, of Equation (4) is the moment m required in Equation (2). Equation (4) is obtained by solving Laplace's equation and involves applying the boundary rules mentioned in the Introduction. The field E_i created within the particle is uniform and directed along the same axis as E_m (but opposes it for $\varepsilon_p > \varepsilon_m$) and is given by [3,28]:

$$E_i = \frac{3\varepsilon_m}{2\varepsilon_m + \varepsilon_p} E_m \tag{5}$$

A central aim of molecular dielectrics is to understand the response of a single dipole moment to an external field. Maxwell's approach is to carve out an evacuated cavity inside a dielectric [20]. The field inside an evacuated spherical cavity (i.e., $\varepsilon_p = 1$) is known as the

Maxwell cavity field. The susceptibility χ_{Mc} of the Maxwell cavity is defined as the ratio of this field and the applied external field, so that from Equation (5):

$$\chi_{Mc} = \frac{3\varepsilon_m}{2\varepsilon_m + 1} \tag{6}$$

The mathematical derivation of P_p in Equation (4) is obtained by solving Laplace's equation, which assumes that no free charges already exist at the interface. Macroscopic particles usually carry a net surface charge, but if uniformly distributed over their surface this charge does not contribute a component to be added to P_p [3]. The time taken for M^{int} to fully develop is typically around 100 nanoseconds, dictated by the relaxation time of the interfacial charges. This, in turn, depends on the permittivity and conductivity of the particle and surrounding medium, but does not depend on particle size [3,16]. For a sphere of radius R, then from Equations (4) and (5):

$$M^{int} = V_p P_p = 3V_p\varepsilon_0\varepsilon_m\left[\frac{\varepsilon_p - \varepsilon_m}{\varepsilon_p + 2\varepsilon_m}\right]E_m = 4\pi R^3\varepsilon_0\varepsilon_m[CM]_{macro}E_m \tag{7}$$

The term in square brackets, $[CM]_{macro}$, is limited to the range $-0.5 < [CM]_{macro} < 1.0$. To accommodate the phase difference between dielectric displacement and ohmic conduction currents, complex dielectric parameters (i.e., with real and imaginary components) are used in this definition. This modification of Maxwell's original DC model is known as Maxwell–Wagner polarization, with the real component of $[CM]_{macro}$ required in Equation (7) [13,14,16]. Displacement currents tend to dominate above ~50 kHz, so that the relative permittivity values, ε_p and ε_m, are used directly in Equation (7), to be replaced with conductivity parameters σ_p and σ_m below ~1 kHz.

Equation (7) describes a spherical particle, but ~75% of globular proteins take the form of prolate spheroids, with ~25% being oblate [29]. The following general result for M^{int} aligned along an axis parallel to the direction of E_m can be applied [15]:

$$M^{int} = \frac{4\pi abc}{3}\varepsilon_0\varepsilon_m\left[\frac{\varepsilon_p - \varepsilon_m}{\varepsilon_m + A(\varepsilon_p - \varepsilon_m)}\right]E_m \tag{8}$$

where a, b, c, are the spheroid's semiaxes. The factor A quantifies the shielding of the particle's interior from E_m by the induced interfacial charges. For a sphere $A = 1/3$, whilst for a prolate spheroid tending towards the shape of a long thin cigar, with its axis parallel to the external field, A tends to zero (i.e., the internal and external fields are similar). For an oblate spheroid tending towards a thin platelet, with its major axis perpendicular to the field, A approaches 1.0 (the internal field is reduced by the factor $\varepsilon_p/\varepsilon_m$).

In modeling the polarization of spheroids, there are advantages to retaining volume V_p of Equation (7) as an independent parameter [30]. The DEP force can then be defined in terms of a susceptibility factor χ_{DEP} that scales linearly with particle volume [1]:

$$F_{DEP} = \varepsilon_0\chi_{DEP}\nabla E_m^2 \tag{9}$$

For a spherical particle ($a = b = c = R$), application of Equations (2) and (6), together with the vector transformation $2(E\cdot\nabla)E = \nabla E^2$, gives:

$$\chi_{DEP} = \frac{3}{2}V_p\varepsilon_m\left\{Re\left[\frac{\varepsilon_p^* - \varepsilon_m^*}{\varepsilon_p^* + 2\varepsilon_m^*}\right]\right\} = \frac{3}{2}V_p\varepsilon_m\,Re[CM]_{macro}^* \tag{10}$$

where the asterisks indicate complex permittivity values and Re the real component. An important application of DEP is the selective separation of a target particle from a mixture based on its characteristic 'cross-over' frequency f_{xo} (Hz). The frequency-dependence of the Clausius–Mossotti factor in a range that encompasses f_{xo} is given by [16]:

$$Re[CM(f)]^*_{macro} \approx \frac{f^2 - f_{xo}^2}{f^2 + 2f_{xo}^2} \tag{11}$$

For frequencies of the applied field less than f_{xo} the DEP force acting on the particle is negative. As the frequency f is increased above f_{xo}, a transition from negative to positive DEP occurs. Different proteins exhibit different values of f_{xo} [16,24], so there is the expectation that this procedure, already proven for the selective separation and manipulation of cells and bacteria, for example, can be used to selectively enrich target proteins from other proteins or biomacromolecules.

Based on common aspect ratios for prolate and oblate proteins [29], the depolarization factor of relevance has limits 0.2 < A < 0.8. Provided that the protein sample is large enough for it to be characterized using macroscopic electrostatics then, from Equation (10), the corresponding DEP susceptibilities related to their induced moments are:

$$\begin{aligned} \chi_{DEP} &= \tfrac{3}{2}V_p\varepsilon_m\left[\tfrac{5(\varepsilon_p-\varepsilon_m)}{\varepsilon_p+4\varepsilon_m}\right] \quad (\text{prolate spheroid}: \; A = 0.2) \\ \chi_{DEP} &= \tfrac{3}{2}V_p\varepsilon_m\left[\tfrac{5(\varepsilon_p-\varepsilon_m)}{4\varepsilon_p+\varepsilon_m}\right] \quad (\text{oblate spheroid}: \; A = 0.8) \end{aligned} \tag{12}$$

These relationships indicate that the free energy per unit volume -(M·E_m), where M is the vector sum of its constituent dipole moments, depends on the shape of a macroscopic particle. This indicates that, even for macroscopic distances, interactions between the constituent dipole moment fields should be taken into account. However, an accurate calculation of the interaction of a specific dipole with all the other dipoles in the sample would be difficult to achieve. The common procedure in dielectrics theory [19] is thus to consider the dipole in question to be at the centre of a sphere containing a finite number of other dipoles, beyond which there is a homogeneous continuum of the dielectric medium. For this model to represent a good approximation, the quantity of the other dipoles in the spherical volume should be sufficient that the 'law of large numbers' [28] is applied to 'smooth-out' the influence of random polarizations and so lend to the sphere an effective permittivity equal to that of a macroscopic sample. This is of relevance to the dielectric models discussed in Section 3.

To overcome Brownian disruptive effects, the DEP force should exceed the thermal energy ($3kT/2$) of the protein's translational degrees of freedom. From Equation (9) the threshold value of ∇E^2_{Th}, above which F_{DEP} can overcome thermal disruption, is given by:

$$\nabla E^2_{Th} = \frac{3kT}{2\varepsilon_0\chi_{DEP}} \tag{13}$$

Threshold values for ∇E^2_{Th} are shown in Figure 2 for T = 295 K as a function of effective particle radius, employing values for χ_{DEP} given by Equation (12) with ε_p = 25, ε_m = 78. The value ε_p = 25 is obtained from a molecular dynamics (MD) simulation of cytochrome-c in a water droplet of 2.4 nm radius [31], and considered to be consistent with hydration-dependent dielectric properties of protein powders [32]. Also shown in Figure 2 are values of the factor ∇E^2_m reported to have been used in protein DEP experiments. It appears [4] that for only two cases [22,33], involving bovine serum albumin (BSA) and immunoglobulin G (IgG), does the experimental magnitude of ∇E^2_m exceed the threshold required to overcome thermal randomization. For two cases involving the prolate proteins BSA and IgG [34,35] the threshold is close to being met. For the other cases [26,27,36–39] shown in Figure 2, experimental values of the field gradient factor ∇E^2_m fall orders of magnitude below that required (according to standard theory) to observe DEP if proteins of monomolecular form had been studied.

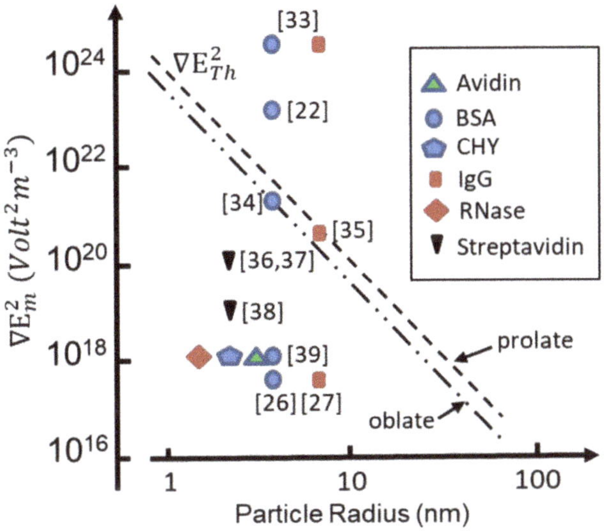

Figure 2. Reported values of ∇E^2_m employed in DEP experiments mapped against the effective (hydrodynamic) radius of the test protein molecule. The straight-line plots show the predicted threshold values of ∇E^2_{Th} required to overcome Brownian forces, derived using Equation (13) for prolate and oblate spheroids. (BSA: bovine serum albumin; CHY: chymotrypsinogen; IgG: immunoglobulin G.)

As discussed in Section 3, solvated proteins are free to tumble and possess an orientational polarizability, $m_p^2/3kT$, associated with its permanent dipole moment m_p [18]. From Equations (2) and (9) the DEP susceptibility, χ_{dDEP}, associated with such a dipole is given as:

$$\chi_{dDEP} = \frac{F_{DEP}}{\varepsilon_o \nabla E_m^2} = \frac{1}{\varepsilon_o \nabla E_m^2}\left(\frac{m_p^2}{3kT}E_m \cdot \nabla\right)E_m = \frac{m_p^2}{6\varepsilon_o kT} \tag{14}$$

The following gives the contribution to the DEP force of a rigid dipole moment relative to that of an induced moment:

$$\frac{F_{dDEP}(rigid\ dipole)}{F_{DEP}(induced)} = \frac{m^2}{6kT\varepsilon_O \chi_{DEP}} \tag{15}$$

As shown in Figure 3, standard DEP theory based on Equation (2) predicts that a rigid dipole moment will experience roughly the same DEP force as an induced moment under the same experimental conditions. This confirms an earlier conclusion [4], based on an analysis only for BSA, that the problem regarding the apparent absence of Brownian disruption for protein DEP is not solved by inserting the magnitude of the permanent dipole moment value into Equation (2). A DEP theory based on molecular dielectrics is required—not one based on macroscopic electrostatics.

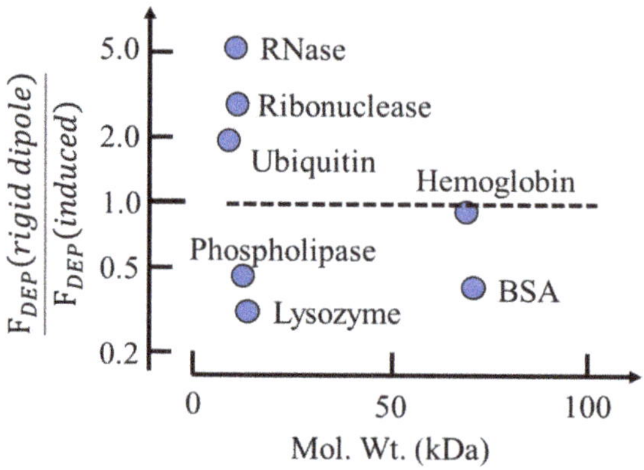

Figure 3. The contribution to the DEP force that a rigid dipole moment makes relative to that of the induced moment is shown for various proteins, as calculated using Equation (15). Values of the hydrodynamic radii were derived using an empirical relationship between protein size and molecular weight (Malvern Panalytical®—Zetasizer Nano ZS) and permanent dipole moment values were derived from the literature [4,24].

Aqueous solutions of proteins, when examined by dielectric or impedance spectroscopy at room temperature, typically exhibit a frequency-dependent relative permittivity as shown in Figure 4. Of particular significance for protein DEP is the β-dispersion, bounded by lower and upper frequency permittivity values of $\varepsilon(\beta lf)$ and $\varepsilon(\beta hf)$, respectively, with a dispersion strength given by $\Delta\varepsilon(\beta) = [\varepsilon(\beta lf) - \varepsilon(\beta hf)]$. The mid-point, $[\varepsilon(\beta lf) - \varepsilon(\beta hf)]/2$, of this dispersion occurs at a frequency $1/(2\pi\tau)$, where τ is the characteristic relaxation time. This is the time required for $1/e$ of the permanent dipoles of an ensemble to return to random orientation after the polarizing field has been removed. For suspensions of cells and other particles of sufficiently low volume concentration, c_v, the magnitude of $\Delta\varepsilon(\beta)$ is given by [3,4,16]:

$$\Delta\varepsilon(\beta) = 3c_v\varepsilon_m[CM]_{macro} \tag{16}$$

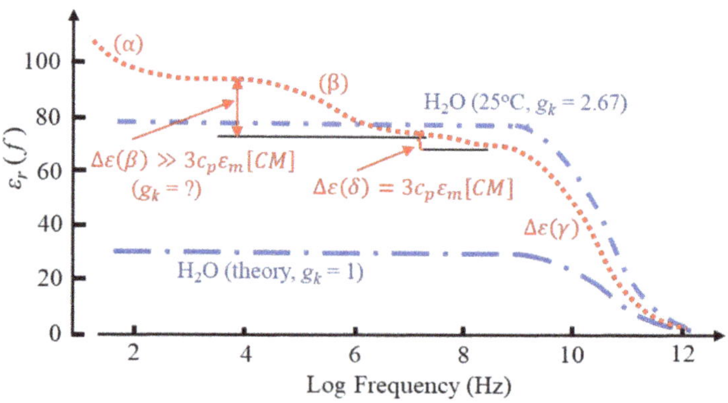

Figure 4. Protein solutions exhibit dielectric dispersions $\Delta\varepsilon$, designated α, β, δ and γ, due to relaxations of the protein's electrical double-layer; permanent dipole; hydration shell; bulk water dipoles, respectively [16]. The magnitude of $\Delta\varepsilon(\beta)$, unlike $\Delta\varepsilon(\delta)$, greatly exceeds that predicted by Equation (16) [4]. $\Delta\varepsilon(\gamma)$ exhibited by pure water arises from correlations of its dipole moments (Kirkwood factor g_k = 2.67), where g_k = 1 corresponds to no correlation.

However, as indicated in Figure 4, whereas the δ-dispersion associated with a protein's hydration shell appears to satisfy this relationship, the β-dispersion does not. Equation (16) is derived from the 'effective medium' theory of dielectric mixtures [24]. This theory assumes that the addition of a small number of 'impurity' particles to a homogeneous dielectric medium can be homogenized, so that after this process the mixture exhibits the same polarization response as the pure solvent. In effect, the ratio of the average electric displacement and average electric field $\langle D\rangle/\langle E\rangle$ is assumed to remain unchanged. The fact that the theory fails for the β-dispersion exhibited by a protein–water mixture implies that it is not a 'passive' mixture—an interaction occurs between the protein and water dipoles. The magnitude of the γ-dispersion observed at ~20 GHz, shown in Figure 4 and due to orientational relaxation of the water dipoles, was understood after introduction into dielectrics theory of the Kirkwood water dipole–dipole correlation factor. An obvious matter to raise is the extent to which the magnitude of the β-dispersion might be influenced by protein–water cross-correlations of dipoles. As discussed here in Sections 3 and 4, this concept had evaded analysis with standard dielectrics theory, but is an important component of the new theory and MD simulations [1].

2.2. The β- and δ-Dispersions Associated with Protein Hydration

It is the custom to identify a dielectric dispersion, using the Greek alphabet, in the order of location on the frequency axis of the applied field frequency. The designation of a particular Greek letter does not specify a specific physico-chemical mechanism responsible for that dispersion. Figure 5 shows the β-dispersion exhibited by solvated BSA [40,41], together with an aqueous suspension of phospholipid nano-vesicles [42].

The dispersion for the vesicles is characterized by a relaxation time τ corresponding to that required for $1/e$ of the total induced interfacial charges to disappear after the polarizing field has been removed. The magnitude of the β-dispersion, as a function of vesicle volume fraction, p, is given by the linear relationship $\Delta\varepsilon(\beta) \approx 143p$ [42]. From Equation (16) this gives $[CM]_{macro}$ = 0.61 [42], a value within the permitted range $-0.5 < [CM]_{macro} < 1.0$. However, an equivalent interpretation is not forthcoming for the β-dispersion shown in Figure 5 for a 12 g/L (0.18 mM) BSA solution [40]. Based on its molecular mass of 66 kD and mass density 1.41 g/cm^3 [43], BSA has a molecular volume of 4.68×10^4 cm^3/mol. This corresponds to the protein occupying 8.42 cm^3/L in a 12 g/L mixture, so that $p = 8.4 \times 10^{-3}$. Thus, the β-dispersion for BSA exhibits a significantly larger dispersion strength than the vesicles, with $\Delta\varepsilon(\beta) \approx 2500p$ and a corresponding value for $[CM]_{macro}$ well above the maximum of 1.0 permitted by the standard theory.

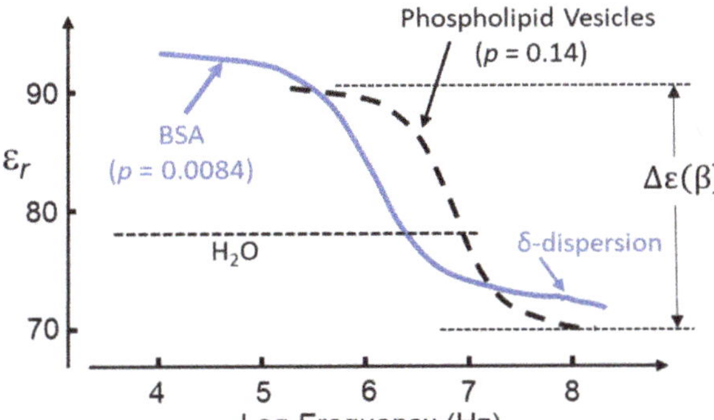

Figure 5. The β-dispersion exhibited by a 0.18 mM (i.e., p = 0.0084) BSA solution [40] is shown, together with that of a suspension of phospholipid vesicles (radius 13.5 nm, volume fraction p = 0.14) [42]. The δ-dispersion is associated with the protein hydration sheath [41,44,45]. Positive DEP can be expected for both particle types in the frequency range where their relative permittivity, ε_r, exceeds that (~78) of pure bulk water.

The δ-dispersion, typically occurring around 70 MHz, is attributed to relaxations of water molecules close to the protein surface [41,44,45]. Two kinds of such water have been identified: (i) water attached directly to the protein surface via H-bonding to oxygen, nitrogen and polar groups; (ii) water, amounting to ~0.36 g water/g protein, that forms a hydration shell around the protein molecule. The magnitude of the δ-dispersion, $\Delta\varepsilon(\delta)$, is found to be directly proportional to the molecular weight and hence surface area of the protein molecule, whereas the value of its characteristic relaxation time (~2 ns) is relatively independent of protein size [45]. The number of bound water molecules can be estimated from $\Delta\varepsilon(\delta)$ using a simple mixture theory [16,44]. For small proteins, such as cytochrome-c or ribonuclease-A, the number of bound water molecules amounts to ~50 per protein, increasing to 135~150 for larger proteins such as hemoglobin and BSA [44]. Based on measurements of the β-, δ- and γ-dispersions exhibited by lysozyme solutions, Wolf et al. conclude that the dynamics of water molecules in the hydration sheath is influenced by interactions with polar residues on the protein surface and less so with the bulk water [45]. This implies that the δ-relaxation is strongly coupled to the β-relaxation of the protein molecule and its H-bonded water molecules but is not strongly influenced by structural fluctuations of the bulk solution (i.e., the γ-dispersion). It is also noteworthy, as indicated in Figure 4, that the magnitude of the dielectric increment $\Delta\varepsilon(\delta)$ corresponds to a realistic value for $[CM]_{macro}$—unlike the case for $\Delta\varepsilon(\beta)$. For example, a 5 wt.% aqueous solution of myoglobin (equivalent to $p = 0.034$) has a measured $\Delta\varepsilon(\delta) = 2.37$ at 25 °C [44]. From Equation (16) this gives $[CM]_{macro} = 0.3$, which lies within the permitted range $-0.5 < [CM]_{macro} < 1.0$. This conclusion also holds for nine other globular proteins of similar concentration investigated by Miura et al., who reported $\Delta\varepsilon(\delta)$ values in the range 1.51 to 5.05 [44]. These dispersion strengths correspond to $[CM]_{macro}$ values in the range 0.19 to 0.63. This implied compliance with classical macroscopic electrostatics is matched by the characteristic relaxation time being independent of protein size. As stated with respect to Equation (5), the relaxation time of induced interfacial charges associated with the Maxwell cavity field depends on the permittivity and conductivity of the particle and surrounding medium, but not on particle size [14,45].

2.3. User-Friendly, Empirical, Theory for Protein DEP

A practical and user-friendly application of Equation (16) arises because $\Delta\varepsilon(\beta)$ contains information for predicting the frequency profile of the DEP response, and where a transition from positive to negative DEP might be expected. For proteins whose β-dispersion is well characterised, and for a protein concentration in the range where a linear relationship between $\Delta\varepsilon(\beta)$ and C_v is found, the variation of permittivity of the aqueous mixture as a function of frequency (Hz) is of the form [18]:

$$\varepsilon_m(f) = \frac{\Delta\varepsilon(\beta)}{1 + (f/f_\beta)^2} + \varepsilon\left(\beta_{hf}\right) \tag{17}$$

where f_β is the frequency marking the inflexion (mid-point) of the β-dispersion and $\varepsilon(\beta_{hf})$ is the high-frequency boundary where the dispersion ends. The dispersion exhibits a dielectric increment in the frequency range where $\varepsilon(\beta_{hf})$ exceeds the experimental value of 78.4 for pure water [3]. This corresponds to where the polarizability (dipole moment per unit volume) of the solvated protein *exceeds* that of the medium it has displaced. The transition to the dielectric decrement occurs at the cross-over frequency, f_{xo}, where ε_m falls below 78.4. The polarizability of the solvated protein is now *less* than that of the medium it has displaced, and a transition from positive to negative DEP can be expected. For the BSA dispersion of Figure 5: $\Delta\varepsilon(\beta) = 21$; $\varepsilon(\beta hf) = 72.4$; $f_\beta \approx 1$ MHz; $f_{xo} \approx 2$ MHz.

Equation (16) is derived from the 'effective medium' theory of dielectric mixtures. For a mixture of protein and water this theory can be applied to analyse the δ-dispersion of Figure 4, but fails for the β-dispersion. This theory assumes that, after the addition of a small number of impurity particles, the mixture can be homogenized so that it exhibits the same polarization response as the pure solution. In effect, the ratio of the average electric displacement and average electric field $\langle D\rangle/\langle E\rangle$ is assumed to remain unchanged. For particles of permittivity ε_p and total partial volume v_p dispersed in a medium of permittivity ε_m, the following relationship is derived [46]:

$$\left(\varepsilon_{eff} - \varepsilon_p\right) v_p \frac{\langle E_i\rangle}{E_m} + \left(\varepsilon_{eff} - \varepsilon_m\right)(1 - v_p)\frac{\langle E_m\rangle}{E_m} = 0 \quad (18)$$

where $\langle E_i\rangle$ and $\langle E_m\rangle$ represent the statistical averages of the electrical field in the interior of a dispersed particle and in the bulk medium, respectively. The field E_i inside each particle is taken to be the Maxwell cavity field given by Equation (5), and substitution of this into Equation (16) gives:

$$\frac{\varepsilon_{eff} - \varepsilon_m}{\varepsilon_{eff} + 2\varepsilon_m} = v_p \frac{\varepsilon_p - \varepsilon_m}{\varepsilon_p + 2\varepsilon_m} = v_p [CM]_{macro} \quad (19)$$

This Equation provides a valuable tool, known as multi-shell modelling, to predict and understand the DEP responses of cells and bacteria [16,47]. It was used by Schwan et al. [42] to derive the lipid bilayer capacitance of the phospholipid vesicles from its β-dispersion shown in Figure 5, and to analyse the δ-dispersion exhibited by solvated BSA [41]. Equation (16) is derived from Equation (19) by assigning $(\varepsilon_{eff} - \varepsilon_m) = \Delta\varepsilon(\beta)$, $v_p = p$, together with the assumption that v_p is small enough to give $\varepsilon_{eff} + 2\varepsilon_m \approx 3\varepsilon_m$. An empirical replacement for $[CM]_{molecular}$ of Figure 1, namely $[CM]^{empirical}$, is derived by assigning v_p as a mole volume fraction C_p and inserting $\varepsilon_{eff} = \kappa\varepsilon_m$ into the denominator of the left-hand side of Equation (18) [4]:

$$(\kappa + 2)[CM]^{empirical} = \left(\frac{\Delta\varepsilon(\beta)}{\varepsilon_m}\right)\left(\frac{C_w \rho_p}{C_p \rho_w}\right) \quad (20)$$

Based on known values for $\Delta\varepsilon(\beta)/C_p$ and protein mass density [4] this relationship produces values for $[CM]^{empirical}$ that vary greatly in magnitude (e.g., ~1110 for BSA; ~37,000 for carboxypeptidase; ~190 for phospholipase) with no apparent dependence on protein molecular weight. Selected examples are given in Figure 6.

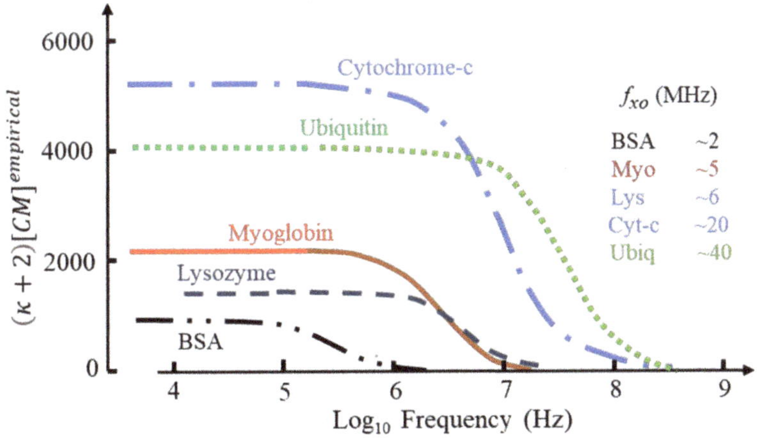

Figure 6. The frequency-dependence of the empirical factor $(\kappa + 2)[CM]^{empirical}$ derived from Equations (17) and (20) for selected proteins. The frequency, f_{xo}, marking the transition between the dielectric increment and decrement of the β-dispersion is also given.

If the values for f_{xo} in Figure 6 indicate where transition from positive to negative DEP occurs, a protocol for separating a mixture of cytochrome-c and the other proteins can be devised. By setting the DEP voltage frequency at ~15 MHz, cytochrome and ubiquitin would be attracted to an electrode array by +ve DEP, with the other proteins repelled by -ve DEP into the (flowing) bulk medium. By adjusting the frequency to ~30 MHz, separation of cytochrome and ubiquitin could result.

The DEP susceptibility profiles shown in Figure 6 also provide a possible explanation for DEP cross-over events reported in the MHz range for BSA [33], avidin [48] and prostate specific antigen [49]. This data is shown in Figure 7.

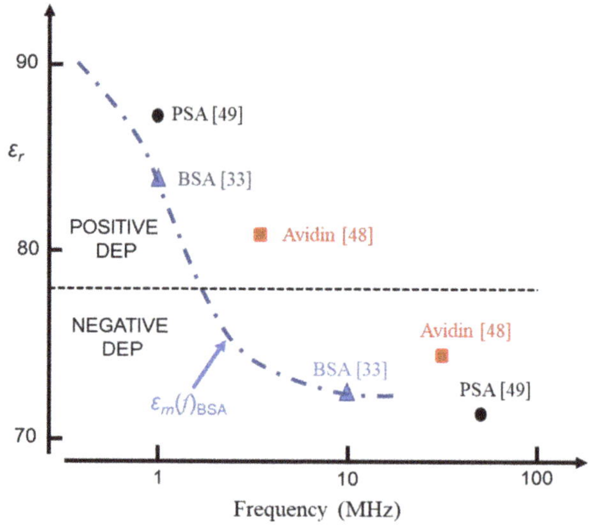

Figure 7. BSA, avidin and prostate specific antigen (PSA) have been reported to exhibit a DEP cross-over at MHz frequencies. This is consistent with the permittivity of the protein solution falling below the value ε_r = 78 for pure water, as shown here for BSA (from data of Figure 6).

The generic expression for the DEP susceptibility is given as

$$\chi_{DEP} = \frac{3}{2} V_p \varepsilon_m \, (\kappa + 2)[CM]^{empirical} \tag{21}$$

The parameter κ can assume two values, either κ = 1 or κ >> 1. The case κ = 1 is given where the particle concerned does not possess a permanent dipole moment, so that χ_{DEP} is given by Equation (10). The β-dispersion it exhibits, when suspended in a fluid, is capable of being analysed in terms of Equation (16) and the DEP force it experiences is given by Equation (9). For protein DEP and other cases of molecular DEP where the sample macromolecule possesses a permanent dipole moment, then χ_{DEP} is given by:

$$\chi_{DEP} = \frac{3}{2} V_p \varepsilon_m \left[(\kappa + 2)[CM]^{empirical} \right] = \frac{3}{2} V_p \varepsilon_m \left[\left(\frac{\Delta\varepsilon(\beta)}{\varepsilon_m} \right) \left(\frac{C_w \rho_p}{C_p \rho_w} \right) \right] \tag{22}$$

The frequency dependence of the DEP force now assumes a modified version of Equation (9), based on Equation (17):

$$F_{DEP}(f) = \frac{3}{2} V_p \varepsilon_0 \varepsilon_m(f) \left[\left(\frac{\Delta\varepsilon(\beta)}{\varepsilon_m} \right) \left(\frac{C_w \rho_p}{C_p \rho_w} \right) \right] \nabla E_m^2 \tag{23}$$

The dielectric relaxation time, τ, for a macromolecule is proportional to its volume [16] and so we can expect the β-dispersion for a protein dimer to be shifted to a frequency that is lower than its monomer counterpart, and also possibly exhibit a larger dispersion strength because of an increased effective dipole moment. Moser et al. [40] investigated the β-dispersion exhibited by monomer and dimer BSA. The characteristic frequency, f_β, of Equation (17) was determined to be ~400 kHz at 25 °C for the dimer, compared to ~1 MHz for the monomer. The dispersion strength $\Delta\varepsilon(\beta)$ exhibited by the dimer was also determined to be ~14% larger than that for the monomer. This offers the possibility that DEP can be used to separate the polymeric forms of a protein.

3. Summary of Dielectric Theory of Relevance to Protein DEP

Although the term 'dielectrophoresis' implies that dielectric theory already underpins the theory and practice of DEP, the extent of this is actually limited to narrow aspects of macroscopic electrostatics. New concepts for molecular DEP, and for protein DEP in particular, have been introduced by Heyden and Matyushov [1], and they highlight the importance of MD simulations as a complimentary tool to dielectric spectroscopy. This summary is to help establish, primarily for members of the DEP community not trained in dielectrics, what is new to established theory and as a guide to the key literature.

Standard DEP theory lacks the ability to adequately deal with particles that possess a permanent dipole moment. The primitive depiction in Figure 8a of a globular protein molecule illustrates that significant contributions to its permanent dipole moment are its ionized acidic and basic peptides located on its surface. An estimate of the moment is calculated from the algebraic summation of the moment vectors, $q_i r_i$, directed towards the protein's centre of mass. X-ray diffraction data provide the 'resting' locations of these charged groups. A more accurate value is obtained by including peptide moments and those formed by α-helixes [16]. A key innovation [1] is to replace the 'boundary problem' of macroscopic electrostatics with calculation of the cross-correlations of the protein dipole and neighbouring water dipoles that, in the inner hydration shell, are located close to the hydrophilic polar and charged side groups on the protein's surface.

For the case shown in Figure 8b, dipole–dipole interactions enhance the effective dipole moment of the solvated protein. As explained in this section, this corresponds to a Kirkwood correlation factor, g_k, larger than unity. If anti-correlation occurs, then g_k is less than unity. The fluctuations of the moment shown in Figure 8c should fit a Gaussian distribution about its mean value. MD simulations for lysozyme and ubiquitin give mean dipole moment values of 145 D and 218 D, respectively, with Gaussian widths of 29 D and 37 D, respectively [1]. Thus, for these proteins, calculations that neglect modulations of the protein dipole moment by conformational fluctuations (i.e., assumes $M_p^2 \approx \left\langle M_p^2 \right\rangle$ does not introduce an error exceeding 4%.

The theory of linear response and susceptibility is important. The susceptibility $\chi(\omega)$ of a system, when a sinusoidal force F of angular frequency ω is applied to it, is given as $\chi(\omega)$ = R/F, where R is the response of the system. The interaction energy between F and R is equal to the negative dot product -F·R. This is the vector notation for -FR cos θ, where θ is the polar angle between F and R. In dielectrics theory, F is the applied field and R is the resulting dipole moment M_T of the system. The dielectric susceptibility of a material is equal to $\left\langle M_T^2 \right\rangle_0/3kT$ where $\left\langle M_T^2 \right\rangle_0$, as shown in Figure 8d, corresponds to the moment's mean square fluctuation *without* application of the force field. Linear response theory connects the *production* of spontaneous thermal fluctuations of a system to the *dissipation* of excess energy produced by externally applied perturbation. Provided that a system of dipoles is at equilibrium when an external (relatively weak) electric field is applied to it, the temporal response of this system to the field can be determined in terms of $\left\langle M_T^2 \right\rangle_0$. Applications of this theory extends to other topics, such as magnetism and elasticity. For example, by examining the spontaneous microscopic fluctuations of a spring-balance with no mass attached to it, a prediction can be made of the extent it will stretch if a specified mass is attached!

Figure 8 depicts important ingredients of protein DEP theory. What follows adds substance to this.

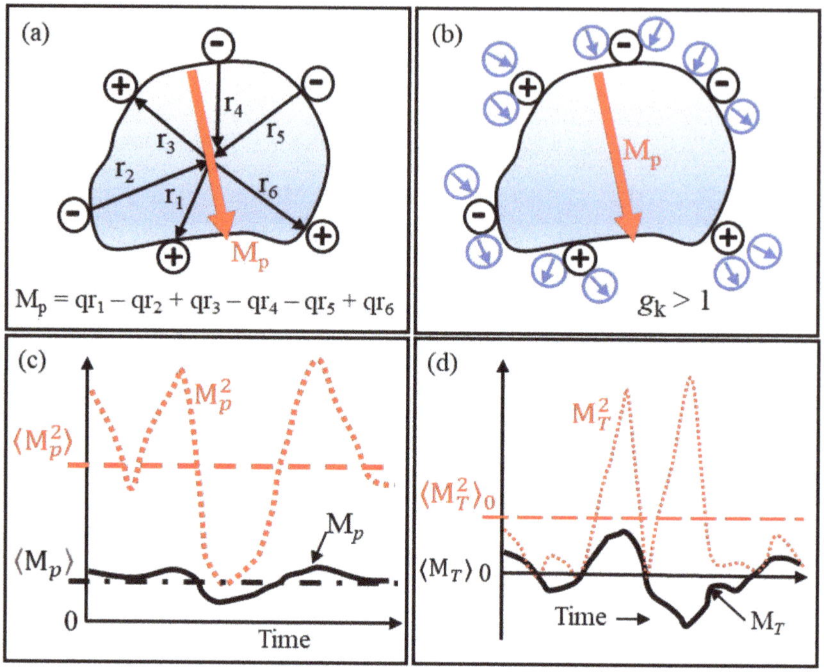

Figure 8. (**a**) Major contributions to the permanent dipole moment M_p of a protein are its peptide groups that carry a charge q. (**b**) H_20 dipoles near charged groups can act to enhance M_p, corresponding to a correlation coefficient $g_k > 1$. Reversal of the polarities shown for the water dipoles gives anti-correlation, with $g_k < 1$. (**c**) Brownian (kT) changes of location and magnitude of q produce fluctuations with time of M_p about its mean value $\langle M_p \rangle$. The dashed red line indicates the mean square $\left\langle M_p^2 \right\rangle$ of these fluctuations. (**d**) Without an applied electric field, orientations of the dipoles in a sample of polar fluid are random. The mean value, $\langle M_T \rangle_0$, of the total moment of these dipoles is thus zero. However, due to random fluctuations, its mean square value $\left\langle M_T^2 \right\rangle_0$ is finite and corresponds to spontaneous polarization.

3.1. The Clausius–Mossotti Law, Lorentz Cavity Field and Debye's Orientational Polarization

Based on Green's electric potential function [7] and Mossotti's hypothesis [9] that the electric fluid residing in each conductive 'corpuscle' of a dielectric is displaced under the action of a local field to form an electric doublet (i.e., dipole), Clausius derives the following relationship (known as the Clausius–Mossotti law) linking Faraday's concept of a specific inductive capacity (i.e., the static dielectric constant) to a simplifying factor g that relates the induced moment of an individual corpuscle to the total polarization of the dielectric [10]:

$$\varepsilon_m = (1 + 2g)/(1 - g) \tag{24}$$

The key steps to the derivation of this relationship are described elsewhere [24]. It is implicit that g is proportional to the packing number density, N, of corpuscles and is thus proportional to the mass density ρ of the dielectric. The following is an alternative and the usual way to express the Clausius–Mossotti law:

$$\left(\frac{\varepsilon_m - 1}{\varepsilon_m + 2} \right) \frac{1}{\rho} = \text{constant} \tag{25}$$

This masks, but does not remove, the implied issue of the factor g approaching a value of unity, where either a small perturbation of temperature or applied field could result in infinite polarization and a ferroelectric transition. This is addressed in Section 3.3. Although Clausius postulates that each corpuscle could possess a permanent dipole moment, he describes their polarization as an induced elastic displacement of internal charges [10]. This is now known as atomic distortion polarizability, comprising the two distinct contributions of electronic and atomic polarizability. The number density N is equal to $N_A\rho/M_W$, with N_A the Avogadro constant, so that the polarization (induced moment per unit volume) P_m is given in terms of the local field E_L as:

$$\mathrm{P}_m = N\alpha \mathrm{E}_L \tag{26}$$

Lorentz evaluates E_L by placing a polarizable particle inside a virtual spherical cavity "whose dimensions are infinitely small in a physical sense" [50]. All the atomic matter, apart from the point charges forming the dipole, are removed from this imagined cavity. The distributed charge induced at the cavity's outer surface creates a cavity field E_L of magnitude $\mathrm{E}_m + \mathrm{P}_m/(3\varepsilon_o)$. The atomic matter that had been removed is now returned, which adds another component $s\mathrm{P}_m/\varepsilon_o$ to the cavity field, to give:

$$\mathrm{E}_L = \mathrm{E}_m + \left(\frac{1}{3} + s\right)\frac{\mathrm{P}_m}{\varepsilon_o} \tag{27}$$

Lorentz shows, for a cubic lattice of polarizable particles, that $s = 0$, and states that otherwise this "is a constant that will be difficult exactly to determine" [50] (pp. 138, 303). This result also holds for an isotropic lattice, provided that no short-range interactions occur between the induced dipole fields. Substituting into Equation (27) the value for P_m given by Equation (4) and assigning $s = 0$, E_L has the value:

$$\mathrm{E}_L = \frac{\varepsilon_m + 2}{3}\mathrm{E}_m \tag{28}$$

This is known as the Lorentz field (sometimes called the Clausius–Mossotti field). The susceptibility of the Lorentz cavity field, equal to the ratio $\mathrm{E}_L/\mathrm{E}_m$, is thus:

$$\chi_{Lc} = \frac{\varepsilon_m + 2}{3}\ (\mathrm{s} = 0)\ \text{ or }\ \chi_{Lc} = \left[\frac{1}{3}(\varepsilon_m + 2) + s(\varepsilon_m - 1)\right] \tag{29}$$

Equation (29) indicates that the Lorentz cavity field, unlike the Maxwell cavity field of Equation (5), is always larger than the applied macroscopic field E_m. This arises because the Lorentz cavity, being virtual and without a physical boundary between it and the surrounding dielectric, does not experience the field-shielding effect of induced interfacial charges as produced for the Maxwell cavity. From Equations (4), (26) and (28) the following relationship is obtained:

$$\frac{\varepsilon_m - 1}{\varepsilon_m + 2} = \frac{N\alpha}{3\varepsilon_o} \tag{30}$$

Lorentz recognizes that this relationship corresponds to the law formulated by Clausius as given by Equation (25). It can be used to derive good estimates of a dielectric's refractive index n ($n^2 = \varepsilon_m$). Debye extends its relevance to polar solutions, by adding the orientational polarization of a molecular dipole of moment m to the atomic distortion polarization [18]:

$$\frac{\varepsilon_m - 1}{\varepsilon_m + 2} = \frac{1}{3\varepsilon_o}\sum_i N\left(\alpha_i + \frac{m_i^2}{3kT}\right) = \frac{1}{3\varepsilon_o}\sum_i \frac{N_A\rho}{M_w}\left(\alpha_i + \frac{m_i^2}{3kT}\right) \tag{31}$$

The polar molecule is regarded as a sphere undergoing rotational Brownian motion, whilst experiencing the Lorentz cavity field inside a spherical cavity within the dielectric.

The summation Σ_i includes all types of polarizable molecule within the dielectric. In the absence of an applied electric field E_m, the direction of orientations of an ensemble of dipoles will, on average, be distributed with the same probability over all directions in space. On applying a field, each dipole gains potential energy $U = -mE_m \cos\theta$, where θ is the polar angle between the direction of the dipole and the applied field. Using classical Boltzmann statistical mechanics and a method resembling that of Langevin for gas molecules carrying a permanent magnetic moment [51], Debye obtains the following value for the average moment per dipole oriented along the applied field's direction [18]:

$$\langle m \rangle_E = m\langle \cos\theta \rangle = \left(\frac{m^2}{3kT}\right) E_m \quad (32)$$

Computer modelling of the dielectric properties of a solvated protein employs a lattice of such Langevin dipoles [52,53]. The value of $\langle \cos\theta \rangle$ is given by the Langevin function, $L(x) = \coth x - 1/x$, and is shown in Figure 9 [18,19].

In dielectric spectroscopy measurements the fields are usually less than 10^5 V/m, so for many kinds of polar molecules (e.g., water) $L(x) \approx x/3$ and values for $\langle \cos\theta \rangle$ are appreciably smaller than unity (typically < 10^{-4}). This indicates that very little change to the directions of the polarized dipoles has occurred compared with their random directions with no field applied. The magnitude of the orientational polarizability for each dipole is thus $\left(\frac{m^2}{3kT}\right)$, and is incorporated as such in Equation (31). For protein iDEP experiments the applied fields are typically of the order ~10^6 to 10^7 V/m [4,5]. For many classes of polar molecule (e.g., water with m = 1.8 D), even with such high fields, the interaction energy, $m \cdot E_m$, is much less than the thermal energy kT. However, globular proteins possess dipole moments several hundred times larger than many other types of polar molecule [16,24]. As shown in Figure 9 for BSA (m = 710 D [54]) the linear relationship given by Equation (32) is no longer applicable and saturation of the orientational polarization is approached. With increasing field strength, $\langle \cos\theta \rangle$ approaches unity, indicating that on average the protein's permanent dipole moment is almost aligned with the field.

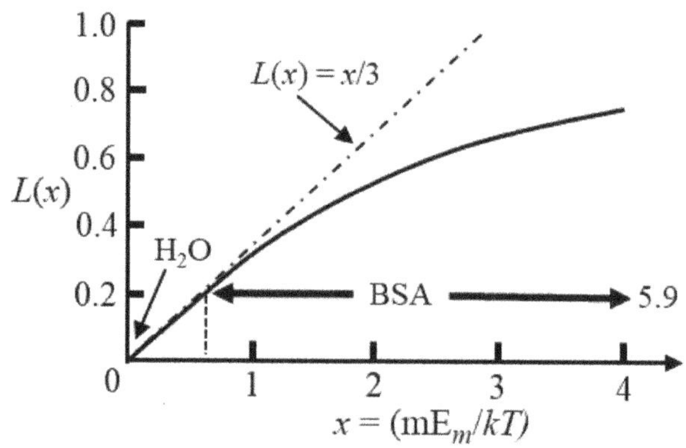

Figure 9. The average orientation of a single dipole along the direction of an applied field is given by the Langevin function ($L(x) = \coth x - 1/x$) [18,19]. Fields of ~10^6 to 10^7 V/m are used in protein DEP experiments [4,5]. Polar molecules typically have dipole moments less than 5 debye units (e.g., 1.8 D for H_2O) so that x << 1. Globular proteins have large dipole moments and as shown for BSA (m = 710 D [54]) saturation of its polarization can occur in DEP studies.

3.2. *Maximum Size of a Lorentz Cavity*

Ramshaw offers a basis for defining the maximum size permitted for a virtual Lorentz cavity [55,56]. He makes the interesting observation that in Equation (4) both P_m (dipole moment per unit volume) and E_m (the macroscopic Maxwell field) can be defined at any location within a dielectric, regardless of whether ε_m exists as a real concept. Thus, the existence of ε_m cannot be guaranteed, by definition. What does this mean within the context of the Lorentz cavity? Ramshaw [55] addresses this by expanding the Clausius–Mossotti function of Equation (25) into powers of the number density, N, of polar molecules, keeping the material's specific density ρ constant. This is equivalent to finding the virial coefficients for a gas [18]. Ramshaw makes no assumption regarding ε_m and also eliminates E_m from his analysis by considering both short-range and long-range orientation correlations of pairs of dipoles. The total correlation between two dipoles, in zero applied field, is regarded as the sum of a direct and an indirect effect that is transmitted by chains of direct effects between intermediate molecules. Beyond a certain total chain length d, where dipole–dipole intermolecular potential energies fall below kT, thermal disruption overcomes coherent correlations of dipole pairs. If a cavity's diameter is smaller than d, ε_m can be considered as well-defined and the cavity field is the Lorentz field, E_L, given by Equation (27). Above this cavity size, no such assumption may be made [55,56].

3.3. Onsager's Reaction Field [57] and Kirkwood's Correlation Factor [58,59]

Onsager, a PhD student of Debye, decided to learn 'about the dielectrics that Debye had done'. He was 'too lazy to go to the libraries and sat down and worked it out and lo and behold it came out quite different!' [60]. Onsager replaces the cavity used by Lorentz and Debye with a much smaller one, of the same radius, R, as the polar molecule of interest. The effect is considered of introducing a rigid point dipole of moment m_o into this cavity, with no external field applied. The thermal average $<m_o>$ of this dipole moment is calculated to seamlessly match it to the known number density of dipoles in the surrounding medium. The point dipole polarises the surrounding homogenous medium, creating interfacial charges and a reaction field inside the cavity. Matyushov considers the physics of this as two charged lobes, at the opposite sides of the void containing the point-dipole, visualized as oriented dipoles of the medium cut through by the dividing surface [28]. Adopting Matyushov's nomenclature, the interface dipole created by the reaction field is:

$$M^{int} = -2\left[\frac{(\varepsilon_m - 1)}{(2\varepsilon_m + 1)}\right] m_o \tag{33}$$

The minus sign signifies that the reaction field acts against the field created by the fixed dipole moment, m_o, and almost cancels it out for polar media of high permittivity [28]. Onsager's introduction of the reaction field removes the potential problem of the factor g approaching a value of unity in Equation (19).

The Kirkwood correlation factor, g_k, quantifies the deviation from randomness of the orientation of a dipole, with respect to its neighbours, in a liquid composed of polar molecules. If N is the number of dipoles, each of moment m, in an ensemble of dipoles, then g_k is given by:

$$g_k = \frac{\left\langle M^2 \right\rangle}{N m^2} \tag{34}$$

where $\left\langle M^2 \right\rangle$ is the mean square value of the ensemble's total dipole moment, M. The significance of g_k is demonstrated in Figure 4, where it addresses the fact that Equation (31) incorrectly predicts water at room temperature to possess a static permittivity, $\varepsilon_m \approx 31$, instead of the experimental value, $\varepsilon_m \approx 78$. In its condensed phase, moment m (corresponding to that of an isolated H_2O molecule) is enhanced due to its orientational fluctuations being amplified through coordinated rotations of dipole neighbours to which it is hydrogen bonded. Kirkwood replaces Σm^2 in Equation (31) with the summation $\Sigma<m_o \cdot m>$ (i.e., $\Sigma\, m_o m \cos\theta$) with θ being the angle between a target dipole, m_o, and a neighbouring dipole, m. The correlation factor g_k is defined as the average of $\cos\theta$ between this target dipole and all the dipoles (including itself) within a volume of polar liquid surrounding it [58,59]. The summation thus includes $<m_o \cdot m_o>$, which is given the value of unity (θ is effectively zero, so cos θ = 1). If there is no correlation at all between the dipoles, then <cos θ> = 0 and g_k = 1.0. Hydrogen bond links between neighbouring dipoles in bulk water leads to correlation. For a perfect and rigid tetrahedral coordination of the H-bonds, about which free rotations are permitted but no bond bending, g_k is given as

$$g_k = 1 + z\cos^2\frac{\theta}{2} \tag{35}$$

where θ is the H-O-H bond angle and z is the coordination number. Based on coordination numbers (z = 4 for a perfect tetrahedron) and inter-molecular distances obtained from X-ray scattering data for water, the value g_k = 2.67 was obtained, to give ε_m = 78.2 [61].

Based on these contributions by Onsager [57] and Kirkwood [58,59], Equation (31) evolves to the general form [16]:

$$\frac{(\varepsilon_m - \varepsilon_\infty)}{\varepsilon_m}\frac{(2\varepsilon_m + \varepsilon_\infty)}{(\varepsilon_\infty + 2)^2} = \frac{1}{3\varepsilon_o}\sum_i N\left(\alpha_i + \frac{g_k m_i^2}{3kT}\right) \tag{36}$$

where ε_m is the static or 'equilibrium' permittivity value (corresponding to where the system of dipoles has attained a constant value of polarization with a constant applied field), and ε_∞ is the square of the refractive index at high frequencies ~10^{13} Hz [20]. Typically, for pure polar liquids, $\varepsilon_m \gg \varepsilon_\infty$ with $\varepsilon_\infty \approx 1$, so that for cases where only the dipole polarization is of interest (i.e., the atomic distortion polarizability α_i is not included) Equation (36) is usually presented as

$$\frac{(\varepsilon_m - 1)(2\varepsilon_m + 1)}{9\varepsilon_m} = N_{dip} g_k \tag{37}$$

where N_{dip} is the dipole polarizability density given by

$$N_{dip} = \frac{Nm^2}{9\varepsilon_0 kT} \tag{38}$$

with N being the number of polar molecules per unit volume. Equation (37) applies to the static case. If we wish to consider an orientational dipole dispersion, $\Delta\varepsilon$, occurring between a low and high frequency limit, the static permittivity, ε_m, is replaced by the complex permittivity, $\varepsilon(\omega) = \varepsilon' - i\varepsilon''$, of the polar liquid at frequency $\omega/2\pi$. On applying a sinusoidal field, $E(\omega) = E_{pk}\exp(i\omega t)$, to the system of dipoles, the variation of the permittivity as a function of frequency is given through the Laplace transform [62]:

$$\frac{[\varepsilon(\omega) - 1][2\varepsilon(\omega) + 1]}{9\varepsilon(\omega)} = \left\{\frac{[\varepsilon_m - 1][2\varepsilon_m + 1]}{9\varepsilon_m}\right\}\mathcal{L}\{-\varphi_M(t)\} \tag{39}$$

The Laplace transform is defined as [63]:

$$\mathcal{L}\{f(t)\} = \int_0^\infty f(t)\exp(-i\omega\tau)dt \tag{40}$$

and in this case $-\varphi_M(t)$ is the autocorrelation (decay) function of the permanent dipoles. This represents a combination of the superposition theorem with Onsager's theory that microscopic fluctuations created by thermal energy obey the same law of decay as macroscopic perturbations produced by an applied field [62]. According to Onsager's proposal [64] the following auto-correlation functions are equivalent:

$$\begin{aligned} \varphi_M(t) &= \tfrac{\langle M(t)\rangle_o}{\langle M(0)\rangle_o} \text{ (Decay function of a macroscopic moment);} \\ \varphi_M(t) &= \tfrac{\langle M(t)\cdot M(0)\rangle_0}{\langle M(0)\cdot M(0)\rangle_0} \text{ (Correlation function of moment fluctuations)} \end{aligned} \tag{41}$$

If these are expressed through a sum of exponential terms $\sum_{i=1}^{n} a_i \exp\left(-\frac{t}{\tau_i}\right)$, the Laplace transform is the sum of equivalent Debye terms:

$$\mathcal{L}\{-\varphi_w(t)\} = \sum_{i=1}^{n} \frac{a_i}{(1 + i\omega\tau_i)} \tag{42}$$

where a_i is the amplitude, $\Delta\varepsilon(i)$, and τ is the characteristic dipole correlation time for each dispersion. $\Delta\varepsilon(i)$ is the difference of real components of $\varepsilon(\omega)$, so that each dispersion is characterized by the frequency variation of N_{dip} of Equation (38), namely:

$$N_{dip}(\omega) = \frac{N_{dip}}{(1 + \omega^2\tau_i^2)} \tag{43}$$

This is the origin of the frequency dependence given for $\varepsilon_m(f)$ in Equation (17).

3.4. *Fröhlich's Theory Relating Permittivity and Spontaneous Polarization*

This theory is used in molecular dynamics (MD) simulations of solvated proteins. Fröhlich [19] (p. 37) considers a macroscopic spherical region of volume V, within an infinite homogeneous dielectric medium. Volume V is large enough to exhibit the same static permittivity, ε_m, as the surrounding homogeneous medium (refer to discussion after Equation (12)). The volume's surface can deviate, so that no molecule is cut by the surface, and encloses an ensemble of dipoles. Each dipole is represented as an elementary charge that can be thermally activated to 'hop' between two adjacent potential energy wells (see [24] for a schematic). The position of each charge is described as the vectorial displacement from the position it would have in the lowest energy level (ground state) of the whole system, and given according to the rules of Boltzmann statistical mechanics. However, the medium outside the sphere is treated as a continuous dielectric described by its macroscopic permittivity. Except at the absolute zero of temperature, the ensemble of charges inside the sphere will not retain a fixed positional configuration—even if macroscopically it is in equilibrium. Owing to thermal fluctuations there is a probability of finding this ensemble with any set of displaced space elements. This means, despite the average moment, $\langle \mathrm{M} \rangle_0$, of the ensemble in the absence of a field being zero, its mean square value, $\left\langle \mathrm{M}^2 \right\rangle_0$, in the absence of a field, has a finite value (see Figure 8d).

Fröhlich adopts the Onsager reaction field as the 'self-field' of the spherical volume and then adds to this the Maxwell cavity field produced by an external field E. The external field is considered to be sufficiently weak that the non-linear (saturation) polarization region shown in Figure 9 is avoided, so that $(\mathrm{M{\cdot}E})/kT \ll 1$. Furthermore, Onsager's 'self-field' is shown by Equation (33) to be the dominant component of the local field. Fröhlich's conclusion that the static permittivity of a sphere of a dielectric embedded in a large specimen of the same material can be expressed in terms of its spontaneous polarization, can thus be anticipated. He derives the following expression for the dielectric's susceptibility:

$$(\varepsilon_m - 1) = \frac{\left\langle \mathrm{M}^2 \right\rangle_o}{3kT\varepsilon_o V}\left[\frac{3\varepsilon_m}{(2\varepsilon_m + 1)}\right] = \frac{\left\langle \mathrm{M}^2 \right\rangle_o}{3kT\varepsilon_o V}\chi_{Mc} \tag{44}$$

where $\left\langle \mathrm{M}^2 \right\rangle_o$ indicates the Boltzmann average in the absence of a field, and from Equation (6) the susceptibility, χ_{Mc}, of the Maxwell cavity field is identified.

Fröhlich generalizes his model by permitting the immersed spherical volume to have an arbitrary static permittivity ε_p. For this 'mixture' model, he obtains the following result [19] (p. 177):

$$(\varepsilon_p - 1) = \frac{\langle M_p^2 \rangle_o}{3kT\varepsilon_o V_p}\left[\frac{(2\varepsilon_m + \varepsilon_p)}{(2\varepsilon_m + 1)}\right] = \frac{\langle M_p^2 \rangle_o}{3kT\varepsilon_o V_p}\chi_{Mcp} \tag{45}$$

where the Maxwell cavity susceptibility, χ_{Mcp}, is given by Equation (5). The fact $\langle M_p^2 \rangle_o$ occurs in the absence of an applied field indicates that the macroscopic sample of volume V_p can be of arbitrary shape. Matyushov expresses this as an 'invariant of rotations of the laboratory frame' [28]. Simonson adopts Equation (45) in a molecular dynamics (MD) simulation at 295 K of a solvated cytochrome-c molecule contained within a 2.4 nm radius sphere of water [31]. His conclusion that $\varepsilon_p \approx 25$ was used to give the ∇E_{Th}^2 values shown in Figure 2.

3.5. South and Grant's Theory of Protein Dipole Relaxation

South and Grant [65] apply Fröhlich's model to a sphere of permittivity $\varepsilon_{\infty w}$ and volume V containing N_w water molecules and N_m protein molecules. Each molecule is represented as a point dipole, m_w for the water and m_p for the protein, suspended in a 'background' medium of effective permittivity, $\varepsilon_{\infty w}$. This 'background' is responsible for spontaneous fluctuations of induced moments. The sphere is immersed in an infinite continuum of permittivity ε_{mix}, i.e., the same as the effective *static* permittivity of the sphere itself. The spontaneous fluctuations of induced and permanent dipoles are assumed to arise from independent mechanisms [66]. The short-range protein–water and water–water correlations are represented by Kirkwood factors g_{kp} and g_{kw}, respectively, and are assumed to be insensitive to protein concentration. Because each molecule is represented as a point dipole, no account is taken of the fact that a globular protein molecule is several thousand-times larger than the water molecules. South and Grant take this into account by representing the water as a continuum, thus removing the water dipoles from their model but retaining their influence by replacing the background medium by one whose permittivity is the static value permittivity, ε_{sw}, of pure water.

The following relationship is derived for the model of a central sphere of effective permittivity $\varepsilon_{\infty w}$ and volume V containing N_p protein molecules, each of effective dipole moment m_p and surrounded by a hydration shell of permittivity ε_{mw}, in an infinite continuum of static permittivity ε_{mix}:

$$(\varepsilon_{mix} - \varepsilon_{sw}) = \frac{N_p g_{kp} m_p^2}{3kT\varepsilon_o V}\left[\frac{3\varepsilon_{mix}}{(2\varepsilon_{mix} + \varepsilon_{sw})}\right] \tag{46}$$

It is assumed that $g_{kp} \approx 1$ for the situation where the overall short-range interactions of a protein with water molecules are nearly spherically symmetrical.

Following the procedure leading to Equation (39), the magnitude of the β-dispersion exhibited by a dilute protein solution is [65]:

$$\Delta\varepsilon(\beta) = \frac{hN_A C_p}{\varepsilon_0 M_w kT} g_{kp} m_p^2 \tag{47}$$

where the number density N_p/V has been replaced by $N_A C_p/M_W$, with N_A the Avogadro constant, M_W the protein's molecular weight and C_p its concentration (g/mL). The value of h is $\frac{1}{2}$ when the solvent is considered to be an assembly of point dipoles, and h = 3/4 when the continuum approximation is made [65]. This result can be compared with the following semiempirical relationship formulated by Oncley [67]

$$\Delta\varepsilon(\beta) = \frac{bN_A C_p}{9\varepsilon_o M_w kT} m_p^2 \tag{48}$$

The parameter b is calibrated by Oncley to be 5.8, based on dipole moment values obtained for simple amino acids by determination of the separation distance of the positive and negative charges carried by their amino and carboxyl groups. However, Takashima and Asami [68] found that a value for b = 4.5 gives better agreement between the calculated and experimental dipole moment for a solvated protein, if a linear relationship between $\Delta\varepsilon(\beta)$ and C_p is found. Based on b = 4.5, from Equations (47) and (48) estimates for g_{kp} of 1.0 (h = 1/2) and 0.67 ($h = \frac{3}{4}$) arise.

Values for g_{kp} less than unity imply anticorrelation of the protein and water dipoles of hydration, and from MD studies this was concluded to be the case for the inner population of water dipoles around ubiquitin and apo-calbindin [69,70]. The inner population of water, less than 0.35 nm from the protein surface, consists of water patches around polar and charged amino side-groups, whereas almost no waters are found this close to the hydrophobic groups [69]. Water molecules in the outer hydration shell (0.35~0.6 nm from the surface) either form hydrogen bonds with the inner bound waters or cover the hydration 'holes' about the non-polar and uncharged areas of the protein surface. In the MD simulations, ubiquitin was found to slow down the dynamics of bulk water molecules located as far away as 1.35 nm from its surface [69].

4. Heyden and Matyushov's Theory and MD Simulations

4.1. The Theory

This builds on foundations laid down by Matyushov and co-workers [71–74]. A key aspect is to replace the dielectric boundary-value problem of macroscopic electrostatics by calculation of the cross correlation of the protein's permanent dipole moment with its polarized hydration shell. The result is replacement of $Re[CM]_{macro}$ with the factor K, so that Equation (10) now takes the form:

$$\chi_{DEP} = \frac{3}{2} V_p \varepsilon_m \, Re[K] \tag{49}$$

As depicted in Figure 1, K holds the key to making the transition from the standard DEP theory to the new one, (from $[CM]_{macroscopic}$ to $[CM]_{molecular}$), and is given by:

$$K = \varepsilon_m \chi_c^{HM} \frac{N_{dip}}{3} + \frac{3\varepsilon_m}{2(\varepsilon_m - 1)} \left(\chi_c^{HM} - \chi_{Lc}\right) \tag{50}$$

The susceptibility factor χ_c^{HM} is new to dielectric theory (where the superscript 'HM' signifies its genesis) and effectively replaces Oncley's empirical parameter, b, of Equation (48). χ_{Lc} is the Lorentz cavity field susceptibility of Equation (6) and N_{dip} is the static dimensionless number density of Equation (36). For a particle without a permanent dipole moment, then N_{dip} = 0 and K is equivalent to $[CM]_{macroscopic}$. For the case where N_{dip} is finite and K effectively acts as $[CM]_{molecular}$, calculation of χ_c^{HM} is required and is given by [1]:

$$\chi_c^{HM} = \chi_c^{HM}(\alpha)\left(\langle \mathrm{M}_p \cdot \mathrm{M}_t \rangle / \left\langle \mathrm{M}_p^2 \right\rangle\right) = \left\{\left(\left\langle \mathrm{M}_p^2 + \langle \mathrm{M}_p \cdot \mathrm{M}_w \rangle\right\rangle\right) / \left\langle \mathrm{M}_p^2 \right\rangle\right\} \tag{51}$$

This calculation, performed in the MD simulation, derives the direct correlation between the dipole moment of the protein and its hydration shell, $\mathrm{M_p}$, and the dipole moment, $\mathrm{M_w}$, induced in the surrounding aqueous medium. This contains the self-variance, $\left\langle \mathrm{M}_p^2 \right\rangle$, and the cross-correlation, $\langle \mathrm{M}_p \cdot \mathrm{M}_w \rangle$, between the hydrated protein and bulk water dipoles. χ_c^{HM} is given by

$$\chi_c^{HM}(\alpha) = \chi_{Lc} - \alpha\left(\frac{2(\varepsilon_m - 1)^2}{3\varepsilon_m(2\varepsilon_m + 1)}\right) \tag{52}$$

The factor α can be assigned two values, namely zero or unity. For α = 0, χ_c^{HM} is equal to the Lorentz cavity susceptibility, χ_{Lc}. The surface between an inner and outer volume of dielectric does not produce the interfacial dipole moment M^{int} of Equation (4).

The assignment of α = 1 brings with it a physical interface and creation of M^{int}, with Equation (52) now given by:

$$\chi_c^{HM}(\alpha = 1) = \frac{3}{(2\varepsilon_m + 1)} \tag{53}$$

In this case, χ_c^{HM} represents the susceptibility of the Maxwell cavity. For reasons clearly explained [1], in deriving Equation (53) from Equation (5), the Maxwell field E_m is E_0/ε_m, where E_0 is the vacuum field. As indicated in Equation (51), calculation of not only the self-variance, $\langle M_p^2 \rangle$, of the protein's moment is involved, but also that of the cross correlations between M_p and the total moment, M_t, of the water dipoles comprising the surrounding dielectric continuum. Of particular importance is the cross correlation involving the moment, M^{int}, associated with polarization of its hydration shell [1]. The standard macroscopic boundary conditions of macroscopic electrostatics, used to derive the DEP susceptibility values given by Equations (6) and (10), apply only for the case where the influence of a particle's induced moment dominates over that of its permanent dipole moment (should it possess one).

The following equation is derived for the collective dielectric dispersions of a protein solution:

$$\Delta\varepsilon_{mix}(\omega) = \left(\frac{9}{2}\right)\nu_p N_{dip}(\omega) + 9\nu_p N_{dip}(\omega)\left[\chi_c^{HM}(\omega) - 1\right] + (1 - \nu_p)\Delta\varepsilon_w(\omega) \tag{54}$$

with v_p the volume fraction of the protein and $N_{dip}(\omega)$ given by Equation (43).

The first term in Equation (54) is the β-dispersion, $\Delta\varepsilon(\beta)$, produced by protein tumbling, followed next by the δ-dispersion, $\Delta\varepsilon(\delta)$, that is assumed to be associated with the protein hydration sheath (see Section 2.2). The magnitude and frequency dependence of $\Delta\varepsilon(\beta)$ is in line with standard dielectric theory, but the expression given for $\Delta\varepsilon(\delta)$ is new. The last term in Equation (54) is the γ-dispersion for the relaxation of bulk water dipoles and is attenuated by the factor (1-v_p) representing the partial volume of low polarizability occupied by the non-relaxing proteins and their hydration shells. It is interesting to note that if $\chi_c^{HM} = 1$, corresponding to no correlations between the protein dipole and water dipoles in its hydration shell, then $\Delta\varepsilon(\delta)$ vanishes. Written in terms of dipole moments of the protein and water, the $\Delta\varepsilon(\delta)$ term in Equation (54) becomes

$$9\nu_p N_{dip}\left(\chi_c^{HM} - 1\right) = \nu_p\left[\frac{(M_p \cdot M_w)}{(V_p\varepsilon_0 kT)}\right] \tag{55}$$

The dispersion strengths $\Delta\varepsilon(\beta)$ and $\Delta\varepsilon(\delta)$ are given by

$$\Delta\varepsilon(\beta) = \frac{9}{2}\nu_p N_{dip} \quad . \tag{56}$$

$$\Delta\varepsilon(\delta) = 9\nu_p N_{dip}\left(\chi_c^{HM} - 1\right) \tag{57}$$

to give

$$\chi_c^{HM} = 1 + \left(\frac{\Delta\varepsilon(\delta)}{2\Delta\varepsilon(\beta)}\right) \tag{58}$$

Based on experimental values reported for the magnitudes of the $\Delta\varepsilon(\beta)$ and $\Delta\varepsilon(\delta)$ dispersions [4,24,44,68], as well as inspection of Figure 5, it is apparent that the term in brackets is small, indicating that $\chi_c^{HM} \approx 1$. In other words, the protein–water Kirkwood correlation factor is close to unity, implying very little correlation, in line with South and Grant's assumption [65].

The DEP response of a solvated protein as a function of frequency, as depicted in Figure 5 for BSA, is primarily related to $\Delta\varepsilon(\beta)$ and thus through Equations (36) and (56) depends on $N_{dip}(\omega)$. The following relationship to connect K and $\Delta\varepsilon(\omega)$ is derived [1]:

$$K(\omega) = \frac{2}{9}\varepsilon(\omega)\left\{\frac{\Delta\varepsilon(\omega)}{v_p} + \Delta\varepsilon_w(\omega)\right\}\left[\frac{\chi_c^{HM}(\omega)}{(2\chi_c^{HM}(\omega - 1))}\right] \quad (59)$$

From the MD simulations, the value of the term in square brackets is estimated to be close to unity. For dilute protein concentrations, as v_p tends to zero, the term $\Delta\varepsilon(\omega)$ can also be neglected, so that to a good approximation:

$$K(\omega) = \frac{2}{9}\varepsilon(\omega)\left\{\frac{\Delta\varepsilon(\omega)}{v_p}\right\} \quad (60)$$

The DEP frequency response is given by Re[K] in Equation (49) and is shown in Figure 10 for lysozyme and ubiquitin as calculated from Equation (60) and based on the MD simulations [1]. Included in this figure are the empirical factors $(\kappa + 2)[CM]^{empirical}$ derived elsewhere for these two proteins [4] (see Figure 6 for estimated DEP cross-over frequencies). When normalized, the calculated and empirical results for ubiquitin are similar, reflecting a close correspondence of the MD simulation [1] and the experimental spectroscopy data [75] used to derive $(\kappa + 2)[CM]^{empirical}$ [4]. The difference shown for the high-frequency tail of lysozyme results from a disparity of the experimental data [44] and the MD simulation—indicating the importance of the latter to molecular DEP studies. The difference in magnitudes shown for ubiquitin is related to the new theory [1] replacing Oncley's empirical factor (b = 4.5) of Equation (48) with the susceptibility factor χ_c^{HM} of Equation (51), together with the fact that the empirical theory [4] employs mass density values for the protein [4] and experimental values for $\Delta\varepsilon(\beta)/v_p$.

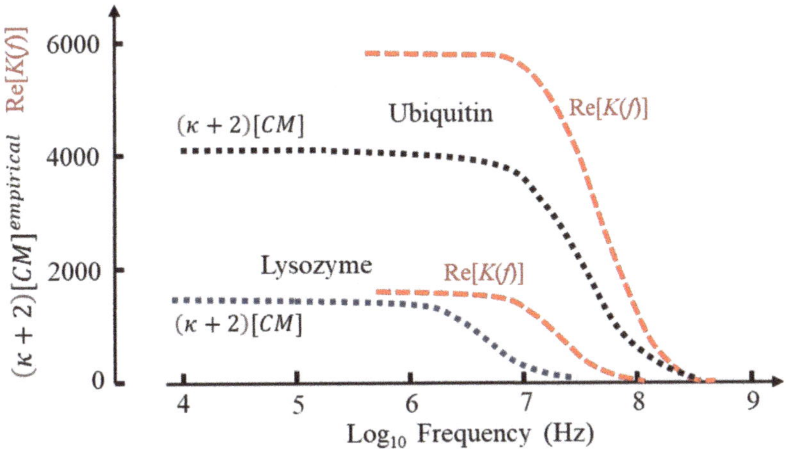

Figure 10. Frequency dependence of the DEP susceptibility factor $Re[K]$ for ubiquitin and lysozyme [1], together with plots for their empirical factors $(\kappa + 2)[CM]$ [4].

4.2. Molecular Dynamics Studies

In brief, the goal of the MD simulations of ubiquitin and lysozyme is to determine the cavity susceptibility given by Equation (51), relating the combined dipole moment, M_p, of the protein and its hydration shell to the dipole moment induced in the surrounding solution. When this equation is applied to the simulation trajectories, M_p is calculated within a sphere of cut-off radius, r_c, drawn around the protein. A correction factor, previously found to be independent of r_c [71], is applied to account for the fact that at each configuration along the simulation trajectory the outside solution is polarized by the dipole M_p.

Two features of these MD simulations are noteworthy and represent clear advances, namely their timescales and the atomic contents of the simulation cells. Simulations with integration steps of either 1 or 2 femtoseconds were performed for run times of 1 and 10 microseconds. In the frequency domain, a simulation run of 10 μs at steps of 1 fs corresponds to the range from 10^5 to 10^{12} radians/sec (i.e., 16 kHz to 160 GHz). This permits analyses to be made of the β, δ and γ relaxations exhibited by medium-sized proteins of molecular weight up to 100 kD (see Figure 4). By comparison, with the computer power available to them at the time, Boresch et al. [76] were limited to simulation runs of 5 ns at 2 fs steps in their pioneering MD simulation of ubiquitin. With such run times, adequate characterization of the δ-dispersion exhibited by globular proteins is possible, but only for proteins of low molecular weight (e.g., ubiquitin, cytochrome-c, lysozyme and myoglobin) is partial investigation of the β-dispersion achievable. Even so, these simulations were able to clarify which mechanisms (self- or cross-correlation) are responsible for the β- and δ-relaxations [76–78]. In subsequent MD simulations extending to 15 ns [69,70], ubiquitin, apo-calbindin D-9K, and the C-terminal SH2 domain of phospholipase C were studied to represent the structural variability found in medium-sized proteins. Two hydration shells were clearly discerned about charged and polar amino acids, and it was concluded that the protein molecule slows down the dynamics of water molecules located as far away as 1.35 nm [70]. The extension from MD run times of 15 nanoseconds to 10 microseconds [1] represents a significant advance. It also permits a significant increase in simulation cell size and content to be made.

The simulation cell (a cube of sides 12.5 nm) contained atomic copies of the proteins, based on high resolution x-ray crystal structures, together with 64,139 and 64,155 water molecules for ubiquitin and lysozyme, respectively, that included those resolved in their crystal structures [1]. The ionization (protonation) states of the titratable amino acid sidechains were determined for pH 7. The resulting total charge of ubiquitin was zero, and the charge (+8) of lysozyme was neutralized by a uniform counter charge. As a starting point for a 1 microsecond MD simulation, the energy of each system was minimized to a local minimum to avoid clashes between the protein and added water molecules, and then equilibrated at 300 K and 1 bar for 0.2 microseconds. These MD simulations require massive computing power. As an estimate, a simulation run of 1 microsecond at steps of 1 femtosecond, where the movements of around 200,000 atoms are determined, requires in excess of 10^{14} calculations to be made!

5. Something Else?

DEP experimentation and protein sample preparation fall under this heading. The published protein DEP data has mainly been obtained using either microfabricated metal electrodes to create the field gradients, referred to as eDEP, or iDEP where insulating microstructures are employed for this purpose. The eDEP experiments have reported results in line with those expected of the empirical [4] and formal theory [1]. However, inconsistent results (even for the same protein) have occurred for the iDEP experiments, and it is here where interesting and potentially exploitable examples of 'Something Else' exist. Some of these inconsistencies may be associated with iDEP departing from experimental guidelines commonly adopted for eDEP.

5.1. eDEP

The first eDEP experiments (and use of the term 'dielectrophoresis') were performed on macroscopic particles and employed metal electrodes in the form of tungsten wire and tinfoil, for example [13]. In early eDEP studies using *microfabricated* metal electrodes to study bacteria and cells [79–83] the following electrokinetic and related phenomena were observed and reported:

- Below ~1 kHz the DEP response of bacteria is dominated by a surface conductivity related to mobile ions in their surface electrical double layer [79,80].
- Electrothermal fluid motion can disrupt DEP at frequencies below ~10 KHz [80].
- Below ~1 kHz electrode polarization should be taken into account [81].
- Below ~100 Hz the electrophoresis and DEP of bacteria are superimposed and enhance positive DEP [81].
- Below ~100 Hz the DEP of mammalian cells rises rapidly with decreasing frequency. Neuraminidase treatment confirms this is associated with cell surface charge [82].
- Below ~500 Hz particles undergoing negative DEP are driven onto the surface of planar electrodes [83]. This is later shown to be caused by AC electroosmosis [84,85].

It is instructive to examine the procedures adopted for the first report of protein DEP by Washizu et al. [2]. They observed positive DEP of fluorescently labelled proteins (avidin, chymotripsinogen, concanavalin and ribonuclease) using microelectrodes fabricated from either aluminium or platinum. Measurements in the frequency range 1 kHz–10 MHz were performed, with the proteins suspended in their dialysis media of conductivity around 0.3 mS/m. Gel chromatography was used to verify that the samples after dialysis were in monomer form, and the similarity of results obtained using aluminium or the less electrochemically active platinum electrodes indicated that generated multivalent positive ions had not enhanced protein aggregation. The fact that DEP of the proteins was observed at a threshold field factor, ∇E^2_{Th}, much lower than that indicated in Figure 2 could possibly have resulted from pearl-chaining of the proteins. Such chaining arises from dipole–dipole interactions, which should be proportional to d^{-4} (where d is the initial spacing between dipoles). This was eliminated as a contributing factor by the similarity of the dynamics of the DEP responses for initial protein concentrations of 0.01 μg/mL and 0.1 μg/mL. However, possible evidence that protein aggregation occurred at the higher initial concentration of 1 μg/mL and for frequencies around 1 kHz was indicated by a 'memory' phenomenon. Samples that had already experienced DEP either exhibited a lower threshold field value for the next measurement, or the DEP force acting on them gradually increased with time of exposure to a constant field. Fluid motion arising from Joule heating was also investigated and eliminated as a possible artifact. In their concluding comments, Washizu et al. [7] state:

> *"although we may be seeing the combination of agglomeration and DEP at the low-frequency region, agglomeration is not a prerequisite for molecular accumulation, and DEP does occur with protein monomers."*

Based on this careful work of Washizu et al. [2] the following can be proposed as guidelines for observing *positive* DEP of protein *monomers*:

(i) Validate (e.g., through gel chromatography) that the samples are protein monomers.
(ii Limit protein concentrations to below ~0.1 μg/mL (i.e., less than 7 μM for many candidate proteins).
(iii) Adopt 1 mS/m as the upper limit for the aqueous solvent conductivity.

In protein eDEP experiments these guidelines have largely been followed. Solution conductivities of 1 mS/m or lower have been employed, with protein sample concentrations mostly below 10 μM. For all proteins investigated a consistent observation of *positive* DEP has been reported for the frequency range 1 kHz–1 MHz, in line with expectations of the empirical [4] and formal [1] theories. For some proteins a cross-over to negative DEP has been reported in the range 1–10 MHz [4], a finding consistent with the proposal that the DEP response for proteins can be predicted in terms of its dielectric β-dispersion [3]. However, the reported protein iDEP data for proteins is not consistent, and so it is here that categories falling under the umbrella of 'Something Else?' can be suggested.

5.2. iDEP

As introduced by Cummings and Singh [86] and reviewed by its practitioners [87–89] the complications that electrophoresis and electroosmosis can bring to eDEP experiments, especially at frequencies below 1 kHz, are translated through iDEP into facilitators of novel microfluidic devices for the selective sorting or concentration of molecular and macroscopic bioparticles. Negative DEP manifests itself as "streaming DEP" where the particles are carried down an array of insulating posts, following stream lines dictated by the spacing and geometry of the posts and the induced electroosmotic fluid flow. Streaming DEP occurs when the DEP force overcomes particle diffusion but is weaker than the combined effects of electrophoresis and electroosmosis. Increasing the magnitude of the externally applied field can change streaming DEP to "trapping DEP", where the DEP force dominates over diffusion as well as electrokinetic flow and target particles can be reversibly immobilized at the insulating posts [86]. This introduces increased difficulties of interpreting particle motions, especially for submicron particles, where a host of potential linear and non-linear electrokinetic and electrothermal effects are available for consideration [90].

Examples can be cited where iDEP of macroscopic particles has been successfully interpreted in terms of the standard DEP theory. Kim et al. [91], in a review of DEP studies of bioparticles below the size of a typical mammalian cell (e.g., microbes, organelles, exosomes, nucleic acids and proteins), provide a comprehensive summary of the 'one core', single- and multi-shell models that have been applied to spherical and non-spherical bioparticles. While microbes can in many cases be described using such classical DEP models, Kim et al. conclude that the theoretical basis of protein- and DNA-iDEP requires further fundamental studies to allow the prediction of biomolecular DEP response and hence their tailored applications. Using DC-iDEP, Lapizco-Encinas et al. [92] separated live from dead *E. coli* and isolated Gram-positive from Gram-negative bacteria. However, the spatial separation of different bacteria resulted from differences of their exhibited negative DEP mobility—the separation by both positive and negative DEP previously reported using eDEP (10–100 kHz) [93,94] to physically separate target bacteria from a mixture was not possible. This result could be consistent with a DEP force determined by polarizations of a bacteria's electrical double-layer, where the Dukhin function now serves as the effective [*CM*] factor [24]. In another study, by controlling the electroosmotic flow through a central outlet port and side outlet channels in a DC-iDEP device, the negative DEP force acting on the larger crystals in a mixture of protein crystals (size range ~80 nm to 20 μm) was sufficient to direct them into a central fluid outlet [95]. The weaker DEP force acting on the smaller crystals (size range ~80–200 nm) resulted in them remaining in the main stream emerging from the side outlets. In another example of controlled iDEP for macroscopic particles, polystyrene beads and yeast cells were focused into a thin stream line using an applied field in the form of a 0.1 Hz square wave. In this situation the negative DEP force remained unchanged in direction but there was a periodic reversal of the electroosmotic and electrophoretic forces [96].

Inconsistent iDEP results for proteins occur when the field is generated using a direct current (DC) voltage. Voltages as high as 4000 V have been used to concentrate PEGylated RNase by positive DEP [97,98]. However, the DEP force acting on the native protein was considered to be too weak to overcome electroosmosis [97]. An interesting situation is shown in Figure 11a where BSA samples, at the same pH and buffer conductivity, exhibit either negative [21] or positive [26,99] DEP. The concentration of the sample exhibiting negative DEP sample is 0.15 mM—much higher than the guideline of less than 10 μM adopted for eDEP studies. Positive DEP was observed for a BSA concentration of 7 nM [26,99]. The tendency of a solvated protein to dimerize or aggregate to form small crystallites is concentration-, temperature-, pH- and buffer-sensitive [4,100,101]. For example, BSA can be driven to a metastable state at a high mass concentration and high values of buffer pH and conductivity. In this respect, it is also of interest to note that in their iDEP experiments with high protein concentrations, Liu and Hayes [23] reported negative DEP responses for α-chymotrypsinogen (4 mM), immunoglobulin G (0.7 mM) and lysozyme (7 mM). Proteins in an iDEP experiment can also be subjected to shear stresses and high electric field strengths, factors known to influence initial protein crystal growth [102,103]. A field driven mechanism (possibly involving pearl chaining) whereby proteins in a metastable state may crystalize is proposed in Figure 11b. As shown in Figure 2, crystallites of diameter 50~100 μm are large enough to exhibit DEP according to standard theory. This is an aspect of protein DEP within the mesoscale between $[CM]_{macroscopic}$ and $[CM]_{molecular}$, where κ or N_{dip}, respectively, may or may not equal unity in either Equation (21) or (50), respectively.

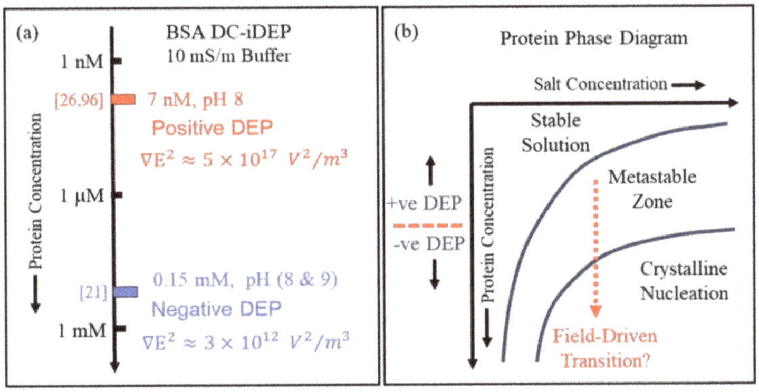

Figure 11. (**a**) BSA samples at the same buffer conductivity and pH exhibit negative iDEP at a concentration of 0.15 mM [21], but positive iDEP at 7 nM [26,98]. The low magnitude of ∇E_m^2 associated with the negative DEP result suggests minimal influence of a non-linear electrokinetic or electrothermal effect. (**b**) A simple protein phase diagram [99,100] to show how crystallization might occur during the DEP of a protein sample in a metastable state.

As reviewed by Hill and Lapizco-Encinas [104], significant efforts have been made to mathematically model iDEP-based microfluidic devices and to identify empirical correction factors that can be added to align model predictions with experimental observations. These correction factors are intended to take account of electrothermally induced fluid flow, Joule heating, particle–particle interactions and temperature gradients, for example. For micron-sized particles (e.g., bacteria, blood cells, polystyrene beads and yeast cells) most correction factors are found to be small (0.3 to ~15), whilst particles of diameter ~1 μm attract larger correction factors—in some cases as large as 500~600, depending on the geometry and layout of the insulating posts, hurdles or restrictions [104].

The electrokinetic forces of electrophoresis and electroosmosis that, together with DEP, act on a particle in an iDEP experiment are traditionally assumed to have a linear dependence on the applied electric field. Their superposition is commonly referred to as 'linear electrokinetics'. However, as recently reviewed [89,91] an increased understanding of how insulators alter the magnitude and non-uniformity of an applied electric field has led to a realization of the importance of non-linear electrokinetic effects [105–109]. For example, a nonlinear field dependence of electrophoresis can give rise to unexpected particle trapping in fluidic channels that is clearly different in origin from DEP trapping [105]. A high magnitude DC uniform electric field can induce nonlinear particle velocities, leading to particle flow reversal beyond a critical field magnitude, referred to as the electrokinetic (EK) equilibrium condition [106,107]. The concept of an amplification factor has been introduced to describe how an insulator constriction can greatly magnify the electric field—information that can be used to significantly reduce (to sub-100 V) the DC voltage required in an iDEP device to manipulate micron-sized particles [108]. This realization that in a so-called iDEP device the DEP force may in fact be largely irrelevant, with non-linear electrophoresis and electroosmosis effects the leading actors, has led to new particle separation strategies. An important example is provided by Quevedo et al. [109], who show that synthetic lysozyme and BSA sub-micron particles, as well as their blends, are separable based on differences in their isoelectric points and as manifested in large differences of the voltage required to selectively trap them. These results mirror to some extent the unique electrokinetic signatures found by Liu and Hayes for α-chymotrypsinogen, immunoglobulin G and lysozyme in their innovative gradient insulator-based (g-iDEP) device operated under DC conditions [23]. Although these proteins exhibited behaviours consistent with negative DEP, non-linear electrokinetic effects may have complicated this interpretation.

Until such time as electrothermally induced fluid flow and electroosmosis is proven to have not been responsible for the DEP cross-over at MHz frequencies shown for avidin, BSA and PSA in Figure 7, this effect should also be included in the "Something Else?" category.

6. Concluding Comments

Practitioners of DEP have backgrounds across the biological, chemical, engineering and physical sciences. Most will appreciate having a simple formula to use when predicting, modelling or analysing experimental data. In the standard DEP theory, for particles of known spheroidal size and shape and without a permanent dipole moment, $[CM]_{\text{macroscopic}}$ serves this purpose very well. However, the only way to predict the DEP response of a solvated globular protein molecule in the frequency range 1 kHz to 100 MHz is to examine either its published dielectric β-dispersion or full spectrum (e.g., Figures 4 and 5), or to obtain these details by means of dielectric spectroscopy and/or MD simulations.

A simple equation that can be derived from Equation (36) for a mixture of two polar molecules is of the form

$$\varepsilon_{mixture}(\omega) \propto \left\{ \frac{v_p g_{kp} m_p^2}{\left(1+\omega^2\tau_p^2\right)} \right\} + \left\{ \frac{\left[(1-v_p) g_{kw} m_w^2\right]}{(1+\omega^2\tau_w^2)} \right\} \tag{61}$$

where v_p is the volume fraction of the component of lowest concentration, with $v_p + v_w = 1$ where v_w is the volume fraction of the other component. For the case of a small concentration of protein (p) dissolved in water (w), the dielectric spectrum should reveal two distinct and separable dispersions. The second term on the right-hand side of this equation represents the γ-dispersion due to relaxation of the water dipoles and is centred near 10 GHz. At ~250 MHz where $(1+\omega^2\tau_w^2) \approx 1$, and $\left(1+\omega^2\tau_p^2\right) \gg 1$, the decrement of the γ-dispersion should equal $(1 - v_p)$ and so provide an indication of the effective volume of the protein with its hydration shell. The magnitude and frequency dependence of the first term on the right-hand side of Equation (61) represents the β-dispersion and provides the information regarding the predicted DEP response (see Figure 10). For a protein of medium molecular weight (e.g., BSA), the small in amplitude but identifiable δ-dispersion near 70 MHz should be observed [41]. For proteins of low molecular weight (e.g., ubiquitin) the δ-dispersion manifests itself as a slight distortion of the high-frequency tail of the β-dispersion [75]. The combined magnitude of $g_{kp} m_p^2$ can be calculated or estimated from classical dielectric theory, but calculation of each component is only now, for the first time, possible as a result of the innovative theory and new microsecond MD simulations presented by Heyden and Matyushov [1].

Equation (9) provides a simple DEP force equation to be used for globular proteins, with the DEP susceptibility χ_{DEP} given by Equation (49). For lysozyme and ubiquitin, the magnitudes and frequency-dependencies of the susceptibility factor $Re[K]$, obtained by Heyden and Matyushov [1], are shown in Figure 10. In the absence, for other proteins, of values for the susceptibility factor χ_c^{HM} of Equation (51) (which effectively replaces the empirical parameter, b, of Equation (48)) empirical values given elsewhere [4,24] can be used because they are based on experimental values for $\Delta\varepsilon(\beta)/v_p$ and provide information regarding the relative magnitudes and frequency profiles to be expected. The new theory [1] and empirical one [4] do not take into account relaxations of electrical double layers and ion diffusion processes, and so cannot be relied upon to predict DEP responses for frequencies below ~1 kHz.

Funding: This research received no external funding.

Acknowledgments: The content of this article was presented (with the same title) at the DEP2020/21 conference held at Flagstaff, Arizona. At that time and here again now, Ralph Hölzel was acknowledged for his past valuable collaborations. Gratitude was also expressed to Matthias Heyden and Dmitry Matyushov for enjoyable and valuable discussions of their new theories and molecular dynamics simulations, and to Victor Perez-Gonzalez for updates on recent advances in the practice and theory of insulator-based electrokinetic studies.

Conflicts of Interest: The author declares no conflict of interest.

References

1. Heyden, M.; Matyushov, D.V. Dielectrophoresis of proteins in solution. *J. Phys. Chem. B* **2020**, *124*, 11634–11647. [CrossRef] [PubMed]
2. Washizu, M.; Suzuki, S.; Kurosawa, O.; Nishizaka, T.; Shinohara, T. Molecular dielectrophoresis of biopolymers. *IEEE Trans. Ind. Appl.* **1994**, *30*, 835–843. [CrossRef]
3. Pethig, R. Limitations of the Clausius-Mossotti function used in dielectrophoresis and electrical impedance studies of biomacromolecules. *Electrophoresis* **2019**, *40*, 2575–2583. [CrossRef] [PubMed]
4. Hölzel, R.; Pethig, R. Protein dielectrophoresis: Status of experiments and an empirical theory. *Micromachines* **2020**, *11*, 533. [CrossRef] [PubMed]
5. Hayes, M.A. Dielectrophoresis of proteins: Experimental data and evolving theory. *Anal. Bioanal. Chem.* **2020**, *412*, 3801–3811. [CrossRef]
6. South, G.P.; Grant, E.H. Dielectric dispersion and dipole moment of myoglobin in water. *Proc. R. Soc. Lond. A* **1972**, *328*, 371–387.
7. Green, G. *An Essay on the Application of Mathematical Analysis to the Theories of Electricity and Magnetism*; Wheelhouse T: Nottingham, UK, 1828.
8. Faraday, M. *Experimental Researches in Electricity*; reprinted in Dover Publications; New York 1965; Taylor and Francis: London, UK, 1855; Volume 3, pp. 1839–1855.
9. Mossotti, O.F. Discussione analitica sull'influenza che l'azione di un mezzo dielettrico ha sulla distribuzione dell'electricitá alla superficie di più corpi elettrici disseminati in esso. *Mem. Mat. Fis. Soc. Ital. Sci.* **1850**, *24*, 49–74.
10. Clausius, R. *Die mechanische Wärmetheorie, Zweiter Band: Die mechanische Behandlung der Electricität*; Vieweg: Braunschweig, Germany, 1879; pp. 62–97.
11. Abraham, M.; Föppl, A. *Einführung in die Maxwellsche Theorie der Elektrizität*, 3rd ed.; Teubner: Leipzig, Germany, 1907; Chapter 3.
12. Becker, R. *The Classical Theory of Electricity and Magnetism*; Hefner Publ. Co.: New York, NY, USA, 1932.
13. Pohl, H.A. *Dielectrophoresis: The Behavior of Neutral Matter in Nonuniform Electric Fields*; Cambridge University Press: Cambridge, UK, 1978.
14. Jones, T.B. *Electromechanics of Particles*; Cambridge University Press: Cambridge, UK, 1995.
15. Stratton, J.A. *Electromagnetic Theory*; McGraw-Hill: New York, NY, USA, 1941.
16. Pethig, R. *Dielectrophoresis: Theory, Methodology and Biological Applications*; John Wiley & Sons: Hoboken, NJ, USA, 2017.
17. Arnold, W.M.; Schwan, H.P.; Zimmermann, U. Surface conductance and other properties of latex particles measured by electrorotation. *J. Phys. Chem.* **1987**, *91*, 5093–5098. [CrossRef]
18. Debye, P. *Polar Molecules*; The Chemical Catalog Co., Dover Publications: New York, NY, USA, 1929.
19. Fröhlich, H. *Theory of Dielectrics*, 2nd ed.; Clarendon Press: Oxford, UK, 1958.
20. Maxwell, J.C. *A Treatise on Electricity and Magnetism*, 3rd ed.; Clarendon Press: Oxford, UK, 1891; Volume 1.
21. Lapizco-Encinas, B.H.; Ozuna-Chacón, S.; Rito-Palomares, M. Protein manipulation with insulator-based dielectrophoresis and direct current electric fields. *J. Chromat. A* **2008**, *1206*, 45–51. [CrossRef]
22. Zhang, P.; Liu, Y. DC biased low-frequency insulating constriction dielectrophoresis for protein biomolecules concentration. *Biofabrication* **2017**, *9*, 045003. [CrossRef]
23. Liu, Y.; Hayes, M.A. Orders of magnitude larger force demonstrated for dielectrophoresis of proteins enabling high-resolution separations based on new mechanisms. *Anal. Chem.* **2021**, *93*, 1352–1359. [CrossRef] [PubMed]
24. Hölzel, R.; Pethig, R. Protein dielectrophoresis: Key dielectric parameters and evolving theory. *Electrophoresis* **2021**, *42*, 513–538. [CrossRef] [PubMed]
25. Kwak, T.J.; Jung, H.; Allen, B.D.; Demirel, M.C.; Chang, W.-J. Dielectrophoretic separation of randomly shaped protein particles. *Sep. Purif. Technol.* **2021**, *262*, 118280. [CrossRef]
26. Camacho-Alanis, F.; Gan, L.; Ros, A. Transitioning streaming to trapping in DC insulator based dielectrophoresis for biomolecules. *Sens. Actuators B Chem.* **2012**, *173*, 668–675. [CrossRef] [PubMed]
27. Nakano, A.; Camacho-Alanis, F.; Ros, A. Insulator-based dielectrophoresis with β-galactosidase in nanostructured devices. *Analyst* **2015**, *140*, 860–868. [CrossRef]
28. Matyushov, D. *Manual for Theoretical Chemistry*; World Scientific: Hackensack, NJ, USA, 2021.
29. Shannon, G.; Marples, C.R.; Toofany, R.D.; Williams, P.M. Evolutionary drivers of protein shape. *Sci. Rep.* **2019**, *9*, 11873. [CrossRef]
30. Gimsa, J.; Wachner, D. A polarization model overcoming the geometric restrictions of the Laplace solution for spheroidal cells: Obtaining new equations for field-induced forces and transmembrane potential. *Biophys. J.* **1999**, *77*, 1316–1326. [CrossRef]

31. Simonson, T. Dielectric constant of cytochrome-c from simulations in a water droplet including all electrostatic interactions. *J. Am. Chem. Soc.* **1998**, *120*, 4875–4876. [CrossRef]
32. Bone, S.; Pethig, R. Dielectric studies of protein hydration and hydration-induced flexibility. *J. Mol. Biol.* **1982**, *157*, 571–575, Erratum in **1985**, *181*, 323–326. [CrossRef]
33. Cao, Z.; Zhu, Y.; Liu, Y.; Dong, S.; Chen, X.; Bai, F.; Song, S.; Fu, J. Dielectrophoresis-based protein enrichment for a highly sensitive immunoassay using Ag/SiO2 nanorod arrays. *Small* **2018**, *14*, 17032265. [CrossRef]
34. Laux, E.M.; Knigge, X.; Bier, F.F.; Wenger, C.; Hölzel, R. Dielectrophoretic immobilization of proteins: Quantification by atomic force microscopy. *Electrophoresis* **2015**, *36*, 2094–2101. [CrossRef] [PubMed]
35. Clarke, R.W.; White, S.S.; Zhou, D.J.; Ying, L.M.; Klenerman, D. Trapping of proteins under physiological conditions in a nanopipette. *Angew. Chem. Int. Ed.* **2005**, *44*, 3747–3750. [CrossRef] [PubMed]
36. Liao, K.T.; Tsegaye, M.; Chaurey, V.; Chou, C.F.; Swami, N.S. Nano-constriction device for rapid protein preconcentration in physiological media through a balance of electrokinetic forces. *Electrophoresis* **2012**, *33*, 1958–1966. [CrossRef] [PubMed]
37. Chaurey, V.; Rohani, A.; Su, Y.H.; Liao, K.T.; Chou, C.F.; Swami, N.S. Scaling down constriction-based (electrodeless) dielectrophoresis devices for trapping nanoscale bioparticles in physiological media of high-conductivity. *Electrophoresis* **2013**, *34*, 1097–1104. [CrossRef] [PubMed]
38. Staton, S.J.R.; Jones, P.V.; Ku, G.; Gilman, S.D.; Kheterpal, I.; Hayes, M.A. Manipulation and capture of A beta amyloid fibrils and monomers by DC insulator gradient dielectrophoresis (DC-iGDEP). *Analyst* **2012**, *137*, 3227–3229. [CrossRef] [PubMed]
39. Zheng, L.; Brody, J.P.; Burke, P.J. Electronic manipulation of DNA, proteins, and nanoparticles for potential circuit assembly. *Biosens. Bioelectron.* **2004**, *20*, 606–619. [CrossRef] [PubMed]
40. Moser, P.; Squire, P.G.; O'Konski, C.T. Electric polarization in proteins—Dielectric dispersion and Kerr effect studies of isoionic bovine serum albumin. *J. Phys. Chem.* **1966**, *70*, 744–756. [CrossRef]
41. Essex, C.G.; Symonds, M.S.; Sheppard, R.J.; Grant, E.H.; Lamote, R.; Soetewey, F.; Rosseneu, M.Y.; Peeters, H. Five-component dielectric dispersion in bovine serum albumin solution. *Phys. Med. Biol.* **1977**, *22*, 1160–1167. [CrossRef]
42. Schwan, H.P.; Takashima, S.; Miyamoto, V.K.; Stoeckenius, W. Electrical properties of phospholipid vesicles. *Biophys. J.* **1970**, *10*, 1102–1119. [CrossRef]
43. Fischer, H.; Polikarpov, I.; Craievich, A.F. Average protein density is a molecular-weight-dependent function. *Protein Sci.* **2004**, *13*, 2825–2827. [CrossRef]
44. Miura, N.; Asaka, N.; Shinyashiki, N.; Mashimo, S. Microwave dielectric study on bound water of globule proteins in aqueous solution. *Biopolymers* **1994**, *34*, 357–364. [CrossRef]
45. Wolf, M.; Gulich, R.; Lunkenheimer, P.; Loidl, A. Relaxation dynamics of a protein solution investigated by dielectric spectroscopy. *Biochim. Biophys. Acta Proteins Proteom.* **2012**, *1824*, 723–730. [CrossRef] [PubMed]
46. Reynolds, J.A.; Hough, J.M. Formulae for dielectric constant of mixtures. *Proc. Phys. Soc. B* **1957**, *70*, 769–775. [CrossRef]
47. Raicu, V. *Dielectric Relaxation in Biological Systems*; Raicu, V., Feldman, Y., Eds.; Oxford University Press: New York, NY, USA, 2015; pp. 60–83.
48. Bakewell, J.G.; Hughes, M.P.; Milner, J.J.; Morgan, H. Dielectrophoretic manipulation of avidin and DNA. In Proceedings of the 20th Annual International Conference of the IEEE Engineering in Medicine and Biology Society, Hong Kong, China, 1 November 1998; Volume 20, pp. 1079–1082.
49. Kim, H.J.; Kim, J.; Yoo, Y.K.; Lee, J.H.; Park, J.H.; Wang, K.S. Sensitivity improvement of an electrical sensor achieved by control of biomolecules based on the negative dielectrophoretic force. *Biosens. Bioelectron.* **2016**, *85*, 977–985. [CrossRef] [PubMed]
50. Lorentz, H.A. *The Theory of Electrons*; Teubner: Leipzig, Germany, 1916.
51. Langevin, P. Sur la théorie du movement brownien. *C. R. Acad. Sci. Paris* **1908**, *146*, 530–533.
52. Warshel, A. *Computer Modeling of Chemical Reactions in Enzymes and Solutions*; Wiley: New York, NY, USA, 1991.
53. Papazyan, A.; Warshel, A. Continuum and dipole-lattice models of solvation. *J. Phys. Chem. B* **1997**, *101*, 11254–11264. [CrossRef]
54. Scheider, W.; Dintzis, H.M.; Oncley, J.L. Changes in the electric dipole vector of human serum albumin due to complexing with fatty acids. *Biophys. J.* **1976**, *16*, 417–431. [CrossRef]
55. Ramshaw, J.D. Existence of dielectric-constant in rigid-dipole fluids—Direct correlation-function. *J. Chem. Phys.* **1972**, *57*, 2684–2690. [CrossRef]
56. Ramshaw, J.D. Existence of the dielectric constant in rigid-dipole fluids: The functional-derivative approach. *J. Chem. Phys.* **1977**, *66*, 3134–3137. [CrossRef]
57. Onsager, L. Electric moments of molecules in liquids. *J. Am. Chem. Soc.* **1936**, *58*, 1486–1493. [CrossRef]
58. Kirkwood, J.G. The dielectric polarization of polar liquids. *J. Chem. Phys.* **1939**, *7*, 911–919. [CrossRef]
59. Kirkwood, J.G. Molecular distribution in liquids. *J. Chem. Phys.* **1939**, *7*, 919–925. [CrossRef]
60. Onsager, L. Autobiographical commentary. In *Critical Phenomena in Alloys, Magnets and Superconductors*; Mills, R.E., Ascher, E., Jaffee, R.I., Eds.; McGraw-Hill: New York, NY, USA, 1971; p. 20.
61. Oster, G.; Kirkwood, J.G. The influence of hindered molecular rotation on the dielectric constants of water, alcohols and other polar liquids. *J. Chem. Phys.* **1943**, *11*, 175–178. [CrossRef]
62. Hill, N.E. The theories of complex permittivity. *J. Phys. C Solid State Phys.* **1972**, *5*, 415–424. [CrossRef]
63. Kreyszig, E. *Advanced Engineering Mathematics*, 10th ed.; John Wiley & Sons: New York, NY, USA, 2015.
64. Onsager, L. Reciprocal relations in irreversible processes. *Phys. Rev.* **1931**, *37*, 405–426. [CrossRef]

65. South, G.P.; Grant, E.H. Theory of dipolar relaxation in aqueous macromolecular solutions. *Biopolymers* **1974**, *13*, 1777–1789. [CrossRef]
66. Scaife, B.K.P. *Complex Permittivity*; English Universities Press: London, UK, 1971.
67. Oncley, J.L. The investigation of proteins by dielectric measurements. *Chem. Rev.* **1942**, *30*, 433–450. [CrossRef]
68. Takashima, S.; Asami, K. Calculation and measurement of the dipole moment of small proteins: Use of protein data base. *Biopolymers* **1993**, *33*, 59–68. [CrossRef]
69. Rudas, T.; Schröder, C.; Boresch, S.; Steinhauser, O. Simulation studies of the protein-water interface. II. Properties at the mesoscopic resolution. *J. Chem. Phys.* **2006**, *124*, 234908. [CrossRef]
70. Schröder, C.; Rudas, T.; Boresch, S.; Steinhauser, O. Simulation studies of the protein-water interface. I. Properties at the molecular resolution. *J. Chem. Phys.* **2006**, *124*, 234907. [CrossRef]
71. Martin, D.R.; Friesen, A.D.; Matyushov, D.V. Electric field inside a "Rossky cavity" in uniformly polarized water. *J. Chem. Phys.* **2011**, *135*, 084514. [CrossRef] [PubMed]
72. Matyushov, D.V. Dipolar response of hydrated proteins. *J. Chem. Phys.* **2012**, *136*, 085102. [CrossRef] [PubMed]
73. Seyedi, S.S.; Matyushov, D.V. Protein dielectrophoresis in solution. *J. Phys. Chem.* **2018**, *122*, 9119–9127. [CrossRef] [PubMed]
74. Matyushov, D.V. Electrostatic solvation and mobility in uniform and non-uniform electric fields: From simple ions to proteins. *Biomicrofluidics* **2019**, *13*, 064106. [CrossRef] [PubMed]
75. Knocks, A.; Weingärtner, H. The dielectric spectrum of ubiquitin in aqueous solution. *J. Phys. Chem. B* **2001**, *105*, 3635–3638. [CrossRef]
76. Boresch, S.; Höchtl, P.; Steinhauser, O. Studying the dielectric properties of a protein solution by computer simulation. *J. Phys. Chem. B* **2000**, *104*, 8743. [CrossRef]
77. Nandi, N.; Bagchi, B. Anomalous dielectric relaxation of aqueous protein solutions. *J. Phys. Chem. A* **1998**, *102*, 8217–8221. [CrossRef]
78. Nandi, N.; Bagchi, B. Reply to the comment by S. Boresch and O. Steiner on the letter by N. Nandi and B. Bagchi entitled "Anomalous dielectric relaxation of aqueous protein solution". *J. Phys. Chem. A* **2001**, *105*, 5509–5510. [CrossRef]
79. Price, J.A.R.; Burt, J.P.H.; Pethig, R. Applications of a new optical technique for measuring the dielectrophoretic behavior of microorganisms. *Biochim. Biophys. Acta* **1988**, *964*, 221–230. [CrossRef]
80. Inoue, T.; Pethig, R.; Al-Ameen, T.A.K.; Burt, J.P.H.; Price, J.A.R. Dielectrophoretic behavior of micrococcus-lysodeikticus and its protoplast. *J. Electrost.* **1988**, *21*, 215–223. [CrossRef]
81. Burt, J.P.H.; Al-Ameen, T.A.K.; Pethig, R. An optical dielectrophoresis spectrometer for low-frequency measurements on colloidal suspensions. *J. Phys. E Sci. Instrum.* **1989**, *22*, 952–957. [CrossRef]
82. Burt, J.P.H.; Pethig, R.; Gascoyne, P.R.C.; Becker, F.F. Dielectrophoretic characterization of Friend murine erythroleukaemic cells as a measure of induced-differentiation. *Biochim. Biophys. Acta* **1990**, *1034*, 93–101. [CrossRef]
83. Pethig, R.; Huang, Y.; Wang, X.-B.; Burt, J.P.H. Positive and negative dielectrophoretic collection of colloidal particles using interdigitated castellated microelectrodes. *J. Phys. D Appl. Phys.* **1992**, *25*, 881–888. [CrossRef]
84. Ramos, A.; Morgan, H.; Green, N.G.; Castellanos, A. AC electric-field-induced fluid flow in microelectrodes. *J. Coll. Interface Sci.* **1999**, *217*, 420–422. [CrossRef]
85. Castellanos, A.; Ramos, A.; Gonzalez, A.; Green, N.G.; Morgan, H. Electrohydrodynamics and dielectrophoresis in microsystems: Scaling laws. *J. Phys. D Appl. Phys.* **2003**, *36*, 2584–2597. [CrossRef]
86. Cummings, E.; Singh, A. Dielectrophoresis in microchips containing arrays of insulating posts: Theoretical and experimental results. *Anal. Chem.* **2003**, *75*, 4724–4731. [CrossRef]
87. Lapizco-Encinas, B.H. Microscale electrokinetic assessments of proteins employing insulating structures. *Curr. Opin. Chem. Eng.* **2020**, *29*, 9–16. [CrossRef]
88. Perez-Gonzalez, V.H. Particle trapping in electrically driven insulator-based microfluidics: Dielectrophoresis and induced charge electrokinetics. *Electrophoresis* **2021**, *42*, 2445–2464. [CrossRef]
89. Lapizco-Encinas, B.H. The latest advances on nonlinear insulator-based electrokinetic microsystems under direct current and low-frequency alternating current fields: A review. *Anal. Bioanal. Chem.* **2022**, *414*, 885–905. [CrossRef]
90. Xuan, X.C. Review of nonlinear electrokinetic flows in insulator-based dielectrophoresis: From induced charge to Joule heating effects. *Electrophoresis* **2022**, *43*, 167–189. [CrossRef]
91. Kim, D.; Sonker, M.; Ros, A. Dielectrophoresis: From Molecular to Micrometer-Scale Analytes. *Anal. Chem.* **2019**, *91*, 277–295. [CrossRef] [PubMed]
92. Lapizco-Encinas, B.H.; Simmons, B.A.; Cummings, E.B.; Fintschenko, Y. Insulator-based dielectrophoresis for the selective concentration and separation of live bacteria in water. *Electrophoresis* **2004**, *25*, 1695–1704. [CrossRef] [PubMed]
93. Pethig, R.; Markx, G.H. Applications of dielectrophoresis in biotechnology. *Trends Biotechnol.* **1997**, *15*, 426–432. [CrossRef]
94. Markx, G.H.; Huang, Y.; Zhou, X.-F.; Pethig, R. Dielectrophoretic characterization and separation of micro-organisms. *Microbiology* **1994**, *140*, 585–591. [CrossRef]
95. Abdallah, B.G.; Roy-Chowdhury, S.; Coe, J.; Fromme, P.; Ros, A. High throughput protein nanocrystal fractionation in a microfluidic sorter. *Anal. Chem.* **2015**, *87*, 4159–4167. [CrossRef] [PubMed]
96. Malekanfard, A.; Beladi-Behbahani, S.; Tzeng, T.-R.; Zhao, H.; Xuan, X. AC insulator-based dielectrophoretic focusing of particles and cells in an "infinite" microchannel. *Anal. Chem.* **2021**, *93*, 5947–5953. [CrossRef]

97. Mata-Gomez, M.A.; Perez-Gonzalez, V.H.; Gallo-Villanueva, R.C.; Gonzalez-Valdez, J.; Rito-Palomares, M.; Martinez-Chapa, S.O. Modelling of electrokinetic phenomena for capture of PEGylated ribonuclease A in a microdevice with insulating structures. *Biomicrofluidics* **2016**, *10*, 033106. [CrossRef]
98. Mata-Gomez, M.A.; Gallo-Villanueva, R.C.; Gonzalez-Valdez, J.; Martinez-Chapa, S.O.; Rito-Palomares, M. Dielectrophoretic behavior of PEGylated RNase A inside a microchannel with diamond-shaped insulating posts. *Electrophoresis* **2016**, *37*, 519–528. [CrossRef]
99. Nakano, A.; Chao, T.-C.; Camacho-Alanis, F.; Ros, A. Immunoglobulin G and bovine serum albumin streaming dielectrophoresis in a microfluidic device. *Electrophoresis* **2011**, *32*, 2314–2322. [CrossRef]
100. Garcia-Ruiz, J.M. Nucleation of protein crystals. *J. Struct. Biol.* **2003**, *142*, 22–31. [CrossRef]
101. Asherie, N. Protein crystallization and phase diagrams. *Methods* **2004**, *34*, 266–272. [CrossRef] [PubMed]
102. Nanev, C.N. Recent Insights into the Crystallization Process; Protein Crystal Nucleation and Growth Peculiarities; Processes in the Presence of Electric Fields. *Crystals* **2017**, *7*, 310. [CrossRef]
103. Pareja-Rivera, C.; Cuellar-Cruz, M.; Esturau-Escofet, N.; Demitri, N.; Polentarutti, M.; Stojanoff, V.; Moreno, A. Recent advances in the understanding of the influence of electric and magnetic fields on protein crystal growth. *Cryst. Growth Des.* **2017**, *1*, 135–145. [CrossRef]
104. Hill, N.; Lapizco-Encinas, B.H. On the use of correction factors for the mathematical modeling of insulator based dielectrophoretic devices. *Electrophoresis* **2019**, *40*, 2541–2552. [CrossRef]
105. Tottori, S.; Misiunas, K.; Keyser, U.F. Nonlinear electrophoresis of highly charged nonpolarizable particles. *Phys. Rev. Lett.* **2019**, *123*, 014502. [CrossRef]
106. Cardenas-Benitez, B.; Jind, B.; Gallo-Villanueva, R.C.; Martinez-Chapa, S.O.; Lapizco-Encinas, B.H.; Perez-Gonzalez, V.H. Direct current electrokinetic particle trapping in insulator-based microfluidics: Theory and experiments. *Anal. Chem.* **2020**, *19*, 12871–12879. [CrossRef]
107. Antunez-Vela, S.; Perez-Gonzalez, V.H.; De Peña, A.; Lentz, C.J.; Lapizco-Encinas, B.H. Simultaneous determination of linear and nonlinear electrophoretic mobilities of cells and microparticles. *Anal. Chem.* **2020**, *22*, 14885–14891. [CrossRef]
108. Ruz-Cuen, R.; de los Santos-Ramirez, J.M.; Cardenas-Benitez, B.; Ramirez-Murillo, C.J.; Miller, A.; Hakim, K.; Lapizco-Encinas, B.H.; Perez-Gonzalez, V.H. Amplification factor in DC insulator-based electrokinetic devices: A theoretical, numerical, and experimental approach to operation voltage reduction for particle trapping. *Lab Chip* **2021**, *21*, 4596–4607. [CrossRef]
109. Quevedo, D.F.; Lentz, C.J.; de Peña, A.C.; Hernandez, Y.; Habibi, N.; Miki, R.; Lahann, J.; Lapizco-Encinas, B.H. Electrokinetic characterization of synthetic protein nanoparticles. *Beilstein J. Nanotechnol.* **2020**, *11*, 1556–1567. [CrossRef]

micromachines

Article

Active, Reactive, and Apparent Power in Dielectrophoresis: Force Corrections from the Capacitive Charging Work on Suspensions Described by Maxwell-Wagner's Mixing Equation

Jan Gimsa

Department of Biophysics, University of Rostock, Gertrudenstr. 11A, 18057 Rostock, Germany; jan.gimsa@uni-rostock.de; Tel.: +49-381-6020; Fax: +49-381-6022

Abstract: A new expression for the dielectrophoresis (DEP) force is derived from the electrical work in a charge-cycle model that allows the field-free transition of a single object between the centers of two adjacent cubic volumes in an inhomogeneous field. The charging work for the capacities of the volumes is calculated in the absence and in the presence of the object using the external permittivity and Maxwell-Wagner's mixing equation, respectively. The model provides additional terms for the Clausius-Mossotti factor, which vanish for the mathematical boundary transition toward zero volume fraction, but which can be interesting for narrow microfluidic systems. The comparison with the classical solution provides a new perspective on the notorious problem of electrostatic modeling of AC electrokinetic effects in lossy media and gives insight into the relationships between active, reactive, and apparent power in DEP force generation. DEP moves more highly polarizable media to locations with a higher field, making a DEP-related increase in the overall polarizability of suspensions intuitive. Calculations of the passage of single objects through a chain of cubic volumes show increased overall effective polarizability in the system for both positive and negative DEP. Therefore, it is proposed that DEP be considered a conditioned polarization mechanism, even if it is slow with respect to the field oscillation. The DEP-induced changes in permittivity and conductivity describe the increase in the overall energy dissipation in the DEP systems consistent with the law of maximum entropy production. Thermodynamics can help explain DEP accumulation of small objects below the limits of Brownian motion.

Keywords: DEP force in narrow volumes; capacitor charge cycle; micro-fluidic volumes; DEP trajectories; 2D and 3D modelling; DEP-induced polarizability increase; conditioned polarization; lossy dispersive materials; thermodynamics; Rayleigh's dissipation function; law of maximum entropy production (LMEP)

Citation: Gimsa, J. Active, Reactive, and Apparent Power in Dielectrophoresis: Force Corrections from the Capacitive Charging Work on Suspensions Described by Maxwell-Wagner's Mixing Equation. *Micromachines* **2021**, *12*, 738. https://doi.org/10.3390/mi12070738

Academic Editor: Antonio Ramos

Received: 18 May 2021
Accepted: 21 June 2021
Published: 23 June 2021

1. Introduction

Up to the high radio frequency range, polarization processes in complex media, such as particle suspensions, are classified either by their electric mechanisms such as polarizations of electric double layers, of structural interfaces (Maxwell-Wagner) and of molecular dipoles through orientation (Debye) or by the frequency range in which they disperse (alpha, beta, and gamma dispersions) [1–3]. At any given frequency, the sum-effect of all polarization processes that can follow the field alternations weakens the field in each half-cycle. With increasing frequency, slower processes disperse and their contributions to medium polarization fade, which is reflected in a frequency-dependent permittivity decrease and the corresponding increase of the complex (specific) conductivity $\underline{\sigma}$ in S/m by up to several orders of magnitude over several frequency decades (complex parameters are underscored). This behavior is described mathematically by the dispersion relation, which considers the field-induced motion of bound charges, which at high frequencies cannot be distinguished from the motion of charges, such as ions, which can already move freely under DC.

In AC fields above approx. 1000 V/m, different AC-electrokinetic movement of freely suspended microscopic objects, such as colloidal particles or biological cells, can be observed when the objects are exposed to homogeneous, inhomogeneous, or rotating fields. While neighboring objects of even polarizability are attracted toward one another, they are attracted or repelled toward electrode surfaces depending on the electrode and object shapes and the object's polarizability relative to that of the suspension medium. The field-induced forces and torques lead to the translation (dielectrophoresis; DEP) and rotation (electrorotation; ROT) of individual objects or their aggregation in "pearl chains", which map the electric field, similar to how iron filings map the magnetic field lines.

Since the slow electrokinetic AC effects, which incidentally also occur in the DC field, are uncoupled from the frequency of the inducing field, they are not usually considered polarization mechanisms. However, in DEP or in electro-thermal pumping it is generally accepted that the lower polarizable medium, which may, depending on the frequency of the external field, be either the suspension medium or the object itself, is displaced by higher polarizable medium at sites of higher field strength. It is intuitive that this process increases the overall polarizability of the suspension. The increase in polarizability by field-induced orientation has previously been shown for suspensions of ellipsoidal objects [4].

The description of the AC electrokinetic behavior of objects in the radio-frequency range is usually based on dipole calculations [3,5,6]. In the classical DEP approach, the object is assumed to be small compared to the characteristic distance of the field inhomogeneity. This allows the DEP force calculation to be based on the assumption of induced dipole moments, i.e., the (weakly) inhomogeneous field induces a symmetric polarization. The interaction of the dipole moment with the inhomogeneous field generates unbalanced forces with the two poles of the dipole, leading to DEP (comparable to the dipolophoresis of a dipole in the field of an ion). However, when objects approach other objects or electrodes, a correct description of, for example deformation, orientation, and DEP forces must include induced multipoles. In addition, multi-body problems such as aggregation and pattern formation play a role [7].

Here, the DEP force is derived from the work difference performed when charging two cubes of equal volume containing either a single suspended object or the pure suspension medium. The work difference resulting from exchanging the positions of the two cubes in an inhomogeneous field provides the DEP force. It is shown that DEP increases the overall polarizability and conductivity in a suspension system. Under the condition of constant voltage, which is generally applied to DEP chambers, the increased conductivity is directly related to the increased dissipation of electrical energy, according to the Rayleigh dissipation function. Two homogeneous spheres with complementary properties are used in model calculations, reflecting the two main structural dispersions observed in biological cells. Comparison with the classical electro-quasistatic dipole approximation, which serves as a reference, suggests that the new approach has a slightly higher accuracy for higher volume fractions, which may be important in narrow microchambers [8–12].

2. Theory

2.1. General Remarks

The overall permittivity or (specific) conductivity of suspensions or emulsions is increased or decreased by introducing objects with higher or lower polarizabilities than the carrier medium. In DEP, media or objects with lower polarizability are replaced by entities with higher polarizability at locations with a high field. The differently polarizable entities can be the suspension media and suspended rigid or emulsified viscous objects in liquid media of different chemical natures or different temperatures (cf. electro-thermal pumps).

The DEP force for an ellipsoidal object is usually derived from the interaction of its induced dipole moment with the inducing field, where the dipole moment is obtained by solving the Laplace equation for electro-quasistatic conditions. Here, the classical DEP force is used as the reference for a different approach, in which the force is derived from the overall electrical work conducted in a charging cycle that permits the field-free transition of the object between two neighboring cuboid boxes in a weakly inhomogeneous field (Figure 1). Without an object, the dielectric properties in the box are given by the external medium. The presence of the object changes the dielectric properties to those of a suspension, which can be calculated by the Maxwell-Wagner mixing equation.

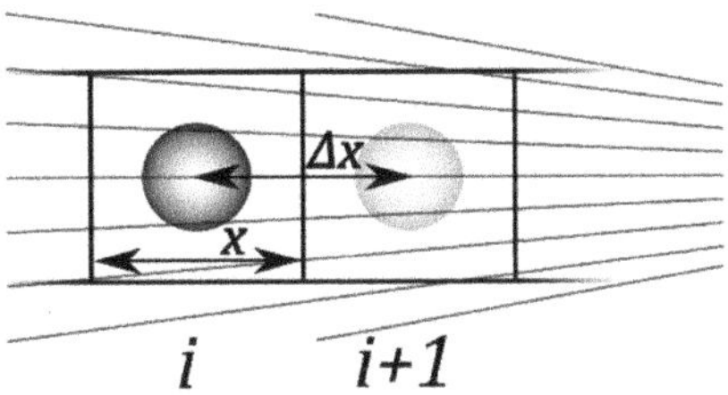

Figure 1. Diagram illustrating the consecutive positions of a spherical object during DEP movement in the inhomogeneous field of a 2D or 3D radial setup. The squares i and i + 1 represent cuboid boxes with x by x geometry ($y = x$) in the sheet plane. The field gradient points in the radial direction. Without any limitation in generality for 3D models of microscopic systems, perpendicular to the sheet plane, a depth of $z = x$ or of $z = 1$ m can be assumed, as is common in 2D models. The distance of travel between the box centers is $\Delta x = x$.

The idea is based on the cycle proposed by Max Born. To circumvent the complicated description of the electrical effects in the transition of charged ions through lipid membranes, Born proposed separating the process into discharging, interface transition, and recharging of the ion.

Version 12 of the Maple software supplication (Maplesoft, Waterloo, ON, Canada) was used to solve and simplify the equations (The main part of the program code is given in the Supplementary Material The data for the figures was exported to Sigma Plot 11.0 (Systat Software, Inc., San Jose, CA, USA).

2.2. Approximation of the Field Gradient

Long, coaxial cylindrical electrodes are a simple means of generating a well-defined inhomogeneous field of radial symmetry. For the symmetry axis in the z-direction, the field component in z-direction is zero. For a more general description, cuboid medium and suspension boxes are initially assumed. This shape allows an easier transition to future numerical, two-dimensional (2D) models. In 2D models, a length of 1 m is usually assumed for the z-dimension to match the physical units. In the subsequent model calculations, cubic boxes are used without restriction of generality for the cuboid shape.

Along the field gradient, the field strength $\underline{E} = \underline{E}_i$ is increased to $\underline{E}_{i+1}$:

$$\underline{E}_{i+1} = (1 + \gamma \Delta x)\underline{E}, \tag{1}$$

when moving $\Delta x = x_{i+1} - x_i$ in radial direction, from the center of box i to the center of box i + 1. For simplicity, vectorial parameters are not marked by arrows. With the inverse length parameter $\gamma > 0\ \mathrm{m}^{-1}$ describing the field inhomogeneity in the x-y field plane, the field gradient in radial (x-) direction is:

$$\frac{\Delta \underline{E}}{\Delta x} = \frac{(1 + \gamma \Delta x)\underline{E} - \underline{E}}{\Delta x} = \gamma \underline{E}. \tag{2}$$

Note that the effects of lateral field components on the induced DEP force are neglected in the dipole approximation by assuming objects that are small in comparison to the characteristic distance of the field inhomogeneity, i.e., for objects with typical semiaxis lengths $< 1/\gamma$. For reasons of symmetry, the induction of a torque by small lateral field components may be excluded at this stage. For $\gamma = 0$, the external field is homogeneous, the DEP force disappears for symmetry reasons, and compression or elongation forces remain.

2.3. DEP Force in the Classical Dipole Approximation

In almost all cases, the electrodes of DEP devices are operated with constant AC voltages that generate the linear field:

$$\underline{E} = E_0 e^{j\omega t}, \tag{3}$$

where $j = \sqrt{-1}$, time t and ω being the circular frequency. E_0 depends on the location in the inhomogeneous field. In $x-$ (radial) direction, the field induces the complex dipole moment:

$$\underline{m} = \varepsilon_0 \varepsilon_e V_0 \underline{f}_{CM} \underline{E} = \varepsilon_0 \varepsilon_e V_0 \left(f_{CM}^{\Re} + j f_{CM}^{\Im} \right) \underline{E}, \tag{4}$$

where ε_0 and ε_e are the permittivity of vacuum and the real part of the relative permittivity of the external medium, which is frequency-independent for aqueous suspension media in the frequency range considered. V_0 is the volume of the ellipsoidal object with principal semiaxes a, b, and c:

$$V_0 = \frac{4\pi}{3} abc. \tag{5}$$

The complex Clausius-Mossotti factor (f_{CM}) has both real ($f_{CM}^{\Re}$) and imaginary ($f_{CM}^{\Im}$) parts. For homogeneous objects of the general ellipsoidal shape in the direction of the (oriented) semiaxis a, it is:

$$\underline{f}_{CM}^{a} = f_{CM}^{a\Re} + j f_{CM}^{a\Im} = \frac{\underline{\varepsilon}_i - \underline{\varepsilon}_e}{\underline{\varepsilon}_e + n^a(\underline{\varepsilon}_i - \underline{\varepsilon}_e)}, \tag{6}$$

where n^a is the depolarizing coefficient along semiaxis a [6,13]. For spherical objects with $n^a = n^b = n^c = 1/3$, Equation (6) reads:

$$\underline{f}_{CM} = f_{CM}^{\Re} + j f_{CM}^{\Im} = 3 \frac{\underline{\varepsilon}_i - \underline{\varepsilon}_e}{\underline{\varepsilon}_i + 2\underline{\varepsilon}_e}. \tag{7}$$

Note that the factors "3" in Equations (5) and (7) are not cancelled out below to avoid confusion for non-spherical objects. The complex relative permittivities of the internal (index i) and external (index e) media are:

$$\underline{\varepsilon}_i = \varepsilon_i - j\frac{\sigma_i}{\omega\varepsilon_0} \qquad \underline{\varepsilon}_e = \varepsilon_e - j\frac{\sigma_e}{\omega\varepsilon_0}, \tag{8}$$

where σ_i and σ_e are the DC conductivities. Note that according to Maxwell's equivalent body notion, Equations (6) or (7) can also describe the Clausius-Mossotti factors of confocal shell models for given field frequency and model parameters [3,6,14–16]. Solutions for shelled spherical, cylindrical (the 2D representation of a sphere) and ellipsoidal objects are readily available [17–22].

Within the dipole approximation, the time-averaged DEP force F_x on an ellipsoidal object is expressed by the real part of the scalar product of the induced dipole moment with the gradient of the complex conjugate field $\underline{E}^* = E_0 e^{-j\omega t}$. Using Equation (2), we get:

$$F_x = \frac{1}{2}\Re\left(\underline{m}\frac{d\underline{E}^*}{dx} \right) \approx \frac{1}{2}\Re\left(\underline{m}\frac{\Delta \underline{E}^*}{\Delta x} \right) = \frac{\Upsilon}{2}\Re(\underline{m}\underline{E}^*). \tag{9}$$

Moments induced by the weak inhomogeneity of the field are dominated by the dipole moment and neglected. Accordingly, the DEP force is generated by the interaction of the real part of the object's dipole moment with the inhomogeneous external field. Introducing Equation (4) into Equation (9) leads to the following well-known result:

$$F_x = \varepsilon_0 \varepsilon_e V_0 \Re\left(f_{CM}^{\Re} + j f_{CM}^{\Im}\right) \underline{E} \frac{\Upsilon}{2} \underline{E}^* = \varepsilon_0 \varepsilon_e V_0 f_{CM}^{\Re} \frac{\Upsilon}{2} E_0^2. \tag{10}$$

2.4. Charging Work for External and Suspension Media Boxes

It is assumed that cuboidal boxes of quadratic cross-section x^2 in the sheet plane and depths z perpendicular to the sheet plane are bounded by equipotential planes ("virtual electrodes") of area xz (Figure 1). The boxes of volume:

$$V_{box} = x^2 z, \tag{11}$$

are flooded with a homogeneous field. They can contain either suspension medium or a suspension with a single spherical or ellipsoidal object located in the center of the box. In the absence of the object (marked by 'e'), box i has the capacitance:

$$\underline{C}_i^e = \varepsilon_0 \underline{\varepsilon}_e \frac{xz}{x} = \varepsilon_0 \underline{\varepsilon}_e z, \tag{12}$$

where x in the denominator stands for the box width, i.e., the distance between the virtual electrodes. In the presence of the object the capacitance is:

$$\underline{C}_i^S = \varepsilon_0 \underline{\varepsilon}_S z. \tag{13}$$

The dielectric is a suspension (marked by capital 'S') with relative permittivity $\underline{\varepsilon}_S$. In the calculation of the electric work, effective (RMS) AC voltages or fields are usually employed to eliminate time averaging. For direct comparison with the dipole approximation, the effective values were substituted by field strength peak values:

$$E_0 = \sqrt{\underline{E}\underline{E}^*} = \sqrt{2} E_{eff}. \tag{14}$$

The electrical work to charge the box with the external medium is:

$$W_e = \frac{x^2 E_{eff}^2}{2} \varepsilon_0 \Re(\underline{\varepsilon}_e) = \frac{x^2 E_0^2}{4} \varepsilon_0 \varepsilon_e = \frac{V_{box} E_0^2}{4z} \varepsilon_0 \varepsilon_e, \tag{15}$$

and for the box with the suspension:

$$W_S = \frac{V_{box} E_0^2}{4z} \varepsilon_0 \Re(\underline{\varepsilon}_S) = \frac{V_{box} E_0^2}{4z} \varepsilon_0 \varepsilon_S^{\Re}, \tag{16}$$

where $\underline{\varepsilon}_S$ is obtained from Maxwell-Wagner's mixing equation for homogeneous ellipsoids [16,23]:

$$\frac{3(\underline{\varepsilon}_S - \underline{\varepsilon}_e)}{\underline{\varepsilon}_S + 2\underline{\varepsilon}_e} = \frac{p}{3} \sum_{k=a,b,c} \underline{f}_{CM}^k. \tag{17}$$

To reduce indexing, homogeneous spherical objects are considered with principal semi-axes of $a = b = c$. With Equations (7) and (8), we obtain:

$$\underline{\varepsilon}_S = \frac{3 + 2p\underline{f}_{CM}}{3 - p\underline{f}_{CM}} \underline{\varepsilon}_e = \frac{3 + 2p\underline{f}_{CM}}{3 - p\underline{f}_{CM}} \varepsilon_e - \frac{3 + 2p\underline{f}_{CM}}{3 - p\underline{f}_{CM}} \left(j \frac{\sigma_e}{\omega \varepsilon_0}\right). \tag{18}$$

Using Equation (18) for $p = 0$, Equation (16) is transformed into Equation (15). Note that the solution of Equation (17) is straightforward for ellipsoidal objects (Equation (6)). Moreover, such objects are always oriented with their longest axis in the field direction if their polarizability is different from that of the suspension medium [4]. The volume fraction of the single object (Figure 1) is:

$$p = \frac{V_0}{V_{box}}. \tag{19}$$

Note that Equation (18) considers the volume fraction but not the number of suspended objects.

2.5. DEP Force Approximation by a Capacitor-Charging Cycle

Consider the work performed to charge the two boxes i and $i + 1$ (Figure 1). Despite the inhomogeneous field, it is assumed that the two boxes i and $i + 1$ are flooded by the homogeneous fields $\underline{E}_i = \underline{E}$ and $\underline{E}_{i+1}$, respectively (Equation (1)). The fields are effective in the centers of the boxes at the considered locations of the object. In the first step of the DEP cycle, both boxes are discharged, box i in the presence of the object and box $i + 1$ in its absence. The conducted work is (Equations (1), (15) and (16)):

$$W_{dis} = -\frac{V_{box}E_0^2}{4z}\left(\Re\left(\underline{C}_i^S\right) + (1+\gamma\Delta x)^2\Re\left(\underline{C}_{i+1}^0\right)\right) = -\frac{\varepsilon_0 V_{box}E_0^2}{4}\left(\varepsilon_S^{\Re} + (1+\gamma\Delta x)^2\varepsilon_e\right). \tag{20}$$

In the second step, the object is transferred from box i to box $i + 1$, without conducting electrical work, before both boxes are recharged in the third step:

$$W_{chrg} = \frac{\varepsilon_0 V_{box}E_0^2}{4}\left(\varepsilon_e + (1+\gamma\Delta x)^2\varepsilon_S^{\Re}\right). \tag{21}$$

The overall work in the cycle is:

$$\Delta W = W_{chrg} + W_{dis} = \varepsilon_0 V_{box}\left(\varepsilon_S^{\Re} - \varepsilon_e\right)\frac{\gamma\Delta x(2+\gamma\Delta x)}{4}E_0^2. \tag{22}$$

The summand $\gamma\Delta x$ in the second parenthesis is neglected, because it is significantly smaller than two for weakly inhomogeneous fields. With Equation (19) we get:

$$\Delta W = \varepsilon_0\frac{V_0}{p}\left(\varepsilon_S^{\Re} - \varepsilon_e\right)\frac{\gamma\Delta x}{2}E_0^2. \tag{23}$$

In positive DEP (positive force), an object that is higher polarizable than the suspension medium travels the distance $\Delta x = x$ between the two box centers in the direction of the field gradient, from the low to the high field. In this case, the sign of the difference in the parenthesis is positive. It is also positive if a low polarizable object starts in the high-field box and travels against the direction of the field gradient in negative DEP. Both positive and negative DEP increase the overall capacitance of the two-box system, which requires increased charging work $(\Delta W > 0)$. The DEP-force is:

$$F_x = \frac{\Delta W}{\Delta x} = \varepsilon_0\frac{V_0}{p}\left(\varepsilon_S^{\Re} - \varepsilon_e\right)\frac{\gamma}{2}E_0^2. \tag{24}$$

Note that the object travels in a 'box-hopping mode'. A more gradual advance could be obtained by assuming a suspension box that is gradually shifted through a long cuboid volume of external medium. This would allow the mathematical transition from the difference quotient to a directional derivative, but would not change the DEP force expression obtained here. After introduction of parameter properties, Equations (23) and (24) can be used to calculate work and force without further simplifications.

However, the relation of Equation (24) to the classical force equation (Equation (10)) is not immediately clear. From considerations of the characteristic properties of suspensions with the mixing equation it is known that the mathematical boundary transition for infinitely small volume fractions $(p \to 0)$ leads to the solutions for single objects [2,24]. For simplicity, the volume fraction p is eliminated in two steps. In the first step, the summands stemming from $\underline{\varepsilon}_e$ in Equation (18) are considered:

$$\varepsilon_S^{\Re} = \Re\left(\frac{3+2p\underline{f}_{CM}}{3-p\underline{f}_{CM}}\underline{\varepsilon}_e\right) = \Re\left(\frac{3+2p\underline{f}_{CM}}{3-p\underline{f}_{CM}}\varepsilon_e\right) - \Re\left(\frac{3+2p\underline{f}_{CM}}{3-p\underline{f}_{CM}}\left(j\frac{\sigma_e}{\omega\varepsilon_0}\right)\right). \quad (25)$$

For $p \to 0$, the parenthesis term of the second, external-conductivity summand becomes purely imaginary and provides only a reactive, but no active contribution, i.e., no effective DEP force (see below). Neglecting the second summand, we get:

$$\varepsilon_S^{\Re} = \Re\left(\frac{3+2p\underline{f}_{CM}}{3-p\underline{f}_{CM}}\right)\varepsilon_e. \quad (26)$$

And after introduction into Equation (24):

$$F_x = \varepsilon_0\varepsilon_e\frac{V_0}{p}\Re\left(\frac{3+2p\underline{f}_{CM}}{3-p\underline{f}_{CM}} - 1\right)\frac{\gamma}{2}E_0^2. \quad (27)$$

By expanding $\underline{f}_{CM}$ with Equations (7) and (8), p is eliminated by the boundary transition $p \to 0$:

$$F_x = 3\varepsilon_0\varepsilon_e V_0\frac{\sigma_i^2+\sigma_e\sigma_i-2\sigma_e^2+\omega^2\varepsilon_0^2\left(\varepsilon_i^2+\varepsilon_e\varepsilon_i-2\varepsilon_e^2\right)}{\sigma_i^2+4\sigma_e\sigma_i+4\sigma_e^2+\omega^2\varepsilon_0^2\left(\varepsilon_i^2+4\varepsilon_e\varepsilon_i+4\varepsilon_e^2\right)}\frac{\gamma}{2}E_0^2 = \varepsilon_0\varepsilon_e V_0 f_{CM}^{\Re}\frac{\gamma}{2}E_0^2. \quad (28)$$

Equation (28) is identical to Equation (10), i.e., the real part of the Clausius-Mossotti factor for spherical homogeneous objects is:

$$f_{CM}^{\Re} = 3\frac{\sigma_i^2+\sigma_e\sigma_i-2\sigma_e^2+\omega^2\varepsilon_0^2\left(\varepsilon_i^2+\varepsilon_e\varepsilon_i-2\varepsilon_e^2\right)}{\sigma_i^2+4\sigma_e\sigma_i+4\sigma_e^2+\omega^2\varepsilon_0^2\left(\varepsilon_i^2+4\varepsilon_e\varepsilon_i+4\varepsilon_e^2\right)}. \quad (29)$$

The limiting frequency cases of Equation (29) are known as DEP plateaus (Figure 2):

$$\underset{\omega\to 0}{f_{CM}^{\Re}} = 3\frac{\sigma_i-\sigma_e}{\sigma_i+2\sigma_e} \text{ and } \underset{\omega\to\infty}{f_{CM}^{\Re}} = 3\frac{\varepsilon_i-\varepsilon_e}{\varepsilon_i+2\varepsilon_e}. \quad (30)$$

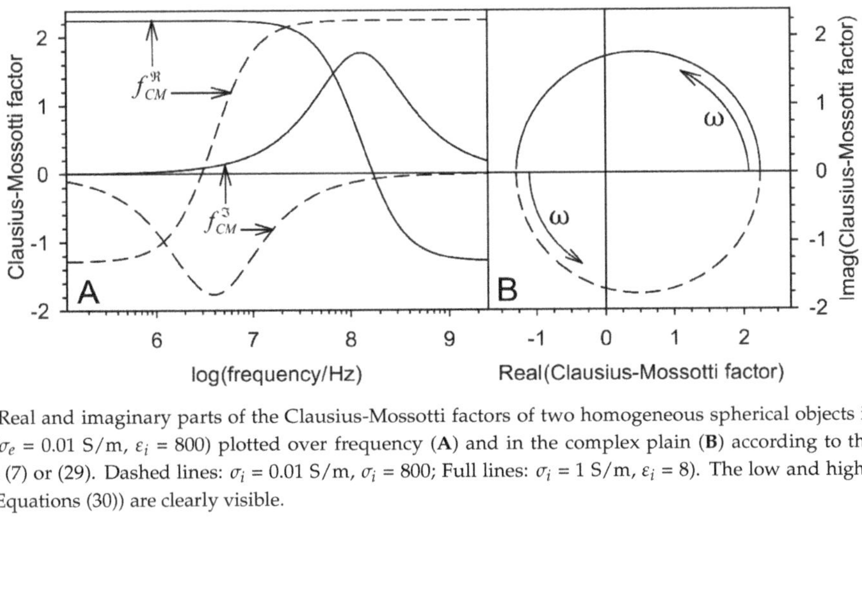

Figure 2. Real and imaginary parts of the Clausius-Mossotti factors of two homogeneous spherical objects in aqueous medium (σ_e = 0.01 S/m, ε_i = 800) plotted over frequency (**A**) and in the complex plain (**B**) according to the identical Equations (7) or (29). Dashed lines: σ_i = 0.01 S/m, σ_i = 800; Full lines: σ_i = 1 S/m, ε_i = 8). The low and high frequency plateaus (Equations (30)) are clearly visible.

2.6. Electrorotation (ROT) Torque

In DEP and ROT, the induced dipole moment (Equation (4)) interacts with a linear inhomogeneous field and with a rotating (circularly polarized) field, respectively [6]. As the frequency-dependent part of the dipole moment, the Clausius-Mossotti factor (Equations (6) and (7)) reflects the DEP force and the ROT torque, which are proportional to the real and imaginary parts of the Clausius-Mossotti factor, respectively.

Even though the ROT torque can also be derived by an appropriate capacitor-charging approach, this is beyond the scope of the present manuscript. Instead, the known relations of the ROT torque to the imaginary part of the Clausius-Mossotti factor are used directly [3,5,6]. In analogy to Equation (10), the ROT torque reads:

$$N_z = \varepsilon_0 \varepsilon_e V_0 f_{CM}^{\Im} E_0^2 k. \tag{31}$$

Torque is induced around axis z, which is oriented in the direction of unit vector k, perpendicular to the plane of rotation of a circular polarized field. Equation (27) is transformed to

$$N_z = \varepsilon_0 \varepsilon_e \frac{V_0}{p} \Im\left(\frac{3 + 2p\underline{f}_{CM}}{3 - p\underline{f}_{CM}}\right) E_0^2 k. \tag{32}$$

Note that the field gradient $\frac{\gamma}{2}$ (unit: m^{-1}) has been dropped from Equations (31) and (32), so that the units of force in Equations (10) and (27) are changing to units of torque.

3. Modelling Results and Discussion

3.1. Model Parameters

Spherical objects with a radius of $a = b = c = 10$ µm (Equation (5)) suspended in cubic boxes (side lengths x = 40 µm) were used for the 3D calculations (Figure 1). These dimensions correspond to a volume fraction of p = 0.0654 (Equation (19)), safely below the 0.1-limit required by Maxwell-Wagner's mixing equation [15]. Here, the lower value can compensate for the different shapes of boxes and objects. Aqueous electrolyte properties with conductivity and relative permittivity of σ_e = 0.1 S/m and ε_e = 80, respectively, were assumed for the external medium. Two complementary parameter settings were chosen for the properties of the homogeneous objects, combining high dielectric contrast at low and high frequencies with strong dispersion:

I σ_i = 0.01 S/m, with ε_i = 800, and
II σ_i = 1 S/m with ε_i = 8.

These parameters were chosen to reflect two strong dispersions that qualitatively correspond to the membrane polarization dispersion (i) and bulk conductance dispersion (ii) of biological cells [3,24] or are found in homogeneous objects [9,10]. Accordingly, each of the two objects sweeps two of the four quadrants of the complex plane swept by biological cells [1,5]. If not stated otherwise, work, forces and dissipation were calculated for normalized field strengths of $E_0 = 1$ V/m and $E_1/E_0 = 1$ (Equation (3)). Note that 10%-field increase per box width corresponds to a field inhomogeneity of $\gamma = 0.1/\Delta x = 0.1/40$ µm $= 2500$ m^{-1} (Equation (2)). For 10 consecutive boxes, the model field strength increases from 1 V/m in box 1 to 2.3579 V/m in box 10. Table 1 lists the values for 10 box centers and 9 box interfaces.

Table 1. Field values used in the calculations.

Box i	Normalized Field E_i/E_0	Normalized Field Squared E_i^2/E_0^2	Transition from Box $i \rightarrow i+1$	Squared Normalized Mean Field at Box Interfaces $\left(\frac{E_{i+1}+E_i}{2E_0}\right)^2$
1	1	1		
2	1.1	1.21	$1 \rightarrow 2$	1.1025
3	1.21	1.4641	$2 \rightarrow 3$	1.3340
4	1.331	1.7716	$3 \rightarrow 4$	1.6142
5	1.4641	2.1436	$4 \rightarrow 5$	1.9531
6	1.6105	2.5937	$5 \rightarrow 6$	2.3633
7	1.7716	3.1384	$6 \rightarrow 7$	2.8596
8	1.9487	3.7975	$7 \rightarrow 8$	3.4601
9	2.1436	4.5950	$8 \rightarrow 9$	4.1867
10	2.3579	5.5599	$9 \rightarrow 10$	5.0660

3.2. *Clausius-Mossotti Factor*

The real and imaginary parts of the Clausius-Mossotti factor represent the normalized DEP force and ROT torque, which are usually plotted over frequency (Figure 2A) or in the complex plain (Figure 2B).

The spectra calculated from Equations (27) and (32) before the boundary transition $p \rightarrow 0$ deviate slightly from the classical model. Presumably, they are more precise in a certain volume fraction range, but this needs to be investigated in more detail. For higher volume fractions, the accuracy is limited by the upper limit of $p < 0.1$ of the mixing equation. For lower volume fractions, it increasingly corresponds to the classical model. Figure 3 shows the relationships for DEP force and ROT torque spectra. Reference spectra were calculated from the spectra of Figure 2 with appropriate prefactors. The visible differences from the reference spectra largely disappear for a box size of $x = 80$ µm, corresponding to $p = 0.0082$.

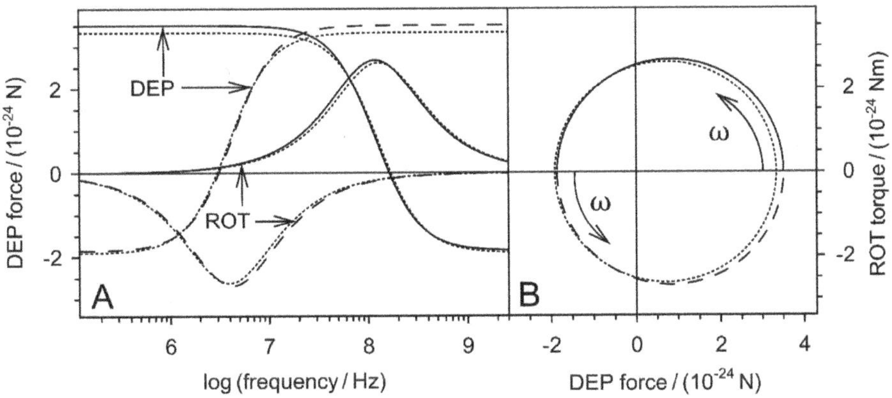

Figure 3. Frequency (**A**) and complex plots (**B**) of DEP-force and ROT-torque spectra of the two spheres of Figure 2 according to Equations (27) and (32) compared with the classical model (dotted lines; Equations (4), (7), (28) and (31)). All spectra were calculated for field strengths of 1 V/m. To obtain corresponding numerical values of forces and moments, the DEP forces were calculated for $\gamma = 1$ m^{-1}.

3.3. Conductivity and Dissipation

The volume-specific field absorption or Rayleigh's dissipation in each box is described by ohmic heating in the absence:

$$P_V = \Re(\underline{\sigma}_e)E_{eff}^2 = \sigma_e \frac{E_0^2}{2}, \tag{33}$$

or presence of the object:

$$P_V = \Re(\underline{\sigma}_S)\frac{E_0^2}{2} = \sigma_S^{\Re}\frac{E_0^2}{2}. \tag{34}$$

The expression for the complex conductivity of a suspension of spherical objects was first derived by Wagner [16]:

$$\underline{\sigma}_S = \frac{3+2p\underline{f}_{CM}}{3-p\underline{f}_{CM}}\underline{\sigma}_e = \frac{3+2p\underline{f}_{CM}}{3-p\underline{f}_{CM}}\sigma_e + \frac{3+2p\underline{f}_{CM}}{3-p\underline{f}_{CM}}j\omega\varepsilon_0\varepsilon_e. \tag{35}$$

This is the conductivity version of Equation (18). Applying the same arguments as to Equation (25), we obtain:

$$\sigma_S^{\Re} = \Re\left(\frac{3+2p\underline{f}_{CM}}{3-p\underline{f}_{CM}}\right)\sigma_e. \tag{36}$$

The DC case (and low frequency limit) of this equation was originally given by Maxwell [15] in his treatise, chapter 9, "Conduction through heterogeneous media":

$$\underset{\omega\to 0}{\underline{\sigma}_S} = \underset{\omega\to 0}{\sigma_S^{\Re}} = \sigma_S = \frac{\sigma_i + 2\sigma_e + 2p(\sigma_i - \sigma_e)}{\sigma_i + 2\sigma_e - p(\sigma_i - \sigma_e)}\sigma_e. \tag{37}$$

Here, it is obtained for $(\omega \to 0)$ or after introducing the DC limit of the Clausius-Mossotti factor (Equation (30)) into Equation (35).

Figure 4 illustrates Equations (26) and (36), i.e., the real parts of the relative permittivity and conductivity in the box with the suspended object. Comparison of the two equations with Equation (27) shows different frequency-independent prefactors and an additional offset. If this is considered, the right ordinate can be rescaled so that the DEP force (Equation (27)) is represented by the same function plots.

Note that the plotted real part of the conductivity can be interpreted as field-normalized volume-specific dissipation (Equation (34)). Obviously, the presence of the object changes the conductivity of the box medium in such a way that the frequency-dependent absorbance of a certain part of the field energy corresponds to the frequency-dependent work conducted in DEP. The perfect correspondence of the curves indicates a physical background.

3.4. Dispersion Relation, Active, Reactive, and Apparent Power

A common representation in impedance research shows the steady decrease of the permittivity of a suspension between frequency plateaus in the so-called dispersion frequency ranges [2,22]. Each decrease in permittivity corresponds to a complementary increase in conductivity. Here, these relationships are described by Equations (18) and (35). The equations can transform into one another using:

$$\underline{\sigma}_S = j\omega\varepsilon_0\underline{\varepsilon}_S. \tag{38}$$

Figure 5 illustrates the relationships of the components of the equations for the permittivity and conductivity for suspension of the two model spheres. The apparent permittivity (Equation (18)) and apparent conductivity (Equation (35)) are the sums of their active and reactive components, which are the first and second summands of each of the two equations.

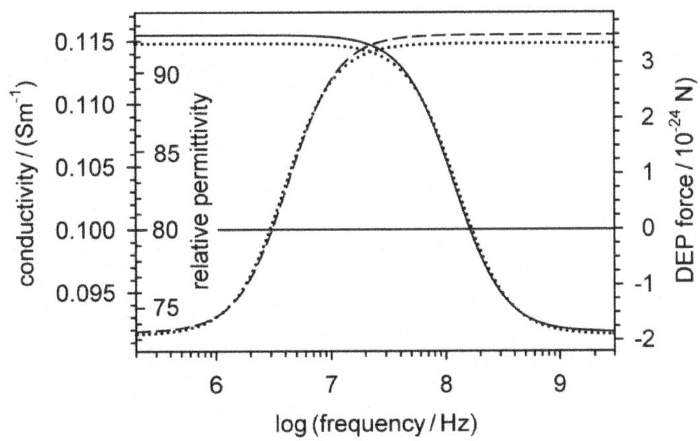

Figure 4. Relationships of the real parts of the relative permittivity and conductivity (Equations (26) and (36)) of the suspension containing two different homogeneous spheres (Dashed lines: σ_i = 0.01 S/m, ε_i = 800; solid lines: σ_i = 1 S/m, ε_i = 8) to the induced DEP force (Equation (27)). The DEP forces are identical to those in Figure 3 when using the relative permittivity and conductivity of the suspension medium (80, 0.1 S/m) as a reference for zero DEP force. The dotted lines show the classical DEP force (Equation (10)).

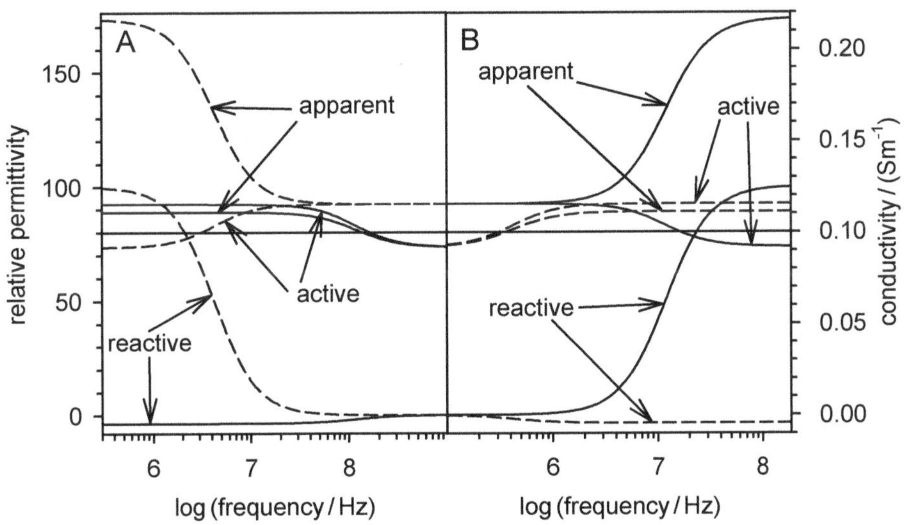

Figure 5. Component plots of the real parts of the relative permittivity ((**A**); Equation (18)) and conductivity ((**B**); Equation (35)) of two suspensions containing a single homogeneous sphere (Dashed lines: σ_i = 0.01 S/m, ε_i = 800; solid lines: σ_i = 1 S/m, ε_i = 8). The apparent permittivity and conductivity are the sums of their active and reactive components.

The active components are proportional to the DEP force (Figure 4), suggesting that the reactive component performs no DEP work, but is capacitively stored at the interface of the object out-of-phase with the active component (cf. ROT) and dissipated in the suspension medium, similarly to the reactive power in the peripheral wiring of electrical machines.

3.5. Dissipation and Charging Work in the Box Chain

Figure 6 shows the effect of DEP translations on the work of charging and energy dissipation in a chain system of 10 cubic boxes subjected to a constant field gradient according to Equation (1). The field-normalized dissipations in each box were calculated using Equation (33) for the external medium conductivity of 0.1 S/m and the field- normalized values squared of Table 1. Without the object, the dissipation in the first box is 0.1 W/m^3, which corresponds to the external conductivity of 0.1 S/m at a field strength of 1 V/m. Without the object, the dissipation along the box chain increases with the square of the field strength (Table 1; Figure 6A, left ordinate). The charge work shows the same field strength dependence (Equations (15) and (16)). It is shown on the right ordinate as "apparent relative permittivity". This parameter assigns an apparent permittivity to each box, so that charging each box with 1 V/m requires the same work as charging the box with the actual field strength in the inhomogeneous field (Table 1). Accordingly, the "apparent relative permittivity" in the first box is 80, the actual permittivity of the suspension medium (cf. left ordinate in Figure 4). For each box, the dissipation and apparent permittivity are plotted for three cases, in the absence and presence of the low (0.01 S/m) or high (1 S/m) conductive sphere. In the calculations, the low frequency limit (Equation (30)) of Equation (36) was used as an example. In the presence of the spheres, the dissipation decreases or increases as indicated for box 6. In DEP, the spheres travel through the box chain against (negative DEP) or in (positive DEP) the direction of the field gradient, depending on their conductivity, i.e., their polarizability relative to that of the suspension medium. The dissipation differences induced by the presence of the spheres increase in the direction of the field gradient, resulting in decreasing and increasing force magnitudes along the DEP trajectories for negative and positive DEP, respectively.

Figure 6B illustrates the DEP effects on the entire box chain considered as a DEP system. Without the object, the field-normalized total dissipation in the chain corresponds to the sum of the 10 boxes (full horizontal line). In the presence of the weakly or strongly conducting sphere, one box contributes to the total dissipation according to Equation (34). The mean chain dissipations (dashed horizontal lines) were calculated assuming a probability 0.1 uniform distribution for the presence of the spheres in each of the boxes. The curves show the dissipation for successive positions of the spheres during DEP translation. In both cases, positive and negative DEP, the field-induced translation leads to an increase in the total dissipation and effective relative polarizability of the DEP system. The latter was calculated by normalizing the sum of the apparent relative permittivities for each sphere position (Figure 6A) to the mean of all positions. The effective relative polarizabilities of one (right ordinate) correspond to the mean dissipations in the entire chain for uniformly distributed (starting) positions of the spheres (dashed horizontal lines). The effective relative polarizabilities of the chain can be used to describe how DEP "conditions" the overall polarizability of the chain system.

In an experimental situation, starting from the random distribution (one), it will increase and rise to a maximum reached when the weakly or strongly conducting (polarizable) object reaches the available positions with the highest or lowest field, respectively. Note that this explanation may be insufficient in real DEP systems, since the presence of the object changes the field distribution, especially near the electrodes, for example by mirror charges. It should also be noted that each field-induced step in or against the direction of the field gradient is directly coupled to the proportional increase or decrease of the active, reactive, and apparent (complex) components plotted in Figure 5.

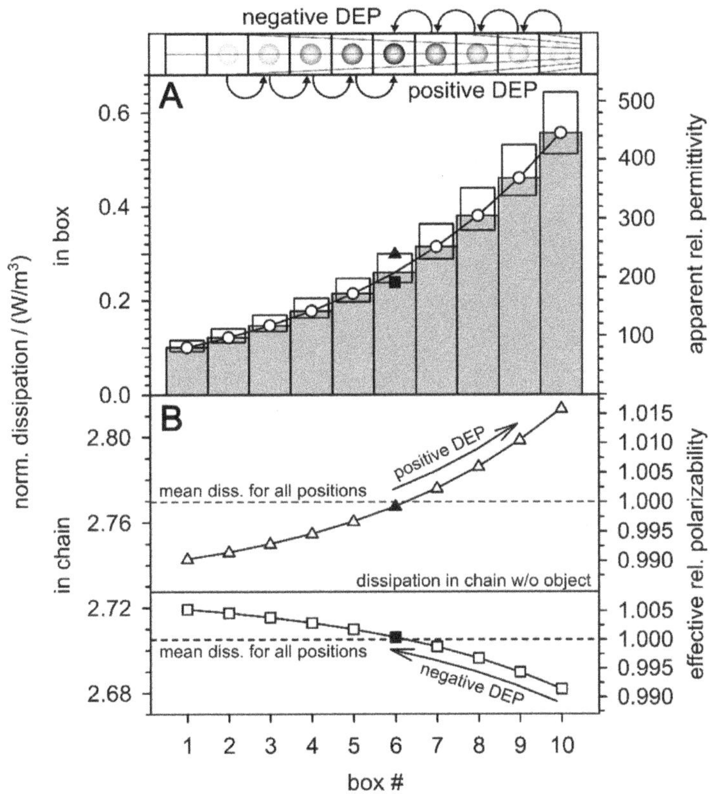

Figure 6. Illustration of DEP-induced changes in field-normalized dissipation in a box chain subjected to an inhomogeneous low-frequency field according to the values of Table 1. (**A**): In the absence of the object, the dissipation in the direction of the field gradient increases with the square of the field strength (circles, gray columns). In the presence of a high (triangles, σ_i = 1 S/m) or low polarizable sphere (squares, σ_i = 0.01 S/m), the active dissipation in the box is increased or decreased, respectively. The work of charging has the same field strength dependence (Equations (15) and (16)). It is plotted as "apparent relative permittivity" above the right ordinate (see text). (**B**): Dependence of the sum of dissipation in the box chain system on the positions of the single objects. Arrows denote DEP "trajectories". The dissipations in the chain system (dashed horizontal lines) correspond to effective relative polarizabilities (right ordinate), which are proportional to the charge work of the whole chain system. Effective relative polarizabilities of one correspond to the average dissipation throughout the chain, for an even distribution of the 10 starting positions for the model spheres (dashed horizontal lines).

3.6. DEP Force in the Box Chain

The original intention of this manuscript was to derive the DEP force from an alternative approach. The new approach actually provided the classical DEP force expression (Equation (28)) after applying the mathematical boundary transition $p \to 0$ to the new force expression Equation (27), which still contains the volume fraction of the object under consideration. Figure 7 compares the two equations. The DEP forces were calculated for the two model spheres at the nine interfaces between the boxes using the DC limit (Equation (30)) of Equation (27).

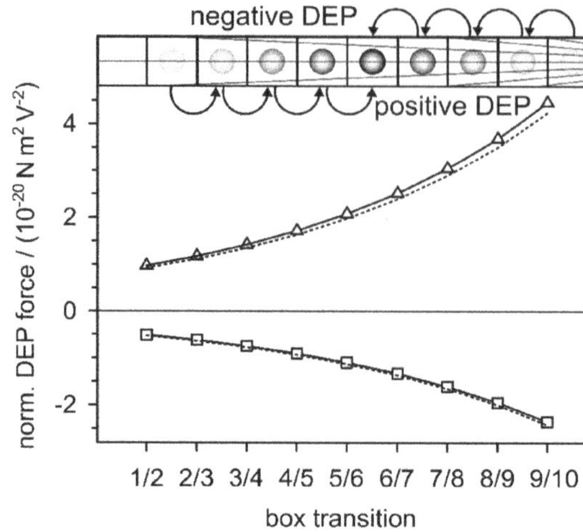

Figure 7. Low frequency plateaus of the DEP forces calculated with Equation (30) for the model spheres with σ_i = 0.01 S/m (squares) and σ_i = 1 S/m (triangles) compared with the classical model of Equation (10) (dotted lines). A field inhomogeneity of $\gamma = 2500\ \text{m}^{-1}$ and normalized field strengths at the box interfaces were assumed (Figure 1, Table 1).

4. General Discussion

4.1. Higher Precision for the DEP Force?

The derivation of Equation (28) and the exact agreement with Equation (10) shed new light on the physical background of the required simplifications and the interrelationships of suspension impedance and AC-electrokinetic effects. The new derivation connects the DEP force and DEP translation directly to the changes in dielectric properties in the system. The new DEP force expression (Equation (27)) is based on the frequency dependencies of the active components of the real part of the permittivity (Equation (26)) or conductivity (Equation (36)) of the suspension of a single object. After the $p \to 0$ boundary transition, the three equations are identical to the classical DEP force (Equations (10) and (28)) when the appropriate prefactors and offsets are considered. Figure 4 shows slight deviations from the classical model due to additional terms, which are eliminated by the boundary transition. This step cancels possible volume-related polarization properties and reduces Equation (27) to the pure dipole effects. It may be worthwhile to consider the lost terms for a more accurate description of the DEP force in narrow environments [9–12]. In the classical electro-quasistatic derivation, it is assumed that the field gradient is undisturbed by the presence of the small objects considered. However, for larger objects, i.e., higher volume fractions, the field distributions with and without objects must diverge. This calls for improvements to the model, also to describe the interaction between objects of similar size [7] or of objects at electrode surfaces with mirror charges.

For testing purposes, it might be worthwhile to compare the new model with multipole models [7]. In the future, it might also be interesting to describe the suspension properties in the charge cycle with mixing equations specifically designed for higher volume fractions, such as Hanai's equation [25,26].

4.2. DEP as Conditioned Polarization Process

DEP dissipates field energy. Figure 6 shows how DEP increases the active power dissipation and polarization in the system, regardless of whether the effective conductivity or polarizability of the object is higher or lower than that of the suspension medium. A similar increase has already been described for the electro-orientation of ellipsoidal objects in homogeneous fields (cf. Appendix B in [4]). In electro-orientation, it is the field-induced orientation of the longest axis of homogeneous objects that maximizes the power dissipation in the system, regardless of whether the objects or the medium are higher polarizable. In DEP and electro-orientation, the higher polarizability of the suspension results from field-induced rearrangements of the suspended objects. Relative to the oscillations of the causative field, these processes can be extremely slow. DEP and electro-orientation are not typical polarization processes where the polarization follows the oscillation of the polarizing field. It is therefore proposed that the DEP and possibly AC electrokinetic processes, such as electro-orientation, electro-deformation or field-induced aggregation as "conditioned polarization processes" be considered.

For an explanation, the positive DEP branch of Figure 6B should be expanded in a Gedankenexperiment. The behavior of the many objects in a real suspension system needs to be reproduced by a large number of box chains in a gradient field at constant electrode voltage. In the inhomogeneous field, all the chains should be aligned along DEP trajectories. Before field-on, a single object in each chain is at a random location. After field-on, all the objects move with an increasing velocity in the direction of the field gradient until each object reaches its final position at locations with the highest field, for example, the electrode surface. This state corresponds to the highest conditioned polarizability of the system. The modeling of such processes is possible taking the Stokes friction into account, but they are beyond the scope of this manuscript. Experimentally, the time course of the increase in polarizability should be detectable in microfluidic systems [11].

4.3. Relations to the Law of Maximum Entropy Production (LMEP)

It has been shown how DEP synchronously increases the total polarizability and dissipation in the suspension system. It is a physical principle in linear systems that forces act along the (energy) field gradient. Despite the quadratic dependence of the DEP force on the field, for the DEP as well, translation along the field gradient is assumed, which is the fastest way to increase the total polarizability and dissipation in the system in agreement with the LMEP. The LMEP states that a system will select the path or assemblage of paths out of the available paths that minimize the potential or maximize the entropy at the fastest rate [27]. In other words, LMEP demands the maximization of entropy production [28–30].

At constant temperature T, the entropy production dS/dt in a box is proportional to dissipation [31] according to Equations (33) and (34) in the absence or presence of the object, respectively. The equations are the electrical version of Rayleigh's dissipation function:

$$\Phi = P_V V_{box} = \frac{dS}{dt} T, \tag{39}$$

which can be expressed in general terms by products of fluxes and their inducing forces [32,33].

In the experimental situation, a dynamic equilibrium with the thermal forces is established after the DEP translation and rearrangement processes of the objects under the influence of the field are completed. In this equilibrium, the electrical energy dissipation should approach a maximum. It is very likely that the DEP systems allow the study of the balance between the stability of the field-induced structures and the entropy production necessary to keep the structures stable [29]. LMEP could provide a thermodynamic explanation for DEP-induced accumulation of viruses [34] and proteins [35], overcoming the dispersive forces associated with Brownian motion and osmotic segregation, which cannot be explained by current DEP theory [36]. Interestingly, the new force expression shows higher values for positive DEP than the classical dipole approach, a tendency that may be enhanced when mixing equations for higher volume fractions are used [25,26].

However, in micro-chamber experiments with a negative DEP, we observed that the expected DEP end positions were not reached because the DEP forces became too weak to overcome sedimentation, surface friction, and subsequent adhesion for increasing distances to the electrodes.

5. Conclusions and Outlook

The DEP force was derived from the electrical charge work of a single-object suspension. Its permittivity has been described by the Maxwell-Wagner mixing equation, which contains the volume fraction but not the number of objects that show the ponderomotive character, i.e., the volume character of the DEP force. The derivation presented can be easily extended to 2D systems and to multishell spheroidal, cylindrical, and ellipsoidal objects with known expressions for their Clausius-Mossotti factors. Along the gradient of an inhomogeneous field, the DEP force was calculated from the differences of the charge work of the system for successive positions of the object. Interestingly, both the increase in capacitive charge work and active energy dissipation provides the frequency dependence of the DEP force when appropriate prefactors and offsets are used. The considerations showed that DEP can be considered as a conditioned polarization process that increases the polarizability, i.e., the total effective permittivity of the DEP system at an extremely low frequency.

The higher polarizability is associated with higher currents, while higher energy dissipation is in accordance with the LMEP. This suggests that DEP movement follows trajectories that increase the polarizability and dissipation at the fastest rate. The application of these principles can simplify the numerical prediction of field-induced forces and object trajectories by simplifying the calculation of the position-dependence of the total conductivity in DEP systems. Moreover, the approach is expected to work not only for modeling DEP trajectories, but also for orientation and aggregation in multi-object systems and for field-induced behavior of objects with irregular shape or internal structures. However, AC electrokinetic forces are induced not only on the suspended objects but also on the suspension media. The superposition of the induced fluid currents on the DEP motion can complicate the prediction of object trajectories in real systems.

Finally, during the completion process of the submission, the author learned of the work of Zheng and Palffy-Muhoray [37], who consider the physical concepts describing electrical energy storage in dissipative materials based on atomic and molecular properties. These authors describe the contradictions between the macroscopic and microscopic approaches and state that the difficulty lies in the partitioning of input power into two distinct components—the dissipation rate and the rate of change of stored energy. The author would like to point to the strong analogy of this description with the contributions of the apparent, active and reactive power components to the DEP presented in this manuscript. Note that the reactive components contributing to the power dissipation can be formally negative.

Supplementary Materials: The following are available online at https://www.mdpi.com/article/10.3390/mi12070738/s1, pdf-file: Derivations on active, reactive and apparent power in dielectrophoresis: Force corrections from the capacitive charging work on suspensions described by Maxwell-Wagner's mixing equation. The Maple code will be provided upon request.

Funding: This research received no external funding.

Acknowledgments: I would like to thank Jessica Schröder for her skillful help with the figures and the reference list. Michal M. Radai (Ra'anana) and Philipp Wysotzki are acknowledged for stimulating discussions. I am grateful to the late Edwin Donath for the discussion on Max Born's ionic cycle at the membrane interface many years ago that inspired this manuscript.

Conflicts of Interest: The author declares no conflict of interest.

References

1. Gimsa, J.; Wachner, D. A Unified Resistor-Capacitor Model for Impedance, Dielectrophoresis, Electrorotation, and Induced Transmembrane Potential. *Biophys. J.* **1998**, *75*, 1107–1116. [CrossRef]
2. Foster, K.R.; Schwan, H.P. Dielectric Properties of Tissues. In *Handbook of Biological Effects of Electromagnetic Fields*, 2nd ed.; Polk, C., Pastow, E., Eds.; CRC Press: Boca Raton, FL, USA, 1996; pp. 25–102.
3. Jones, T.B. *Electromechanics of Particles;* Cambridge University Press: Cambridge, UK; New York, NY, USA, 1995; ISBN 9780521431965.

4. Gimsa, J. Can the law of maximum entropy production describe the field-induced orientation of ellipsoids of rotation? *J. Phys. Commun.* **2020**, *4*, 085017. [CrossRef]
5. Fuhr, G.R.; Gimsa, J.; Glaser, R. Interpretation of electrorotation of protoplasts. 1. Theoretical considerations. *Studia Biophys.* **1985**, *108*, 149–164.
6. Gimsa, J. A comprehensive approach to electro-orientation, electrodeformation, dielectrophoresis, and electrorotation of ellipsoidal particles and biological cells. B1101. *Bioelectrochemistry* **2001**, *54*, 23–31. [CrossRef]
7. Huang, J.P.; Karttunen, M.; Yu, K.W.; Dong, L.; Gu, G.Q. Electrokinetic behavior of two touching inhomogeneous biological cells and colloidal particles: Effects of multipolar interactions. *Phys. Rev. E* **2004**, *69*, 51402. [CrossRef]
8. Barat, D.; Spencer, D.; Benazzi, G.; Mowlem, M.C.; Morgan, H. Simultaneous high speed optical and impedance analysis of single particles with a microfluidic cytometer. *Lab. Chip* **2012**, *12*, 118–126. [CrossRef]
9. Chen, Q.; Yuan, Y.J. A review of polystyrene bead manipulation by dielectrophoresis. *RSC Adv.* **2019**, *9*, 4963–4981. [CrossRef]
10. Jiang, T.; Jia, Y.; Sun, H.; Deng, X.; Tang, D.; Ren, Y. Dielectrophoresis Response of Water-in-Oil-in-Water Double Emulsion Droplets with Singular or Dual Cores. *Micromachines* **2020**, *11*, 1121. [CrossRef] [PubMed]
11. Morgan, H.; Holmes, D.; Green, N.G. High speed simultaneous single particle impedance and fluorescence analysis on a chip. *Curr. Appl. Phys.* **2006**, *6*, 367–370. [CrossRef]
12. Ramos, A.; Morgan, H.; Green, N.G.; Castellanos, A. AC electrokinetics: A review of forces in microelectrode structures. *J. Phys. D Appl. Phys.* **1998**, *31*, 2338–2353. [CrossRef]
13. Landau, L.D.; Lifšic, E.M.; Pitaevskij, L.P. *Electrodynamics of Continuous Media*, 2nd ed.; Pergamon Press: Oxford, UK, 1984; ISBN 0080302750.
14. Gimsa, J.; Wachner, D. A Polarization Model Overcoming the Geometric Restrictions of the Laplace Solution for Spheroidal Cells: Obtaining New Equations for Field-Induced Forces and Transmembrane Potential. B1115. *Biophys. J.* **1999**, *77*, 1316–1326. [CrossRef]
15. Maxwell, J.C. *A Treatise on Electricity and Magnetism*; Clarendon Press: Oxford, UK, 1873; ISBN 0198503733.
16. Wagner, K.W. Erklärung der dielektrischen Nachwirkungsvorgänge auf Grund Maxwellscher Vorstellungen. *Archiv. F. Elektrotechnik* **1914**, *2*, 371–387. [CrossRef]
17. Pastushenko, V.P.; Kuzmin, P.I.; Chizmadshev, Y.A. Dielectrophoresis and electrorotation: A unified theory of spherically symmetrical cells. *Studia Biophys.* **1985**, *110*, 51–57.
18. Asami, K.; Hanai, T.; Koizumi, N. Dielectric Approach to Suspensions of Ellipsoidal Particles Covered with a Shell in Particular Reference to Biological Cells. *Jpn. J. Appl. Phys.* **1980**, *19*, 359–365. [CrossRef]
19. Bohren, C.F.; Huffman, D.R. *Absorption and Scattering of Light by Small Particles*; Wiley: New York, NY, USA, 1998; ISBN 9780471293408.
20. Kakutani, T.; Shibatani, S.; Sugai, M. Electrorotation of non-spherical cells: Theory for ellipsoidal cells with an arbitrary number of shells. *Bioelectrochem. Bioenerg.* **1993**, *31*, 131–145. [CrossRef]
21. Sokirko, A.V. The electrorotation of axisymmetrical cell. *Biol. Mem.* **1992**, *6*, 587–600.
22. Stubbe, M.; Gimsa, J. Maxwell's Mixing Equation Revisited: Characteristic Impedance Equations for Ellipsoidal Cells. *Biophys. J.* **2015**, *109*, 194–208. [CrossRef]
23. Gimsa, J.; Gimsa, U. The influence of insulating and conductive ellipsoidal objects on the impedance and permittivity of media. *J. Electrostat.* **2017**, *90*, 131–138. [CrossRef]
24. Gimsa, J.; Stubbe, M.; Gimsa, U. A short tutorial contribution to impedance and AC-electrokinetic characterization and manipulation of cells and media: Are electric methods more versatile than acoustic and laser methods? *J. Electr. Bioimpedance* **2014**, *5*, 74–91. [CrossRef]
25. Hanai, T. Electrical properties of emulsions. In *Emulsion Science*; Sherman, P., Ed.; Academic Press: London, UK; New York, NY, USA, 1968; pp. 354–477.
26. Tuncer, E.; Gubański, S.M.; Nettelblad, B. Dielectric relaxation in dielectric mixtures: Application of the finite element method and its comparison with dielectric mixture formulas. *J. Appl. Phys.* **2001**, *89*, 8092–8100. [CrossRef]
27. Swenson, R. The fourth law of thermodynamics or the law of maximum entropy production (LMEP). *Chemistry* **2009**, *18*, 333–339.
28. Stenholm, S. On entropy production. *Ann. Phys.* **2008**, *323*, 2892–2904. [CrossRef]
29. Niven, R.K. Steady state of a dissipative flow-controlled system and the maximum entropy production principle. *Phys. Rev. E* **2009**, *80*, 21113. [CrossRef] [PubMed]
30. Grinstein, G.; Linsker, R. Comments on a derivation and application of the 'maximum entropy production' principle. *J. Phys. A Math. Theor.* **2007**, *40*, 9717–9720. [CrossRef]
31. Atkins, P.W. *Physical Chemistry*, 5th ed.; Oxford University Press: Oxford, UK, 1994.
32. Glaser, R. *Biophysics: An Introduction*, 2nd ed.; Springer: Berlin/Heidelberg, Germany, 2012; ISBN 9783642252129.
33. Scaife, B. On the Rayleigh dissipation function for dielectric media. *J. Mol. Liq.* **1989**, *43*, 101–107. [CrossRef]
34. Gimsa, J. New Light-Scattering and Field-Trapping Methods Access the Internal Electric Structure of Submicron Particles, like Influenza Viruses. In *Electrical Bioimpedance Methods: Applications to Medicine and Biotechnology*; Riu, P.J., Ed.; New York Academy of Sciences: New York, NY, USA, 1999; pp. 287–298. ISBN 1-57331-191-X.
35. Hölzel, R.; Pethig, R. Protein Dielectrophoresis: I. Status of Experiments and an Empirical Theory. *Micromachines* **2020**, *11*, 533. [CrossRef] [PubMed]

36. Green, N.G.; Ramos, A.; Morgan, H. AC electrokinetics: A survey of sub-micrometre particle dynamics. *J. Phys. D Appl. Phys.* **2000**, *33*, 632–641. [CrossRef]
37. Zheng, X.; Palffy-Muhoray, P. Electrical energy storage and dissipation in materials. *Phys. Lett. A* **2015**, *379*, 1853–1856. [CrossRef]

micromachines

Article

Dielectrophoresis from the System's Point of View: A Tale of Inhomogeneous Object Polarization, Mirror Charges, High Repelling and Snap-to-Surface Forces and Complex Trajectories Featuring Bifurcation Points and Watersheds

Jan Gimsa [1,*] and Michal M. Radai [2]

[1] Department of Biophysics, University of Rostock, Gertrudenstr. 11A, 18057 Rostock, Germany
[2] Independent Researcher, HaPrachim 19, Ra'anana 4339963, Israel; michal.radai@gmail.com
* Correspondence: jan.gimsa@uni-rostock.de; Tel.: +49-381-6020; Fax: +49-381-6022

Abstract: Microscopic objects change the apparent permittivity and conductivity of aqueous systems and thus their overall polarizability. In inhomogeneous fields, dielectrophoresis (DEP) increases the overall polarizability of the system by moving more highly polarizable objects or media to locations with a higher field. The DEP force is usually calculated from the object's point of view using the interaction of the object's induced dipole or multipole moments with the inducing field. Recently, we were able to derive the DEP force from the work required to charge suspension volumes with a single object moving in an inhomogeneous field. The capacitance of the volumes was described using Maxwell–Wagner's mixing equation. Here, we generalize this system's-point-of-view approach describing the overall polarizability of the whole DEP system as a function of the position of the object with a numerical "conductance field". As an example, we consider high- and low conductive 200 µm 2D spheres in a square 1 × 1 mm chamber with plain-versus-pointed electrode configuration. For given starting points, the trajectories of the sphere and the corresponding DEP forces were calculated from the conductance gradients. The model describes watersheds; saddle points; attractive and repulsive forces in front of the pointed electrode, increased by factors >600 compared to forces in the chamber volume where the classical dipole approach remains applicable; and DEP motions with and against the field gradient under "positive DEP" conditions. We believe that our approach can explain experimental findings such as the accumulation of viruses and proteins, where the dipole approach cannot account for sufficiently high holding forces to defeat Brownian motion.

Keywords: system's perspective; MatLab® model; microfluidics; DEP trajectory; LMEP; protein dielectrophoresis; virus trapping; LOC; µTAS; force spectroscopy

Citation: Gimsa, J.; Radai, M.M. Dielectrophoresis from the System's Point of View: A Tale of Inhomogeneous Object Polarization, Mirror Charges, High Repelling and Snap-to-Surface Forces and Complex Trajectories Featuring Bifurcation Points and Watersheds. *Micromachines* **2022**, *13*, 1002. https://doi.org/10.3390/mi13071002

Academic Editor: Rodrigo Martinez-Duarte

Received: 31 May 2022
Accepted: 24 June 2022
Published: 26 June 2022

1. Introduction

Analytically, dielectrophoresis (DEP) is usually modeled using the electroquasistatic dipole approach [1]. There are few descriptions with the free energy approach or Maxwell's stress tensor [2]. Almost every approach is from the object's point of view. Recently, we presented a new analytical model from the system's perspective. It is based on the capacitive charge work to suspend a single spherical object [3]. For DEP and the electro-orientation of ellipsoidal objects, our results have shown a steady increase in the overall polarizability of the suspension systems [4]. Even though this increase is slow with regard to the field oscillation, we propose considering electro-orientation and DEP as "conditioned polarization mechanisms." Moreover, our results suggest that the law of maximum entropy production (LMEP) [5–8] provides a powerful phenomenological criterion for AC–electrokinetic effects. Our derivations have shown the importance of distinguishing between active and reactive components in DEP [3]. In the case of suspensions, the reactive components of the impedance result in extraordinarily high permittivities and conductivities at low and high frequencies, respectively. We suspect that this stealth effect has prevented—for over a century—any discussion of AC–electrokinetic forces in terms of the electrical work performed on suspensions.

In the dipole approach, the object is assumed to be small compared to the characteristic length of the field inhomogeneity. This allows the assumption of a homogeneous effective polarization described by the induced dipole moment with two equal dipole charges. Their interaction with the slightly inhomogeneous external field produces unequal forces at the two poles of the object, leading to DEP. The dependencies of the dipole moment on the field frequency and media parameters are summarized in the Clausius–Mossotti factor (CMF) [1,9]. Objects that are more and less polarizable than the suspension medium are assumed to move in (positive DEP) and against (negative DEP) the field gradient direction, respectively. This view is mainly correct for small objects. Objects of small size can "sense" the field gradient very locally and with negligible distortion of the external field. Consequently, their DEP trajectories "track" the steepest field gradient at each point.

However, in microchambers, complicated field distribution and inhomogeneous object polarization is typical, because the objects are relatively large with respect to the chamber [10–15]. The simple CMF description becomes problematic because the total force results from the superposition of polarization contributions from the entire volume of the inhomogeneously polarized object with the inhomogeneous field [16,17]. Examples include individual objects inducing mirror charges at the electrode surface or the attraction of two adjacent objects of the same size, where each object is subject to the field resulting from the inhomogeneous polarization of the respective other object. Analytically, these relationships are described by multipole models [18–20].

Our system's approach shows how the DEP force can be derived from the charge work with an object moving between suspension volumes in an inhomogeneous field [3]. Maxwell–Wagner's mixing equation described the apparent (or complex) specific conductivity of the volumes [21,22]. After separating the reactive and active components of the capacitive charge work, it could be shown that the active component drives the DEP [3]. At a given field frequency, the advancement of the object within the field gradient increases the overall polarizability of the DEP system through positive and negative DEPs in unison with its effective overall conductivity and, in turn, the dissipation of the electric field energy in ohmic heat.

In this paper, we generalize this approach by introducing a numerical "conductance matrix" to describe the overall (DC) polarizability of the system as a function of the position of the object in the DEP chamber. The assumption of DC properties for the object and the external medium prevents problems in separating reactive contributions in the electric work conducted on the DEP system [3]. However, it does not reduce the complexity of the field-induced object behavior since DEP is determined by the real (in-phase) part of the object polarization, even in the presence of complex media properties.

As one example, we considered high- and low-conductive 200 μm 2D spheres in a square 1 × 1 mm DEP chamber with 199 by 199 "2D voxels" using the classical-plain versus pointed-electrode configuration. The conductance matrix contains the overall chamber conductances calculated for each position that was geometrically accessible to the sphere center. The matrix values were used as interpolation points for the MatLab® quiver line function to calculate "conductance fields", which completely describe the DEP behavior of the sphere. For a given start position, the complex trajectories of the sphere's center follow the conductance gradient for the whole sphere, i.e., each step increases the overall conductance of the DEP system, and hence the dissipation of electric field energy at the fastest rate according to the LMEP.

2. Theory

2.1. General Remarks

The specific apparent, i.e., complex conductivity of aqueous media, is reduced by objects made of material with low conductivity or permittivity and increased in the presence of objects with high conductivity or permittivity. The actual effect is frequency-dependent. While the effective conductivity of suspensions increases with frequency, their effective permittivity drops [23]. Analytically, the conductivity of a suspension of monodisperse objects can be described by mixing equations [21]. For a given volume fraction, the effect of the objects on the conductivity of the suspension depends on their shape, orientation, and arrangement in relation to the external electric field [24–28].

The shape and frequency dependence of the induced dipole moment for objects confined by closed surfaces of the second degree (ellipsoids, spheroids, spheres, and cylinders) is generally summarized by the unitless, complex CMF, which has real (in-phase) and imaginary (out-of-phase) parts. For a homogeneous, general ellipsoid, it is described by the complex conductivities of the external ($\underline{\sigma}_e$) and object ($\underline{\sigma}_i$) media [29]:

$$\underline{f}_{CM} = f_{CM}^{\Re} + jf_{CM}^{\Im} = \frac{\underline{\sigma}_i - \underline{\sigma}_e}{\underline{\sigma}_e + n(\underline{\sigma}_i - \underline{\sigma}_e)} \tag{1}$$

Complex parameters are underscored. j being $\sqrt{-1}$. The ellipsoid's shape is coded in the depolarizing coefficient n along the axis oriented in the field direction. For 3D and 2D spheres, it is $1/3$ and $1/2$, respectively. Note that the circle representing the sphere in 2D has the polarizability of a cylinder oriented perpendicular to the field in 3D [30].

The CMF is generally derived for a homogeneous external field, which induces a dipole moment. The DEP force is proportional to the real part of the CMF $f_{CM}^{\Re} = \Re(\underline{f}_{CM})$, and any real polarization ratio of an object and external medium occurring for frequency-dependent properties can be modeled by combining appropriate DC conductivities for the external and object media. The imaginary part $f_{CM}^{\Im} = \Im(\underline{f}_{CM})$ vanishes for the low- ($\omega \to 0$) and high- ($\omega \to \infty$) frequency limits, and the CMF is described by the real parts of the media's conductivities (σ_i, σ_e) or permittivities (ε_i,ε_e). With $n = 1/2$ for the 2D sphere:

$$f_{CM}^{\Re} = \underset{\omega \to 0}{\underline{f}_{CM}} = 2\frac{\sigma_i - \sigma_e}{\sigma_i + \sigma_e} \text{ or } f_{CM}^{\Re} = \underset{\omega \to \infty}{\underline{f}_{CM}} = 2\frac{\varepsilon_i - \varepsilon_e}{\varepsilon_i + \varepsilon_e}, \text{ respectively.} \tag{2}$$

The same CMFs are obtained at the frequency limits for the same conductivity and permittivity ratios. The factors sweep the range between -2.0 and 2.0 (-1.5 and 3.0 for the 3D sphere) with the limiting values reached for $\sigma_i << \sigma_e$ or $\varepsilon_i << \varepsilon_e$, and $\sigma_i >> \sigma_e$ or $\varepsilon_i >> \varepsilon_e$, respectively.

The CMFs of Equation (2) are three times larger than the usual expressions because the depolarization coefficient of $1/3$ of the 3D sphere has not been separated and truncated against the $1/3$ in the volume term; a step that is historically justified but is a simplification only for 3D spheres [9,17]. This allowed us to retain the full volume term in Equation (17), which reflects the ponderomotive (bodily) nature of the DEP force.

In the model below, we use the low-frequency limit by combining a tenfold ratio of external conductivity and object conductivity (1.0 S/m with 0.1 S/m and vice versa) corresponding to sphere and external medium conductances of 1.0 S and 0.1 S in 2D. These parameters yield CMFs of -1.64 and 1.64 for 2D spheres.

2.2. Charge Work and Conductance Change

A chamber of cuboid shape with two plain-parallel rectangular y by z electrodes of distance x is to be filled with a medium of complex specific conductivity $\underline{\sigma}_e$. The complex conductance of the chamber is:

$$\underline{L}_e = \underline{\sigma}_e \frac{yz}{x} = (\sigma_e + j\omega\varepsilon_0\varepsilon_e)k \tag{3}$$

σ_e and ε_e are the real parts of the conductivity and permittivity. ω and ε_0 being the circular frequency and the permittivity of vacuum. The cell constant k is the generalized geometry factor relating the conductance for chambers of any geometry to the conductivity of the measured medium. With a single object suspended at location i, the suspension's effective conductivity is $\sigma_{S(i)}$ [3]. The chamber conductance is:

$$\underline{L}_{S(i)} = \underline{\sigma}_{S(i)} k = \left(\sigma_{S(i)} + j\omega\varepsilon_0\varepsilon_{S(i)}\right) k \tag{4}$$

Neglecting the stray capacitance, the same cell constant relates the suspension chamber's capacitance to the suspension's permittivity. The chamber can be described as a lossy capacitor, using the relation between complex conductance and capacitance:

$$\underline{C}_{S(i)} = -j\frac{\underline{L}_{S(i)}}{\omega} = -j\frac{\underline{\sigma}_{S(i)}}{\omega} k = \left(\varepsilon_0\varepsilon_{S(i)} - j\frac{\sigma_{S(i)}}{\omega}\right) k \tag{5}$$

The charge work conducted on the capacitor by the rms AC-voltage V_{eff} applied to the chamber is:

$$W^C_{S(i)} = \frac{\Re\left(\underline{C}_{S(i)}\right)}{2} V^2_{eff} = \frac{C_{S(i)}}{2} V^2_{eff} = \frac{\varepsilon_0\varepsilon_{S(i)} k}{2} V^2_{eff} \tag{6}$$

The energy (heat) dissipation in the chamber is:

$$\underline{P}_{S(i)} = \underline{L}_{S(i)} V^2_{eff} = \underline{\sigma}_{S(i)} k V^2_{eff} \tag{7}$$

If the field at the object's location is inhomogeneous, the capacitive charge work can induce DEP. In principle, the DEP work conducted in moving the object to location $i+1$ can be obtained from:

$$\Delta W^C = W^C_{S(i+1)} - W^C_{S(i)} = \frac{C_{S(i+1)} - C_{S(i)}}{2} V^2_{eff} = \varepsilon_0 \frac{\varepsilon_{S(i+1)} - \varepsilon_{S(i)}}{2} k V^2_{eff} \tag{8}$$

However, the description of the suspension properties, e.g., by mixing equations, may introduce reactive components and requires the identification of the active component of the apparent charge work, which drives DEP [3]:

$$\Delta W^C_{DEP} = \frac{\varepsilon_0}{2}\left(\varepsilon^{active}_{S(i+1)} - \varepsilon^{active}_{S(i)}\right) k V^2_{eff} = \frac{\varepsilon_0 \Delta\varepsilon_{DEP}}{2} k V^2_{eff} \tag{9}$$

A first, a necessary but not sufficient condition for identifying $\Delta\varepsilon_{DEP}$ is that $\varepsilon^{active}_{S(i)}$ and $\varepsilon^{active}_{S(i+1)}$ are strictly the real components of permittivity. There are three ways to ensure that only active components enter Equation (9): analysis of the expression to eliminate reactive components and using the high-frequency limit of permittivity expressions or the low-frequency limit of conductivity expressions. For the two limiting cases, the reactive components of the DEP force vanish, similar to the imaginary components for $\omega \to 0$ and $\omega \to \infty$ in Equations (4) and (5), respectively [3]. For an illustration of the correspondence of the real, i.e., active parts of the suspension's limiting permittivity and conductivity cases with the induced DEP force, see Figure 4 in [3]. DEP movement changes the dissipation by:

$$\Delta\underline{P} = \left(\underline{L}_{S(i+1)} - \underline{L}_{S(i)}\right) V^2_{eff} = \left(\underline{\sigma}_{S(i+1)} - \underline{\sigma}_{S(i)}\right) k V^2_{eff} \tag{10}$$

By analogy with Equation (9), the active component of dissipation is generated in the immediate vicinity of the object (compare with "influential-radius") in phase with the external field. In contrast, the reactive component is generated by out-of-phase field components, mainly in the volume of the suspension medium. For $\omega \to 0$, the (out-of-phase) reactive components vanish, and the apparent and active components of dissipation become identical [3]:

$$\Delta P_{DEP} = \left(\sigma^{active}_{S(i+1)} - \sigma^{active}_{S(i)}\right) k V^2_{eff} = \left(L^{active}_{S(i+1)} - L^{active}_{S(i)}\right) V^2_{eff} = \Delta L_{DEP} V^2_{eff} = \Delta L V^2_{eff} \tag{11}$$

The total dissipation of the DEP system becomes proportional to its DC conductance, simplifying the analysis of the DEP behavior.

2.3. DEP Force

For the fastest increase in the overall polarizability of the system and its active components, the DEP step from location i to $i+1$ must be oriented in the direction of the maximum differential quotient of the capacitive charge work or, more generally, the direction of the charge work gradient [3]. Using Equation (9) and the step width $\Delta r = \left|\vec{r}_{i+1} - \vec{r}_i\right| = r_{i+1} - r_i$ calculated from the location vectors $\vec{r}_i$ and $\vec{r}_{i+1}$, we obtain the DEP force:

$$\vec{F}^C_{DEP} = grad\left(W^C_{DEP}\right) \approx MAX\left(\frac{\Delta W^C_{DEP}}{\left|\vec{r}_{i+1} - \vec{r}_i\right|}\right)\frac{\vec{r}_{i+1} - \vec{r}_i}{\left|\vec{r}_{i+1} - \vec{r}_i\right|} = \frac{\varepsilon_0 k}{2} MAX\left(\frac{\Delta\varepsilon_{DEP}}{\Delta r}\right)V^2_{eff}\frac{\vec{r}_{i+1} - \vec{r}_i}{\Delta r} \tag{12}$$

$\frac{\vec{r}_{i+1} - \vec{r}_i}{\Delta r}$ defines the unit vector pointing in the direction of DEP translation. The DEP-induced differences in the active components of the charge work are always positive. At low frequency or DC, the increase in polarizability is strictly proportional to an increase in the conductance of the system and consequently dissipation. Accordingly, the DEP trajectory of a single object can be calculated from the maxima of the differential quotients of the DC conductance. In analogy to Equation (12), from Equation (11), we obtain:

$$\vec{F}_{DEP} \sim grad(L_{DEP})V^2_{eff} \approx MAX\left(\frac{\Delta L_{DEP}}{\Delta r}\right)V^2_{eff}\frac{\vec{r}_{i+1} - \vec{r}_i}{\Delta r} \tag{13}$$

Note that this relation between the DEP force and system polarizability coincides with the DC limit of Maxwell–Wagner's mixture equation [22] (cf. [3]). To compare forces between different chamber and electrode setups, Equation (13) was normalized to the square of the chamber voltage and the basic conductance L_{Basic} of the chamber without an object. In 3D, we obtain the normalized force:

$$\vec{F}_{DEP} \sim \frac{1}{L_{Basic}} MAX\left(\frac{\Delta L_{DEP}}{\Delta r_i}\right)\frac{\vec{r}_{i+1} - \vec{r}_i}{\Delta r} \tag{14}$$

In the 2D description, fields, currents, etc., have no z-component, as in the case of thin films of uniform thickness, e.g., a metal layer on glass. Here, we use a 2D model geometry with a thickness of $z = 1m$ perpendicular to the sheet plane, neglecting the z-components. Combining the 3D suspension conductivity with thickness yields the DC-sheet conductances $L^{2D}_{S(i)} = \sigma^{active}_{S(i)} z$ and $L^{2D}_{S(i+1)} = \sigma^{active}_{S(i+1)} z$. Equation (11) reads:

$$\Delta P^{2D}_{DEP} = \Delta\sigma_{DEP} z k^{2D} V^2_{eff} = \Delta L^{2D}_{DEP} V^2_{eff} \tag{15}$$

with k^{2D} being the 2D-cell constant. Note that both conductance differences, ΔL_{DEP} and ΔL^{2D}_{DEP} have the unit Siemens. In 2D, Equation (14) reads:

$$\vec{F}^{2D}_{DEP} \sim \frac{1}{L^{2D}_{Basic}} MAX\left(\frac{\Delta L^{2D}_{DEP}}{\Delta r_i}\right)\frac{\vec{r}_{i+1} - \vec{r}_i}{\Delta r} \tag{16}$$

L^{2D}_{Basic} is the system's sheet conductance without an object. Note that the right-hand side of Equation (16) has unit "m". The straightforward approach using the system's capacitive-charging work (Equation (12)), similar to the derivation in [3], provides the "Newton" for the DEP force. Probably, from the object's point of view, the correct proportionality factor in a 3D model includes the magnetic field constant. In the system approach used here, a normalized force is obtained, which can be converted into an exact force for a given location (see below).

3. Materials and Methods

3.1. *Software*

A 2D numerical solver based on the finite-volume method was implemented in MatLab® (version R2018b). It was developed to simulate the potential distributions, current paths, and total conductance for arbitrary geometries and conductivity distributions with current sources (electrodes) [31].

The total conductance data for the 2D system with 199 × 199 2D voxels were stored in a matrix and used as interpolation points for the MatLab® quiver line function to calculate the conductance field.

SigmaPlot 11.0 (Systat Software GmbH, Erkrath, Germany) was used for postprocessing and plotting data in line graphs. Inkscape 1.1.2 (GNU General Public License, version 3) was used to create graphical images and overlays of graphs with matrix images.

The data points of the Figures are given in the Supplementary Materials.

3.2. *Numerical 2D Model*

Without an object, a square chamber of $x = y = 1m$ confined by plain-parallel electrodes with a depth of 1 m perpendicular to the sheet plane has a (sheet) conductance of 0.1 and 1 S for volume conductivities of 0.1 and 1 S/m, respectively. The same sheet conductance occurs for square cm- or µm-size chambers with a depth of 1 m. Since only conductance and no size-related, frequency-dependent polarizabilities are considered, the 2D model is independent of a specific dimension in the x-y plane. To recognize microfluidic geometries, we assume an area of 1 × 1-mm^2 for the DEP chamber, which is formed by 199 × 199 square elements. Each of the elements was assigned a homogeneous area conductance. Due to the assumed thickness, we refer to these elements as "2D voxels" ("2D volume pixels").

The electrodes are located outside the chamber volume. The pointed and plain electrodes were formed by a single and row of 199 highly conductive 500-S voxels. The sheet conductance of the chamber was calculated for all positions accessible to a single 200 µm 2D sphere with a diameter of 39 voxels (Figure 1). The odd number symmetry defines a single central voxel and allows precise localization with respect to the pointed one-voxel electrode. Using 95 (19 voxels) and 497 µm (99 voxels) spheres, we showed that the results in this range do not qualitatively depend on the size of the sphere.

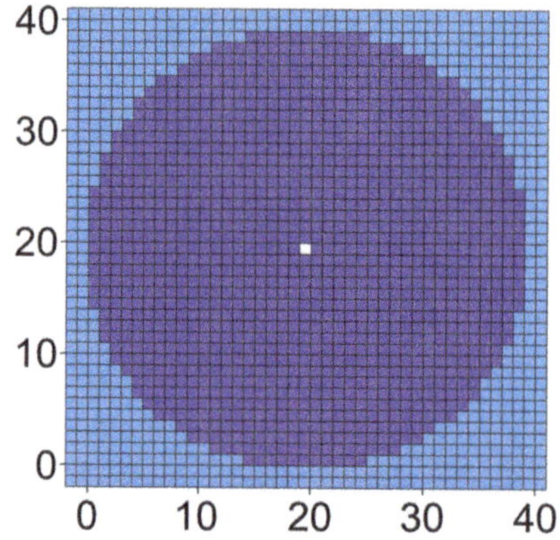

Figure 1. 200 µm 2D sphere approximated with a diameter of 39 voxels in the horizontal and vertical directions. The central voxel is marked in white. Linear rows of 11 voxels form the horizontal and vertical edges.

4. Result and Discussion

4.1. DEP Chamber Characterization

The classical setup with one pointed and one plain electrode was chosen to demonstrate the capability of our system approach. Figure 2A shows the field distribution without the sphere. Benign DEP behavior was observed in the range marked by the double arrow, which mainly corresponds to the dipole model. Figure 2B,C show the potential, field strength, and field gradient along the symmetry line. In Figures 2–4, current lines were used instead of field lines to more clearly show the polarization of the sphere.

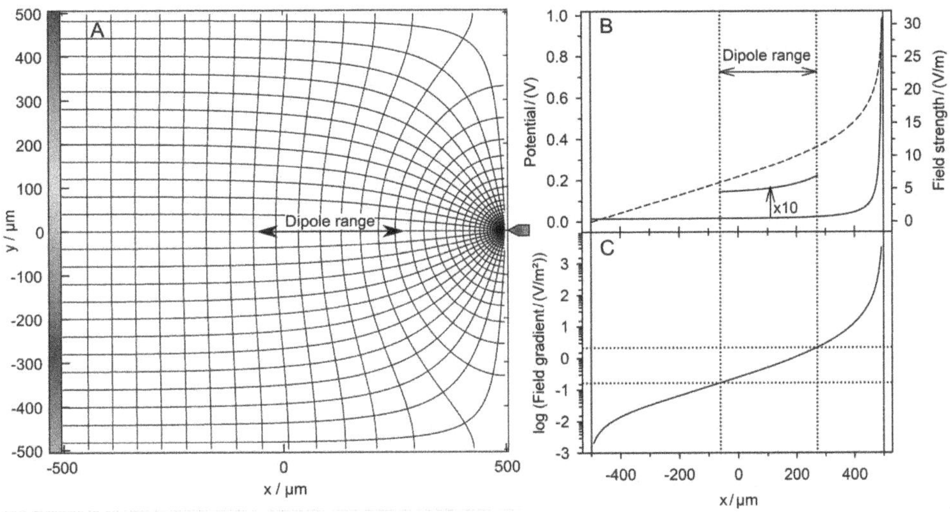

Figure 2. (**A**) Potential and current line distributions in the 1 × 1-mm^2 chamber without the sphere energized with 1 V at the pointed electrode (center right) versus 0 V at the plain electrode (vertical gray bar on the left). At the symmetry line, the dipole range is marked. (**B**) Potential (dashed) and field strength (full) along the symmetry line of the chamber (500 μm $\leq$ x $\leq$ 500 μm, y = 0 μm). Vertical lines mark the limits of the chamber volume. The curve was enlarged by multiplication with a factor of 10 to show the field behavior in the dipole region more clearly. (**C**): Field gradient along the symmetry line.

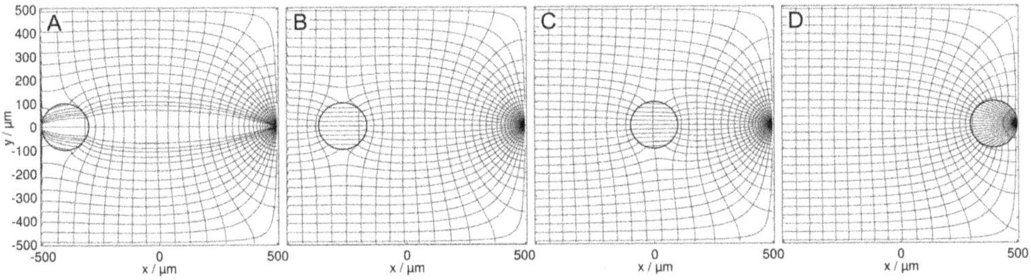

Figure 3. Potential and current line distributions for different positions of the 1.0 S sphere in 0.1 S medium, in front of the plain electrode (**A**), on the watershed (**B**), in a largely homogeneous field region (**C**), and at the pointed electrode (**D**). The conductances are (**A**): 35.744 mS, (**B**): 35.563 mS, (**C**): 35.647 mS, and (**D**): 83.912 mS. The basic sheet conductance L^{2D}_{Basic} of 34.908 mS without a sphere corresponds to a cell constant of $k^{2D} = 0.34908$.

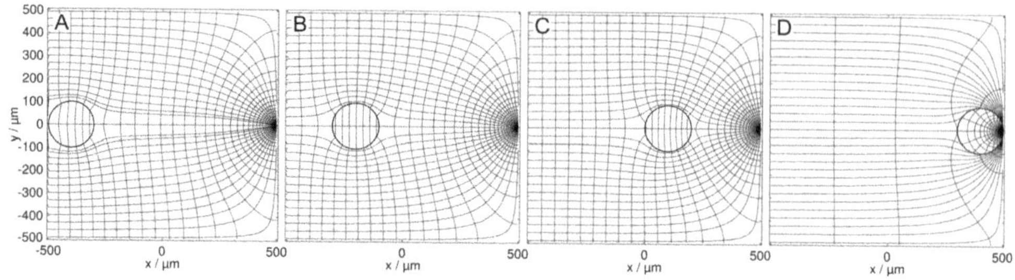

Figure 4. Potential and current line distributions for different positions of the 0.1 S sphere in the 1.0 S medium, at the plain electrode (**A**), in a largely homogeneous field region (**B**,**C**), and in front of the pointed electrode (**D**). The conductances are (**A**): 343.29 mS, (**B**): 342.28 mS, (**C**): 340.10 mS, and (**D**): 60.682 mS. The basic sheet conductance L^{2D}_{Basic} of 348.97 mS without sphere corresponds to a cell constant of $k^{2D} = 0.34897$.

Theoretically, all plots in Figure 2 and the cell constant of the chamber k^{2D} calculated from Equation (2) are independent of the medium conductance. The basic conductance values L^{2D}_{Basic} of the chamber without an object were calculated with media of 0.1 S and 1.0 S from voltage and current using a MatLab® routine. The two obtained cell constants showed negligible numerical differences.

4.2. DEP System with a Homogeneous Sphere

Figures 3 and 4 show the field distributions in the DEP system for the two complementary conductance ratios of sphere and suspension medium for different sphere positions. According to theory, more favorable positions result in a higher overall conductance of the system.

4.3. Calculation of Trajectories and Forces

For a single-object suspension, the electric work conducted in the inhomogeneous field of a linear DEP system leads to the system's overall permittivity and conductivity increase [3]. For objects with effective conductivities that are higher ("positive DEP") and lower ("negative DEP") than those of the external medium, this is true, even though the total conductance of the systems without the object is always lower or higher, respectively, than with the object (Figures 3 and 4).

The DEP behavior of the sphere was modeled using the "conductance matrix". The 160×160 matrix elements were calculated as the overall sheet conductances of the system, with the sphere's center located at each of the 160×160 accessible voxel coordinates. The basic sheet conductance determines the upper and lower boundary of the overall conductance of the DEP systems with the low- and high-conductance sphere, respectively. As a reference, the mean chamber conductance $\overline{L}^{2D}$ was calculated from all values in the conductance matrix. It corresponds to the average start conductance obtained in a field-free, thermally relaxed DEP system for infinitely many starting positions of the sphere.

It was insufficient to consider DEP steps of the sphere's center voxel to one of the up to eight neighboring voxels in the rectangular and diagonal directions to construct DEP trajectories from a given start voxel. We found that this "eight-neighbors" approach did not, for example, prevent the incorrect crossing of a bent watershed, i.e., bifurcation-boundary lines separating the catchment areas of different endpoints. Thus, we applied the quiver line function of MatLab® to generate a "conductance field" using the elements of the conductance matrix as interpolation points. The conductance field provided smooth and more precise trajectories, watersheds, saddle points (bifurcation points), and normalized DEP forces. The program shifted the object stepwise along a quiver line in the direction of the maximum overall conductance increase to construct a trajectory. Positions with object voxels located outside the chamber area were excluded, i.e., the sphere was deflected by the chamber walls moving along the interface until reaching a point of attraction (endpoint). Clearly, the trajectories do not influence the endpoint conductance, while the DEP work conducted depends on the trajectory. Along each trajectory, the normalized DEP force was calculated with Equation (16).

One may ask whether the pointed electrode is very sharp compared with the experimental situation. However, there are at least two arguments against this assumption. First, in 3D, the 2D-pointed electrode corresponds to a 1 m vertical blade, and second, a voxel ratio of electrode to object of 1:39 (Figure 1) is similar to the size ratio of 110 nm-thick glass-chip electrodes and 4 μm cells [32].

4.4. Trajectories and Forces

Figures 5 and 6 show the results for the two complementary conductance combinations. The 19-voxels-wide, white frames in Figures 5A and 6A are geometrically inaccessible to the center of the sphere. In both conductance scenarios, the system's sheet conductance increases steadily along each trajectory toward a specific endpoint (Figures 5B and 6B). In Figures 5B,C and 6B,C, sheet conductance and normalized DEP force, respectively, are plotted over the same abscissas.

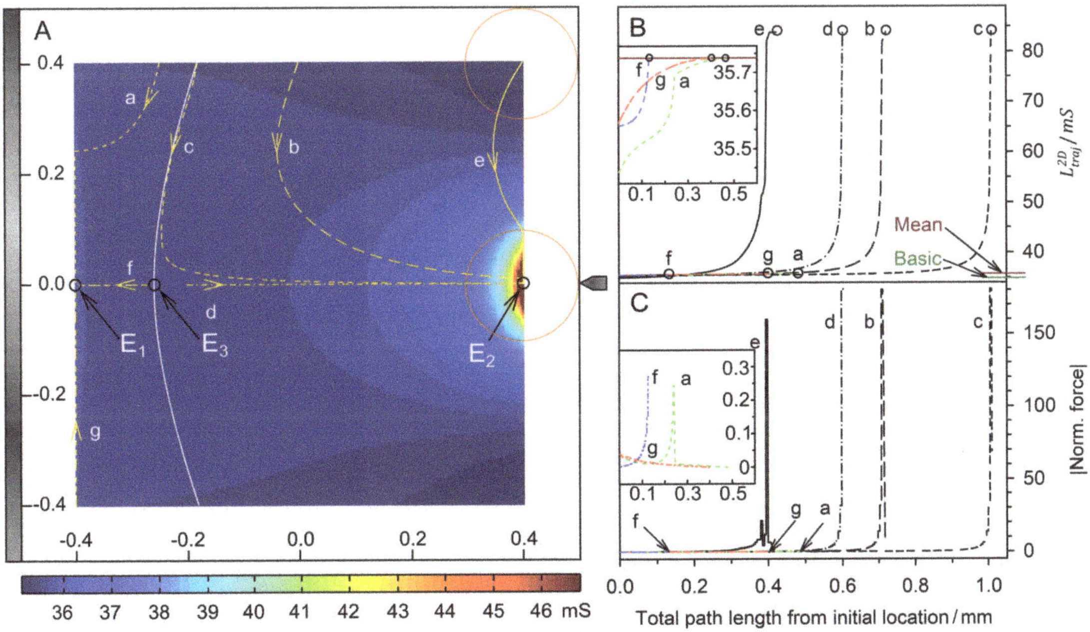

Figure 5. Single 200 μm, 2D sphere of 1.0 S (reddish circles in (**A**)) in the chamber of Figure 2 with 0.1 S medium. The mean conductance is $\bar{L}^{2D}$ = 35.739 mS. (**A**) Conductance field plot with trajectories (a–g). A watershed (bent white line) separates the two caption areas of the stable endpoints E_1 and E_2. E_3 is an instable saddle point in the middle of the watershed. (**B**) Sheet conductance along the trajectories. Basic and mean conductance are marked. The system's sheet conductance increases steadily along each trajectory, reaching moderate and high peak values at the endpoints E_2 and E_1, respectively. Trajectories b, c, d, and e end at E_2. Trajectories a, f, and g end at E_1, reaching $\bar{L}^{2D}$ by coincidence (insert). (**C**) Normalized DEP forces calculated with Equation (16) from the conductance values in (**B**).

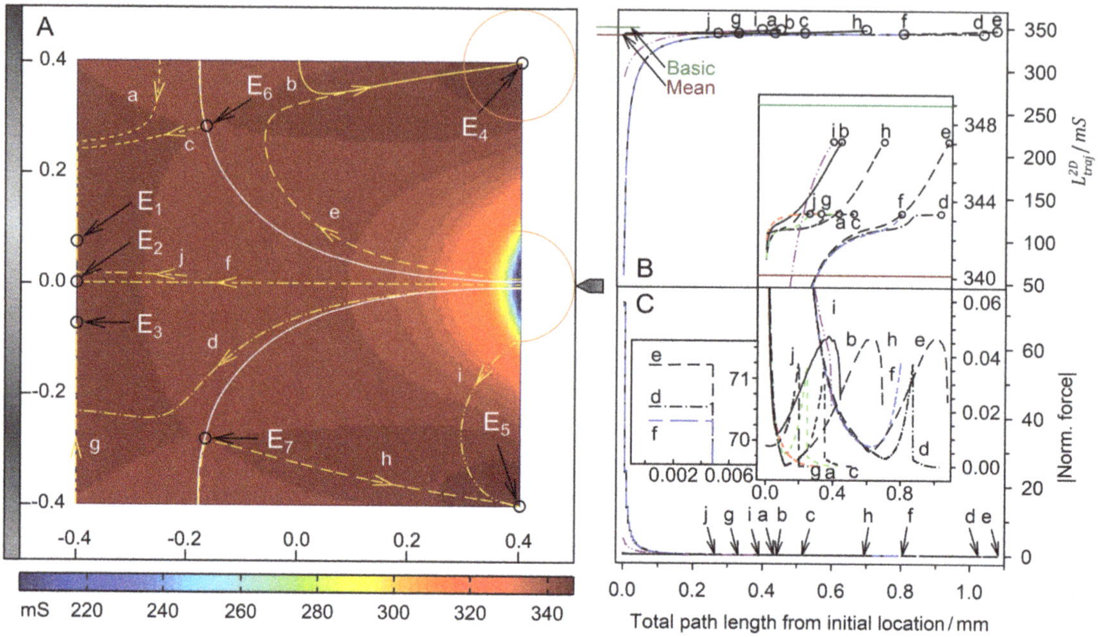

Figure 6. Single 200 µm 2D sphere of 0.1 S (reddish circles in (**A**)) in the chamber of Figure 2 with 1.0 S medium. The mean conductance is $\overline{L}^{2D}$ = 340.14 mS. (**A**) Conductance field plot with trajectories (a–j). Two watersheds (bent white lines) and one symmetry line (trajectory f) separate four catchment areas with the four stable endpoints (E_4, E_1, E_3, and E_5). E_6, and E_7 are instable saddle points. E_2 is an instable minimum at the end of the symmetry line. Trajectories close to f, such as j, are diverted to E_1 or E_3. (**B**) Sheet conductance along the trajectories. Basic and mean conductance are marked. Trajectories b and e end at E_4; trajectories a, c, and j at E_1; trajectory f at E_2; trajectories d and g at E_3; trajectories i and h at E_5. (**C**) Normalized DEP force calculated with Equation (16). Forces along trajectories parallel to the plain electrode toward the endpoints E_1, E_2 and E_3 are very low. The constant forces observed at the start of trajectories e, d, and f (Figure 6C, left insert) may be due to the reversal of the effect discussed above.

At endpoint E_1, the sheet conductance reaches a value close to $\overline{L}^{2D}$ (Figure 5B, insert) but more than twice that value at E_2 (Figure 5B). At the instable saddle point E_3, the DEP force vanishes. The sphere should theoretically travel along the watershed toward E_3 for start points on the watershed. However, a stable trajectory along the watershed could not be established for numerical reasons.

In Figure 5A (trajectories b and e), geometric restriction by the chamber wall or plain electrode (trajectory *a*) causes deflections in the sphere's trajectories. These are visible in the plots of the sheet conductance (Figure 5B) and, in particular, the normalized force (Figure 5C). The deflections occur before the sphere travels a longer distance along the plain electrode (trajectory *a*) or the chamber wall (trajectory e), or after it hits the restriction near an endpoint (E_2, trajectories *b* and *c*). In the latter case, the maximum force is observed when the sphere touches the restriction. The final "correction" steps toward the endpoint generate a lower force. A similar process, although with a long "correction distance," can be seen in trajectory *a* (Figure 5B,C, insert: green dashed curve). In the case of a "direct hit", i.e., when the sphere reaches the endpoint directly (trajectories *d* and *f*), the force curve ends in the peak value. The peak forces reach very high values at the pointed electrode (E_2) and moderate values when the sphere reaches the plain electrode (E_1).

The final steps along the projected conductance gradient before the vertical edge of the 2D sphere is attached to the plain electrode or chamber wall cause a greater increase in conductance and produce a higher force than the deflected motion in the attached state. However, minor movement in parallel to the restriction in the vicinity of an endpoint does not significantly change the sphere's center distance to the tip of the pointed or center of the plain electrode. Thus, the overall conductance of the system does not change dramatically, and the resulting forces are not exceptionally high when the center or one of the neighboring voxels of the sphere's edge touches the pointed electrode (E_2) or the center voxel of the plain electrode (E_1). Accordingly, the peaks with finally decreasing force (trajectories *b*, *c*, and *e*) can be explained by minor corrections of the position near an endpoint (Figure 5C). We suppose that such a force reduction after the final peak is also observed in high-resolution 3D models.

However, the force curve here is probably additionally modulated by the shape approximation of the 2D sphere with straight vertical edges (Figure 1). Another effect that may play a role in this behavior was observed in 3D COMSOL Multiphysics® (www.comsol.com) simulations with a highly conductive sphere in a coaxial DEP chamber. In this system, we found the conductivity minimum not in the attached state of the sphere but at a very short distance from the center electrode (results not published). For the model geometry used here, a comparable distance would be in the voxel size order of magnitude, preventing further investigation in this work.

Trajectories *d* and *f* run along the symmetry line between the pointed and plain electrodes. At saddle point E_3, they start in opposite directions to different endpoints, although the common view would predict an attraction by the pointed electrode (cf. Figure 2). The existence of the watershed separating two regions of attraction along the symmetry line contradicts the dipole view.

In Figure 6, the sheet conductance of the system reaches slightly higher peak values at the "hidden" endpoints E_4 and E_5 than at the endpoints at the plain electrode. We suppose that the sphere's size determines the shape of the watersheds and whether low-conductance spheres can "hide" away from the electrodes. The peak forces calculated for the 0.1 S sphere are generally lower than for the 1.0 S sphere.

4.5. Mirror Charge Effects

In the dipole model view, the 1.0 S sphere moves along the field gradient (Figure 2C), from the plain electrode to the pointed electrode, from areas with low field to those with high field. In our model, the sphere is initially attracted to the plain electrode (Figure 5A, trajectory *f*) and to the pointed electrode only beyond a watershed (Figure 5A, trajectory *d*). We suppose that the attraction toward the plain electrode is caused by mirror charges that exceed the dipole effect in the weak gradient in front of the plain electrode.

In the dipole model view, the 0.1 S sphere moves against the field gradient (Figure 2C), from the pointed to the plain electrode, from high field to low field areas (Figure 6A, trajectory *f*). Above the first 100 μm, Figure 6C shows a very high, steadily decreasing repulsive force. However, the force increases again approx. 200 μm away from the plain electrode, despite the decreasing field gradient, until the sphere attaches to the electrode (insert of Figure 6C, trajectory *f*). The increase in force is related to the increase in the chamber conductance due to an overall reduced screening of the plain electrode by the low-conductive sphere (Figure 4A). This view is consistent with the interaction with mirror charges induced at the plain electrode.

While the mirror charge effect clearly dominates for the 1.0 S sphere at distances of about 150 μm from the plain electrode, the observability of the 0.1 S sphere is blurred by the synchronous action of two equidirectional forces. We suggest that the mirror charge's contribution to the DEP force depends strongly on the sizes and curvatures of the object and electrode. The electrode areas must be large relative to the object size for a high mirror charge to be induced. This effect is negligible at the pointed electrode.

4.6. DEP Force Reversibility in the Dipole Range

For 2D spheres or infinitely long 3D cylinders whose axis of symmetry is perpendicular to the field, the exchange of the conductivities of the external medium and the object reverses the sign of the CMF at constant magnitude (Equation (2)).

Accordingly, the direction of the dipole force is inverted for any position in the DEP chamber. If dipole forces prevailed, every trajectory would have to be exactly reversed, and the quotient of the DEP force magnitudes would have to be (minus) one. However, Figure 7 already shows a more complex picture along the symmetry line of the DEP chamber.

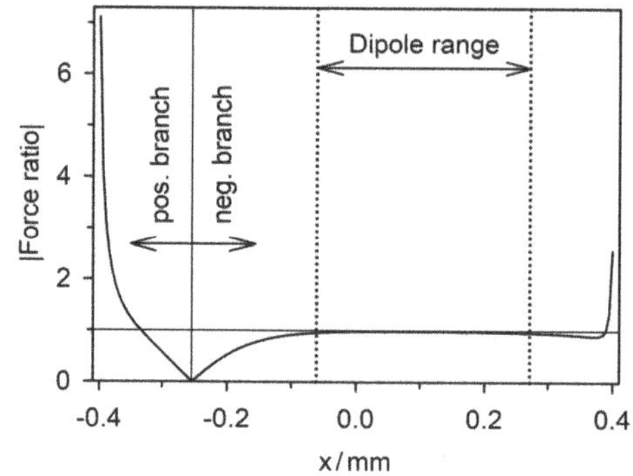

Figure 7. Absolute value of the quotient of the normalized DEP forces acting on the 1.0 S and 0.1 S spheres plotted along the symmetry line of the chambers (trajectories d and f of Figure 5 and trajectory f of Figure 6).

The plot of the quotient has characteristic regions separated by a zero point representing the vanishing force for the 1.0 S sphere at the watershed. The positive branch results from the attraction of the 1.0 S and 0.1 S spheres to the plain electrode. While the force magnitudes remain low in the positive branch, the quotient of seven at the plain electrode indicates the high-conductance sphere's more efficient induction of mirror charges. Moreover, at the pointed electrode, this sphere experiences a force magnitude of around twice as high as the low-conductance sphere. From zero to the right, the force ratio reaches the expected -1 plateau, indicating dipole-like behavior. The plateau ranges from approx. -70 to 270 μm, i.e., over about 42% of the electrode distance along the symmetry axis accessible to the sphere.

We suggest that reversibility is a criterion for the applicability of the dipole approach in certain regions of the DEP chamber. Nonreversibility indicates the presence of higher-order moments, mirror charges, and so on. Interestingly, the total conductance of the DEP chamber corresponds to the average conductance, roughly in the middle of the dipole region.

4.7. Relating Normalized to Actual DEP Forces

In the dipole region, the DEP forces only reach moderate magnitudes compared to the forces in front of the pointed electrode, where they reach magnitudes up to thousands of times higher than in the dipole region. Figure 8 considers the "dipole range" where the forces of the classical dipole model can be quantified and directly compared with the normalized DEP forces from Equation (16).

Stokes friction limits DEP velocities to the linear range in aqueous media, i.e., the velocities are proportional to the driving forces. Accordingly, experimentally observed accelerated DEP motion near electrodes is caused by changes in the DEP force. In the dipole model, the DEP force is:

$$\vec{F}_{DEP} = \Re\left(\underline{\vec{m}}\right) \cdot grad\left(\vec{E}\right) = \varepsilon_0 \varepsilon_e V_0 f_{CM}^{\Re} \vec{E} \cdot grad\left(\vec{E}\right) \tag{17}$$

where ε_0 and V_0 are the permittivity of vacuum and the volume of an ellipsoidal object:

$$V_0 = \frac{4\pi}{3}abc \tag{18}$$

with the principal semiaxes a, b, and c (2D sphere: $a = b$, $c = 1$ m). For an electrode voltage of 1 V and the field parameters of Figure 8, we obtain $\vec{E} \cdot grad\left(\vec{E}\right) = 0.5602\ \mathrm{V^2/m^3}$, which can easily be rescaled to any experimental electrode voltage. Experiments or calculations can provide values for the actual DEP force at the chamber position x = 186 µm, y = 0 µm, where our model yields normalized forces of approx. 0.12 for the 2D sphere. Clearly, the 2D inhomogeneity in the polarization of the 2D sphere causes the very high force magnitudes at the pointed electrode (Figures 3D, 4D, 5C and 6C). They are about 1500 (1.0 S sphere) and 580 (0.1 S sphere) times higher than the forces in the dipole region at the chamber position considered.

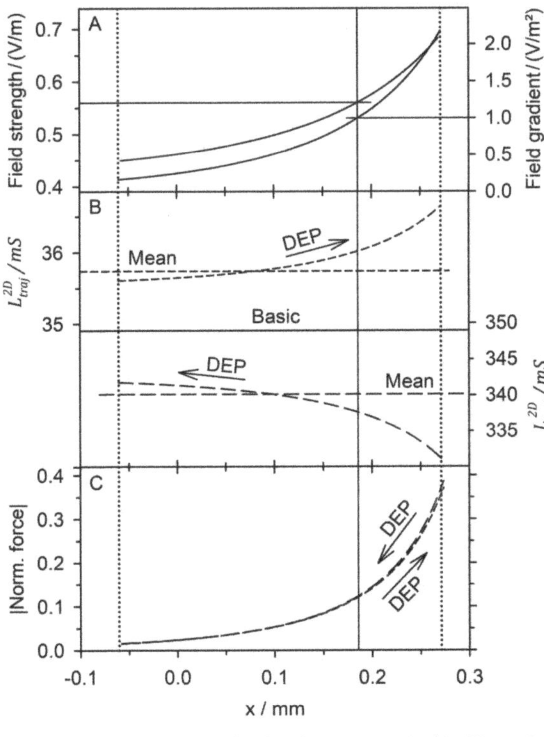

Figure 8. DEP behavior in the dipole range marked in Figure 2. (**A**) Field strength and field gradient along the symmetry line for 1 V potential difference at the electrodes. The vertical auxiliary line at x = 186 µm is perpendicular at a field gradient of 1 V/m^2 (field strength of 0.5602 V/m). Horizontal auxiliary lines run out to the respective ordinates from the intersections with field strength and field gradient plots. (**B**) DEP of the sphere increases the overall conductance of the chamber. Top: 1.0 S sphere, 0.1 S medium (short, dashed line, left ordinate, positive DEP), bottom: 0.1 S sphere 1.0 S medium (long dashed line, right ordinate, negative DEP). At x = 186 µm, the conductances are 36.02 mS and 337.6 mS. The common baseline corresponds to the basic conductances of the two setups. (**C**) At x = 186 µm, the normalized forces from Equation (16) are 0.1211 (1.0 S sphere, 0.1 S medium) and 0.1234 (0.1 S sphere, 1.0 S medium).

The exact force conversion with Equation (17) indeed calls for calculating the conductance field for a 3D object. However, in dipole theory, the force magnitude experienced by a 3D sphere in positive DEP can be twice that experienced in negative DEPs (Equations (1) and (17)). Moreover, we believe that 3D inhomogeneity would induce even higher forces at the pointed electrode when polarizing a 3D sphere.

4.8. Remark on "Positive" and "Negative" DEP

Trajectories with a changing orientation within an inhomogeneous field indicate a problem with defining the sign of the DEP force in relation to the external field gradient. While "positive" and "negative" DEPs correspond to the common understanding in the dipole region, the definition becomes fuzzy elsewhere in the chamber, especially when the force direction reverses because the object itself changes its "field environment,", e.g., due to mirror charges.

4.9. Thermodynamic Aspects

The theoretical description of AC–electrokinetic effects, such as electro-orientation, DEP, electrorotation, or mutual attraction, usually relies on electrostatic approaches. However, for lossy media, the validity of the approach is not clear per se, since electrostatic systems are generally in a state of equilibrium without energy dissipation and entropy production by resistive and displacement currents. Moreover, the electrokinetic effects induced must themselves lead to energy dissipation. Despite these seemingly severe problems, our LMEP approach and experimental observations in the dipole region agree very closely with object-oriented electrostatic models.

Assuming that the DEP system is near equilibrium in its linear range and the application of the electrode voltage causes only minor deflection of the system from equilibrium, which remains in the linear range, it should approach a new "voltage-on equilibrium" through the minimization of entropy production according to Prigogine's principle [33–35]. Then, after voltage-off, the system should return to its previous state. Nevertheless, our theoretical and experimental findings suggest that electrokinetic phenomena increase the overall energy dissipation and are in contradiction with Prigogine's principle. In fact, the problem had already been addressed in 1912 and discussed by [36] (see also the references contained therein). In light of this work, one may wonder how a Kirchhoff network, which is generally the electrotechnical basis of our approach, would react if it could rearrange itself. Perhaps a kind of "Kirchhoff Rearrangement" is the effect we observe with the DEP?

From the point of view of the system, the work on a volume of material can be stored or dissipated (i) as electric field energy, (ii) as magnetic field energy, or (iii) as Joule heat [29]. In our model, we use the degree of the DEP system's overall increase in conductance as a criterion for the DEP force induced. At constant electrode voltage, dissipation of electrical energy is proportional to the square of the applied voltage according to Rayleigh's dissipation function, and it increases with the total conductance of the system [37]. While a small proportion of this energy is "dissipated" in DEP translation, DEP increases the total energy dissipation and the electrical work that must be done while DEP progresses.

In particular, it has been shown how DEP is related to the complex, i.e., apparent permittivity and conductivity of the suspension, both of which consist of an active and a reactive part. Like electrical machines, the reactive part (capacitively stored at the objects) is out of phase with the active component and performs no DEP work. While the DEP force is proportional to permittivity and conductivity's active components, the reactive components are dissipated [3]. A related discussion on the contributions of electronic polarization to the total field energy in lossy dielectrics seems to be underway [38].

4.10. Nonspherical Objects

An applied field can change the impedance of a suspension by inducing electro-orientation, DEP translation, or electrodeformation of objects [11,39]. As shown in theory and experiments, the (frequency-dependent) axis with the highest CMF is aligned in linear fields [32,40]. In homogeneous ellipsoids, the longest axis always has the highest CMF and is aligned [27]. This leads to a reduced suspension impedance and, at constant field strength, increased electrical power dissipation. For homogeneous spheroids, it has been theoretically demonstrated that the field-induced orientation moments are proportional to the increase in the conductivity of the suspension they induce [4].

A nonspherical object in an inhomogeneous field experiences force and torque, simultaneously resulting in both DEP and electro-orientation. Furthermore, the object's movement modifies both force and torque. Friction opposes both types of motion and can, for example, prevent a complete alignment at a particular location before the object moves to another location where a different alignment is induced. The situation is further complicated by the different nature of the friction opposing the translational and rotational motion, as can be seen, for example, from the different radius dependencies for the translation (~R: Stokes friction) and rotation of spheres (~R^3) [41].

5. Conclusions and Outlook

Recently, we derived the classical DEP force expression from the capacitive charge work gradient on a suspension of a single object in an inhomogeneous field, but abstracting from the actual chamber and electrode geometries. Here, we extended this approach to the entire DEP chamber by introducing a conductance field, the low-frequency equivalent of the capacitance field. The fields fully describe the object's DEP behavior and inherently account for inhomogeneous object polarization, mirror charges, electrode shielding effects, and so on.

Our model simplifies the computation of DEP forces in complex field environments. However, if the approach is applied to nonspherical objects or multibody systems, for example, to compute aggregation patterns, this comes at the expense of high computational effort, especially in 3D systems. Appropriate methods, such as Monte Carlo simulations, would likely reduce the computation time.

Objects with an effective conductivity lower or higher than that of the suspension medium usually show negative or positive DEPs, in other words, they move counter to or in the direction of the field gradient. Here, we reveal some exceptions to this rule outside regions where dipole effects dominate, something that may call for a conceptual rethink. A manuscript is in preparation extending the present results to pointed-versus-pointed, plain-versus-plain electrodes, and to four-pointed-electrode arrangements in one-versus-three, side-by-side, and field-cage drive modes.

We believe that our model can explain experimental findings such as the paradoxical accumulation of viruses and proteins in field cages or at electrode edges, where the dipole approach cannot account for sufficiently high trapping forces to withstand Brownian motion [16,17,42–44]. Forces large enough to trap small objects can result from inhomogeneous object polarization at electrodes or other surfaces and near to other objects. Although these forces act over larger distances than the Van der Waals forces do, the distances appear to be too small to trap objects from the entire suspension volume. This is why we propose the "sticky-fly-trap model". Viruses or molecules that approach electrode surfaces or other viruses or molecules by media flow or diffusion, to the point of becoming inhomogeneously polarized, snap to the surface or can form aggregates. This mechanism is suggested by experiments, which show that the aggregation of objects takes longer than redispersion after the field has been turned off. The aggregation is limited by the "undirected", random motion, while the dissolution of the aggregate is achieved by a "directed" diffusion away from the high concentration.

One final point is worth mentioning: the striking similarities between the snap-to-surface behavior in the DEP model and in our earlier force spectroscopy experiments with like-charged glass spheres and mica surfaces [45]. In these experiments, snap to surface has been observed over distances of up to 800 nm, clearly too large to be explained by the Van der Waals attraction, although this may lock the bead at the surface once it is reached. Here, we propose a "self-DEP" mechanism to explain the large snap-to-surface distances. Self-DEP can result from the oscillating dipole induced in a vibrating bead with fixed surface charges, e.g., in the mid-kHz range. In the case of dispersion, the vibration leads to the induction of a dipole by separating the center of the bead's fixed charges, which move with the object, and the center of the countercharges in the external medium. In "self-DEP", the interaction of the oscillating dipole with mirror charges induced on conducting or polarizable surfaces leads to attractive forces. The contribution of mirror charges to DEP forces has been described by Pethig; chapter 5.4 in [46]. Here, we found the attraction by mirror charges for both high- and low-polarizable objects.

The systems perspective allowed us to identify new approaches and perhaps even new fields of work for DEP research. These are (i) the modeling of the high repulsion and snap-to-surface forces induced by the inhomogeneous polarization of the objects; (ii) the calculation of DEP forces in complex field environments and multibody systems; and (iii) the role of active and reactive contributions in frequency-dependent DEP and other electrokinetic AC models in relation to the total work done on the systems [3].

Supplementary Materials: The following supporting information can be downloaded at: https://www.mdpi.com/article/10.3390/mi13071002/s1, conductance matrix with trajectory coordinates for Figures 5 and 6: L_Matrix_Fig.5 and L_Matrix_Fig.6; Data points for Figures 2B,C, 7 and 8: Data_Figs.2_7_8.

Author Contributions: Conceptualization, J.G.; methodology, J.G. and M.M.R.; software, M.M.R.; validation, J.G. and M.M.R.; formal analysis, J.G. and M.M.R.; writing—original draft preparation, J.G.; writing—review and editing, J.G.; visualization, J.G. and M.M.R. All authors have read and agreed to the published version of the manuscript.

Funding: This research received no external funding.

Institutional Review Board Statement: Not applicable.

Informed Consent Statement: Not applicable.

Data Availability Statement: Not applicable.

Acknowledgments: We would like to thank Jessica Schröder for her expert work on the diagrams and help with the references.

Conflicts of Interest: The authors declare no conflict of interest.

References

1. Jones, T.B. *Electromechanics of Particles*; Cambridge University Press: Cambridge, NY, USA, 1995; ISBN 9780521431965.
2. Wang, X.; Wang, X.-B.; Gascoyne, P.R. General expressions for dielectrophoretic force and electrorotational torque derived using the Maxwell stress tensor method. *J. Electrostat.* **1997**, *39*, 277–295. [CrossRef]
3. Gimsa, J. Active, reactive, and apparent power in dielectrophoresis: Force corrections from the capacitive charging work on suspensions described by Maxwell-Wagner's mixing equation. *Micromachines* **2021**, *12*, 738. [CrossRef] [PubMed]
4. Gimsa, J. Can the law of maximum entropy production describe the field-induced orientation of ellipsoids of rotation? *J. Phys. Commun.* **2020**, *4*, 085017. [CrossRef]
5. Swenson, R. Emergent attractors and the law of maximum entropy production: Foundations to a theory of general evolution. *Syst. Res.* **1989**, *6*, 187–198. [CrossRef]
6. Grinstein, G.; Linsker, R. Comments on a derivation and application of the 'maximum entropy production' principle. *J. Phys. A Math. Theor.* **2007**, *40*, 9717–9720. [CrossRef]
7. Niven, R.K. Steady state of a dissipative flow-controlled system and the maximum entropy production principle. *Phys. Rev. E* **2009**, *80*, 21113. [CrossRef]
8. Martyushev, L.M.; Seleznev, V.D. Maximum entropy production principle in physics, chemistry and biology. *Phys. Rep.* **2006**, *426*, 1–45. [CrossRef]
9. Gimsa, J.; Wachner, D. A polarization model overcoming the geometric restrictions of the Laplace solution for spheroidal cells: Obtaining new equations for field-induced forces and transmembrane potential. *Biophys. J.* **1999**, *77*, 1316–1326. [CrossRef]

10. Reichle, C.; Müller, T.; Schnelle, T.; Fuhr, G. Electro-rotation in octopole micro cages. *J. Phys. D Appl. Phys.* **1999**, *32*, 2128–2135. [CrossRef]
11. Ramos, A.; Morgan, H.; Green, N.G.; Castellanos, A. AC electrokinetics: A review of forces in microelectrode structures. *J. Phys. D Appl. Phys.* **1998**, *31*, 2338–2353. [CrossRef]
12. Kang, Y.; Li, D. Electrokinetic motion of particles and cells in microchannels. *Microfluid. Nanofluid.* **2009**, *6*, 431–460. [CrossRef]
13. Fritzsch, F.S.O.; Blank, L.M.; Dusny, C.; Schmid, A. Miniaturized octupole cytometry for cell type independent trapping and analysis. *Microfluid. Nanofluid.* **2017**, *21*, 130. [CrossRef]
14. Dürr, M.; Kentsch, J.; Müller, T.; Schnelle, T.; Stelzle, M. Microdevices for manipulation and accumulation of micro- and nanoparticles by dielectrophoresis. *Electrophoresis* **2003**, *24*, 722–731. [CrossRef] [PubMed]
15. Broche, L.M.; Labeed, F.H.; Hughes, M.P. Extraction of dielectric properties of multiple populations from dielectrophoretic collection spectrum data. *Phys. Med. Biol.* **2005**, *50*, 2267–2274. [CrossRef]
16. Waskasi, M.M.; Lazaric, A.; Heyden, M. Solvent-mediated forces in protein dielectrophoresis. *Electrophoresis* **2021**, *42*, 2060–2069. [CrossRef]
17. Pethig, R. Protein Dielectrophoresis: A tale of two Clausius-Mossottis-or something else? *Micromachines* **2022**, *13*, 261. [CrossRef]
18. Huang, J.P.; Karttunen, M.; Yu, K.W.; Dong, L.; Gu, G.Q. Electrokinetic behavior of two touching inhomogeneous biological cells and colloidal particles: Effects of multipolar interactions. *Phys. Rev. E* **2004**, *69*, 51402. [CrossRef]
19. Washizu, M.; Jones, T.B. Multipolar dielectrophoretic force calculation. *J. Electrostat.* **1994**, *33*, 187–198. [CrossRef]
20. Hossan, M.R.; Dillon, R.; Roy, A.K.; Dutta, P. Modeling and simulation of dielectrophoretic particle-particle interactions and assembly. *J. Colloid Interface Sci.* **2013**, *394*, 619–629. [CrossRef]
21. Wagner, K.W. Erklärung der dielektrischen Nachwirkungsvorgänge auf Grund Maxwellscher Vorstellungen. *Arch. Elektrotech.* **1914**, *2*, 371–387. [CrossRef]
22. Maxwell, J.C. *A Treatise on Electricity and Magnetism*; Clarendon Press: Oxford, UK, 1873; ISBN 0198503733.
23. Foster, K.R.; Schwan, H.P. Dielectric properties of tissues. In *Handbook of Biological Effects of Electromagnetic Fields*, 2nd ed.; Polk, C., Pastow, E., Eds.; CRC Press: Boca Raton, NY, USA, 1996; pp. 25–102.
24. Stubbe, M.; Gimsa, J. Maxwell's mixing equation revisited: Characteristic impedance equations for ellipsoidal cells. *Biophys. J.* **2015**, *109*, 194–208. [CrossRef] [PubMed]
25. Schwan, H.P. Electrical properties of tissue and cell suspensions. *Adv. Biol. Med. Phys.* **1957**, *5*, 147–209. [CrossRef] [PubMed]
26. Jiang, T.; Jia, Y.; Sun, H.; Deng, X.; Tang, D.; Ren, Y. Dielectrophoresis response of water-in-oil-in-water double emulsion droplets with singular or dual cores. *Micromachines* **2020**, *11*, 1121. [CrossRef] [PubMed]
27. Gimsa, J.; Gimsa, U. The influence of insulating and conductive ellipsoidal objects on the impedance and permittivity of media. *J. Electrostat.* **2017**, *90*, 131–138. [CrossRef]
28. Asami, K.; Hanai, T.; Koizumi, N. Dielectric approach to suspensions of ellipsoidal particles covered with a shell in particular reference to biological cells. *Jpn. J. Appl. Phys.* **1980**, *19*, 359–365. [CrossRef]
29. Landau, L.D.; Lifšic, E.M.; Pitaevskij, L.P. *Electrodynamics of Continuous Media*, 2nd ed.; Pergamon Press: Oxford, UK, 1984; ISBN 0080302750.
30. Stubbe, M.; Gimsa, J. Furthering the state of knowledge on the electric properties of hemi-ellipsoidal single cells and cell patches on electrodes. *Biosens. Bioelectron.* **2018**, *105*, 166–172. [CrossRef]
31. Rosenfeld, M.; Tanami, R.; Abboud, S. Numerical solution of the potential due to dipole sources in volume conductors with arbitrary geometry and conductivity. *IEEE Trans. Biomed. Eng.* **1996**, *43*, 679–689. [CrossRef]
32. Gimsa, J.; Titipornpun, K.; Stubbe, M.; Gimsa, U. Combined detection of AC-electrokinetic effects: Experiments with three-axial chicken red blood cells. *Electrophoresis* **2018**, *39*, 2253–2261. [CrossRef]
33. Stenholm, S. On entropy production. *Ann. Phys.* **2008**, *323*, 2892–2904. [CrossRef]
34. Onsager, L.; Machlup, S. Fluctuations and irreversible processes. *Phys. Rev.* **1953**, *91*, 1505–1512. [CrossRef]
35. Prigogine, I. *Introduction to Thermodynamics of Irreversible Processes*, 3rd ed.; Interscience Publishers: New York, NY, USA, 1967; ISBN 0470699280.
36. Zupanović, P.; Juretić, D.; Botrić, S. Kirchhoff's loop law and the maximum entropy production principle. *Phys. Rev. E* **2004**, *70*, 56108. [CrossRef] [PubMed]
37. Scaife, B. On the Rayleigh dissipation function for dielectric media. *J. Mol. Liq.* **1989**, *43*, 101–107. [CrossRef]
38. Zheng, X.; Palffy-Muhoray, P. Electrical energy storage and dissipation in materials. *Phys. Lett. A* **2015**, *379*, 1853–1856. [CrossRef]
39. Green, N.G.; Ramos, A.; Morgan, H. AC electrokinetics: A survey of sub-micrometre particle dynamics. *J. Phys. D Appl. Phys.* **2000**, *33*, 632–641. [CrossRef]
40. Gimsa, J. Combined AC-electrokinetic effects: Theoretical considerations on a three-axial ellipsoidal model. *Electrophoresis* **2018**, *39*, 1339–1348. [CrossRef] [PubMed]
41. Sokirko, A.V. The electrorotation of axisymmetrical cell. *Biol. Mem.* **1992**, *6*, 587–600.
42. Hölzel, R.; Pethig, R. Protein Dielectrophoresis: I. Status of experiments and an empirical theory. *Micromachines* **2020**, *11*, 533. [CrossRef]
43. Hölzel, R.; Calander, N.; Chiragwandi, Z.; Willander, M.; Bier, F.F. Trapping single molecules by dielectrophoresis. *Phys. Rev. Lett.* **2005**, *95*, 128102. [CrossRef]

44. Gimsa, J. New light-scattering and field-trapping methods access the internal electric structure of submicron particles, like Influenza viruses. In *Electrical Bioimpedance Methods: Applications to Medicine and Biotechnology*; Riu, P.J., Ed.; New York Academy of Sciences: New York, NY, USA, 1999; pp. 287–298, ISBN 1-57331-191-X.
45. Gimsa, J.; Wysotzki, P.; Perutkova, S.; Weihe, T.; Elter, P.; Marszałek, P.; Kralj-Iglic, V.; Müller, T.; Iglic, A. Spermidine-induced attraction of like-charged surfaces is correlated with the pH-dependent spermidine charge: Force spectroscopy characterization. *Langmuir* **2018**, *34*, 2725–2733. [CrossRef]
46. Pethig, R. *Dielectrophoresis: Theory, Methodology and Biological Applications*; John Wiley & Sons: Hoboken, NJ, USA, 2017; ISBN 9781118671450.

micromachines

Article

Modeling Brownian Microparticle Trajectories in Lab-on-a-Chip Devices with Time Varying Dielectrophoretic or Optical Forces

Mohammad Asif Zaman [1,*], Mo Wu [1], Punnag Padhy [1], Michael A. Jensen [2], Lambertus Hesselink [1] and Ronald W. Davis [2,3]

[1] Department of Electrical Engineering, Stanford University, Stanford, CA 94305, USA; mowu@stanford.edu (M.W.); punnag@stanford.edu (P.P.); hesselink@ee.stanford.edu (L.H.)
[2] Stanford Genome Technology Center, Department of Biochemistry, Stanford University, Palo Alto, CA 94304, USA; m.a.jensen@stanford.edu (M.A.J.); dnamarkr@stanford.edu (R.W.D.)
[3] Department of Genetics, Stanford University, Palo Alto, CA 94304, USA
* Correspondence: zaman@stanford.edu

Abstract: Lab-on-a-chip (LOC) devices capable of manipulating micro/nano-sized samples have spurred advances in biotechnology and chemistry. Designing and analyzing new and more advanced LOCs require accurate modeling and simulation of sample/particle dynamics inside such devices. In this work, we present a generalized computational physics model to simulate particle/sample trajectories under the influence of dielectrophoretic or optical forces inside LOC devices. The model takes into account time varying applied forces, Brownian motion, fluid flow, collision mechanics, and hindered diffusion caused by hydrodynamic interactions. We develop a numerical solver incorporating the aforementioned physics and use it to simulate two example cases: first, an optical trapping experiment, and second, a dielectrophoretic cell sorter device. In both cases, the numerical results are found to be consistent with experimental observations, thus proving the generality of the model. The numerical solver can simulate time evolution of the positions and velocities of an arbitrarily large number of particles simultaneously. This allows us to characterize and optimize a wide range of LOCs. The developed numerical solver is made freely available through a GitHub repository so that researchers can use it to develop and simulate new designs.

Keywords: Brownian dynamics; Lab-on-a-chip; Langevin equation; dielectrophoresis; optical trap

Citation: Zaman, M.A.; Wu, M.; Padhy, P.; Jensen, M.A.; Hesselink, L.; Davis, R.W. Modeling Brownian Microparticle Trajectories in Lab-on-a-Chip Devices with Time Varying Dielectrophoretic or Optical Forces. *Micromachines* **2021**, *12*, 1265. https://doi.org/10.3390/mi12101265

Academic Editor: Rodrigo Martinez-Duarte

Received: 26 September 2021
Accepted: 15 October 2021
Published: 18 October 2021

1. Introduction

Recent decades have seen rapid advances in the field of biotechnology. Lab-on-a-chip (LOC) devices capable of manipulating micro- and nano-sized bio-samples have played a significant role in fueling this progress. The research on designing novel and complex LOC devices is ongoing. A computational-physics framework capable of simulating the trajectory of a micro- or nano-sample can be a very useful tool for designing and analyzing LOCs [1,2].

Although the main application of LOCs is in the field of biology, the design process of the devices requires modeling several physical phenomena. Trapping and manipulating small-sized samples are often a key feature of LOCs. On-chip manipulation of micro-sized samples (or microparticles) is usually accomplished by using viscous drag forces applied through fluid flow along with dielectrophoresis (DEP) [3–5] or optical [6–8]/optoelectronic [9–11] techniques (DEP is more widely used). LOCs can have integrated micro-electrodes and microfluidic channels to carry out these functions [12]. Designing a device to achieve a specific functionality requires accurately predicting how a micro-object behaves under the influence of relevant forces. For example, microfluidic particle/cell sorter devices often utilize material-selective DEP forces to create separate trajectories for different types of particles (e.g., dead vs live cells) [13–18]. The design process hinges on engineering these trajectories. Similar arguments can be made for LOC devices used in other applications as well. A mathematical simulation model capable of predicting particle trajectories can help designers identify potential issues and optimize the device before proceeding to the fabrication step; consequently, saving time and cost.

Along with the effects of external forces, colloidal particles (we use the term *particle* throughout the paper to refer to any colloidal micro- or nano-specimen of roughly spherical shape) in LOCs exhibit Brownian motion which can affect their trajectories [19]. Brownian motion arises from random collisions between the particle and the molecules of the suspension fluid [20,21]. Although the phenomenon is usually not prominent for large particle sizes (tens of microns) [18], it can be notable when sizes are in sub-micron range. Therefore, it should be included in a generalized model. The Langevin equation, which takes into account the effects of external forces and Brownian motion, can be used for this purpose [20,22]. The viscous drag force experienced by the moving particle in the suspension fluid (which itself can be in motion) is integrated within the equation. This makes it ideal for modeling colloidal particle trajectories in a static or moving fluid.

Unlike colloids in a boundless isotropic fluid, particles in LOC devices are bounded within a finite-sized liquid chamber. As such, additional mechanics need to be taken into account for modeling particle behavior. Firstly, a particle close to the walls of the fluid chamber or a microfluidic channel would experience *hindered* diffusion instead of free diffusion [23–25]. This is due to the hydrodynamic interaction between the particle and the no-slip boundary layer [2] of the fluid near a solid surface. Another factor to consider is particle collision with the solid surfaces of the chip which alters its motion. In multi-particle systems, particle–particle collisions are also possible. The effect of collisions cannot be directly included within the Langevin equation and therefore, must be modeled separately. In this work, we present a numerical model that takes into account all the aforementioned physics to calculate particle trajectories inside LOCs. The numerical solver uses a finite-difference scheme to solve the Langevin partial-differential equation and obtain the time evolution of particle positions. We model hindered diffusion by using a particle-position dependent diffusion tensor in the Langevin equation. A collision detection algorithm is used at each time-step. If detected, an elastic collision model is used to modify the results of the Langevin mechanics. Our generalized model is capable of simulating heterogeneous multi-particle systems, where each particle can differ in size, mass, and material properties. In addition, a large number of particles can be simulated simultaneously. The numerical solver computes the particle positions in parallel and, therefore, does not require long computation time even for large number of particles. Thus, it is ideal for simulating and analyzing various different LOC environments.

The numerical solver is implemented in python and is made available publicly through a GitHub repository [26]. We chose python as it is open source and has many freely available scientific-computing libraries. The software and the mathematical libraries required to run the solver are freely available. Along with the numerical engine, we utilized the graphical libraries to create built-in visualization schemes for analyzing relevant data. This includes the ability to save the animation of the particle motion in the form of a video file and logging the position and velocity values of each particle.

To demonstrate the usefulness of the platform, we present simulation results of two LOC systems. The first example shows optical trapping of multiple microparticles. A dynamic nature is incorporated by activating and deactivating the trap at specific time instances. The corresponding simulation results show the trapping mechanics, as well as the dynamics of the particles when the trap is released. The second example covers a more complex LOC that uses DEP to sort and separate cells. Such LOCs have many applications [12,14,15]. A microfluidic device with flowing liquid is considered for this case. Material selective DEP force is used to direct viable (live) and non-viable (dead) yeast cells along different paths. These two distinctly different examples show the versatility of our solver for simulating various LOC configurations. Moreover, the solver is developed in modular form. Thus, switching from one device geometry to another requires modifying only a few modules and plugging in external force terms.

The rest of the paper is organized as follows: Section 2 introduces the Langevin equation and its discretization to the finite difference form. The particle position dependent diffusion tensor and collision mechanics are discussed in Sections 3 and 4, respectively. Section 5 describes how all the physics were integrated in the numerical solver. The simulation results for the two different LOCs are discussed in Section 6. Finally, concluding remarks are made in Section 7.

2. Langevin Equation

The motion of a colloidal micro- or nano-particle under the influence of an external force-field can be modeled using the Langevin equation. For a colloidal particle in a low Reynolds number environment with shear-flow, the Langevin equation has the form [20,27–29]:

$$m\frac{\partial \mathbf{v}(\mathbf{r},t)}{\partial t} = \frac{k_B T}{\overleftrightarrow{\mathbf{D}}(\mathbf{r})}\left[\mathbf{v}_f(\mathbf{r},t) - \mathbf{v}(\mathbf{r},t) + \sqrt{2}\overleftrightarrow{\mathbf{D}}_{\frac{1}{2}}(\mathbf{r})\mathbf{W}(t)\right] + \mathbf{F}_{\text{ext}}(\mathbf{r},t)\,. \tag{1}$$

Here, t is the time variable, m, $\mathbf{r}$ and $\mathbf{v}(\mathbf{r},t)$ are the mass, position, and velocity of the particle, respectively, $\mathbf{v}_f(\mathbf{r},t)$ is the velocity of the fluid, k_B is the Boltzmann constant, T is the temperature, $\mathbf{F}_{\text{ext}}$ is the net external applied force acting on the particle, $\overleftrightarrow{\mathbf{D}}$ is the diffusion tensor, and $\mathbf{W}(t)$ is a vector white noise term. The tensor $\overleftrightarrow{\mathbf{D}}_{\frac{1}{2}}$ is defined as the element-wise square root of $\overleftrightarrow{\mathbf{D}}$. Each Cartesian component of $\mathbf{W}(t)$ is a random process with zero mean and unit variance. The $m\dot{\mathbf{v}}(\mathbf{r},t)$ term represents the inertia of the particle. This term can often be dropped for small particles where the mass is negligible [20]. The friction/drag force term, $[\mathbf{v}_f(\mathbf{r},t) - \mathbf{v}(\mathbf{r},t)]k_BT/\overleftrightarrow{\mathbf{D}}$, depends on the viscous properties of the fluid [17]. The white noise term $\sqrt{2}\overleftrightarrow{\mathbf{D}}_{\frac{1}{2}}(\mathbf{r})\mathbf{W}(t)$ is the fluctuating force resulting from random collisions between the particle and fluid molecules. The $\mathbf{F}_{\text{ext}}$ term encapsulates all external forces acting on the particle (e.g., optical trapping force, dielectrophoretic force, gravitational force, etc.). Whereas most research articles use the derivatives of the position vector, we express Equation (1) in terms of the velocity vector and its derivative. We do so for easier subsequent integration of collision physics into the model which depends on the velocity vectors. Thus, directly solving the velocity vectors simplifies the analysis.

Equation (1) is a stochastic differential equation. The presence of the discontinuous white noise term makes solving a stochastic differential equation more complex than solving an ordinary differential equation [30]. Our approach to numerically solve Equation (1) involves using the Euler–Maruyama method to convert it into a finite difference equation [30,31]. We use the following discretization scheme:

$$m\frac{\mathbf{v}_{i+1} - \mathbf{v}_i}{\Delta t} = \frac{k_B T}{\overleftrightarrow{\mathbf{D}}(\mathbf{r}_i)}\left[\mathbf{v}_{f,i} - \mathbf{v}_{i+1} + \sqrt{\frac{2}{\Delta t}}\overleftrightarrow{\mathbf{D}}_{\frac{1}{2}}(\mathbf{r}_i)\mathbf{w}_i\right] + \mathbf{F}_{\text{ext},i}(\mathbf{r}_i)\,. \tag{2}$$

Here, $i = 0, 1, \cdots N_t$ represents the time index and Δt is the time step. The set of discrete time where the numerical solution is calculated is given by $t_i = i\Delta t$. The term $\mathbf{w}_i$ is a random vector whose Cartesian components are Gaussian random numbers with zero mean and unit variance. Solving for $\mathbf{v}_{i+1}$ from Equation (2) gives:

$$\mathbf{v}_{i+1} = \frac{\overleftrightarrow{\mathbf{\Lambda}}}{1+\overleftrightarrow{\mathbf{\Lambda}}}\mathbf{v}_i + \frac{1}{1+\overleftrightarrow{\mathbf{\Lambda}}}\left[\mathbf{v}_{f,i} + \sqrt{\frac{2}{\Delta t}}\overleftrightarrow{\mathbf{D}}_{\frac{1}{2}}(\mathbf{r}_i)\mathbf{w}_i + \frac{\overleftrightarrow{\mathbf{D}}(\mathbf{r}_i)}{k_B T}\mathbf{F}_{\text{ext},i}(\mathbf{r}_i)\right]. \tag{3}$$

Here, $\overleftrightarrow{\mathbf{\Lambda}} \triangleq \frac{m\overleftrightarrow{\mathbf{D}}}{k_B T \Delta t}$ is a unit-less tensor quantity defined to make the equation compact. It should be noted that we use $\mathbf{v}_{i+1}$ instead of $\mathbf{v}_i$ in the right hand side of Equation (2). This ensures the stability of the numerical scheme for arbitrary Δt values. This fact can be verified by setting the fluid velocity, the white noise term, and the external force to be zero and observing the simplified velocity update equation: $\mathbf{v}_{i+1} = \frac{\overleftrightarrow{\mathbf{\Lambda}}}{1+\overleftrightarrow{\mathbf{\Lambda}}}\mathbf{v}_i$. As the multiplicative factor is less than unity (every component of $\overleftrightarrow{\mathbf{\Lambda}}$ is positive), the velocity will remain bounded over successive iterations and thus, stay stable.

After calculating the velocity vector at a given time step i, the position vector can be easily calculated using the Euler–Cromer method [32]:

$$\mathbf{r}_{i+1} = \mathbf{r}_i + \mathbf{v}_{i+1}\Delta t\,. \tag{4}$$

The position and velocity data of each particle is calculated using the same equations. The process is repeated for the next discrete time instance until the simulation end time is reached.

3. Diffusion Tensor and Hydrodynamic Interactions

The Langevin equation contains the diffusion tensor which is a representation of the fluid–particle interaction. The diffusion tensor in an isotropic homogeneous medium, $\overleftrightarrow{\mathbf{D}}_0$, depends on the material properties of the suspension medium, particle size, and the temperature. It is given by [23,24]:

$$\overleftrightarrow{\mathbf{D}}_0 = \frac{k_B T}{6\pi\eta r_o}\overleftrightarrow{\mathbf{I}}\,, \tag{5}$$

where $\overleftrightarrow{\mathbf{I}}$ is a 3×3 unit tensor, r_o is the radius of the particle, and η is the dynamic viscosity of the medium. Equation (5) is equivalent to using a scalar diffusion coefficient. For LOC devices, however, the fluid chamber or microfluidic channels are not large enough compared to the particle size for assuming isotropic conditions. In addition, particles are often located near the bottom surface during manipulation in such devices. So, the particle dynamics are different from the isotropic case. The hydrodynamic interactions between the particle and the static fluid layers near a surface hinder the diffusion process. Thus, the diffusion tensor becomes dependent on the relative position of the particle with respect to the solid surfaces of the device. The generalized diffusion tensor for such cases is given by [23,24]:

$$\overleftrightarrow{\mathbf{D}}(\mathbf{r}) = \frac{k_B T}{6\pi\eta r_o}\overleftrightarrow{\mathbf{H}}(\mathbf{r})\,. \tag{6}$$

$\overleftrightarrow{\mathbf{H}}(\mathbf{r})$ is a 3×3 unitless tensor that depends on the distance of the particle from the solid surfaces. For a geometry aligned with the coordinate system, only the diagonal elements of the tensor will be non-zero, i.e., $\overleftrightarrow{\mathbf{H}}(\mathbf{r}) = [H_{jk}(\mathbf{r})]$, $H_{jk}(\mathbf{r}) = 0\ \forall j \neq k$.

The majority of the hydrodynamic interactions in LOCs arise from the bottom/top surface along which particle manipulation takes place. Thus, we consider a case with a single solid surface located at the $z = 0$ plane. The components of $\overleftrightarrow{\mathbf{H}}(\mathbf{r})$ can be expressed as $H_{11}(\mathbf{r}) = H_{22}(\mathbf{r}) = H_{\parallel}(\mathbf{r})$ and $H_{33}(\mathbf{r}) = H_{\perp}(\mathbf{r})$ where [23–25]:

$$H_{\parallel}(\mathbf{r}) = 1 - \frac{9r_o}{16z} + \frac{r_o^3}{8z^3} - \frac{45r_o^4}{256z^4} - \frac{r_o^5}{16z^5}, \tag{7}$$

$$H_{\perp}(\mathbf{r}) = \frac{6z^2 + 2r_o z}{6z^2 + 9r_o z + 2r_o^2}\,. \tag{8}$$

Here, z is the z-position of the particle. In Figure 1, $H_{\parallel}(\mathbf{r})$ and $H_{\perp}(\mathbf{r})$ are plotted as functions of the normalized variable z/r_o. It can be observed that for particle position far from the $z = 0$ surface (i.e., high z/r_o values), the value of both terms approach unity and the diffusion tensor approaches the free space diffusion tensor. The closer the particle is to the surface, the larger the variation is from the free space case. Values less than unity suggests that the diffusion process is hindered near solid surfaces.

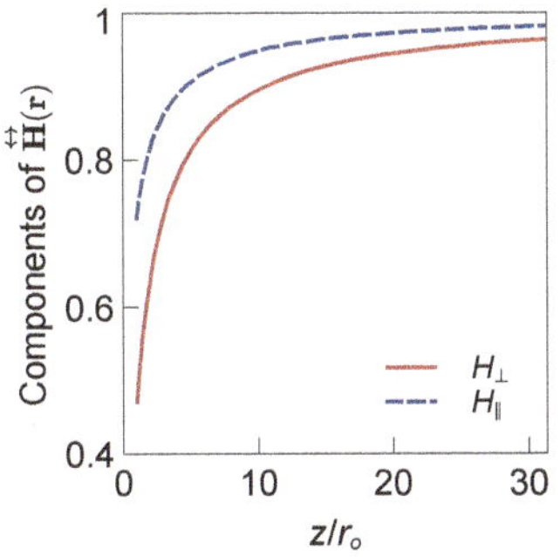

Figure 1. Components of $\overleftrightarrow{\mathbf{H}}(\mathbf{r})$ as functions of z/r_o considering a solid surface located at the $z = 0$ plane.

4. Collision Mechanics

In a multi-particle system, particle-particle collisions need to be taken into account when simulating particle trajectories. We consider a system with N_p particles where particle p and particle q collide at time instance i. Their mass, position, and initial velocity are m_p, m_q, $\mathbf{r}_{p,i}$, $\mathbf{r}_{q,i}$, and $\mathbf{v}_{p,i}$, $\mathbf{v}_{q,i}$, respectively. The velocity vectors after the collision, $\mathbf{v}_{p,i}^{\mathrm{col,pp}}$ and $\mathbf{v}_{q,i}^{\mathrm{col,pp}}$, can be calculated using the theory of elastic collisions where the kinetic energy and momentum of the particles are conserved.

The collision mechanics in three-dimensional systems can be treated by considering the velocities along the normal and tangential directions separately. We note that the velocity along the normal direction of the colliding surfaces are altered after collision, whereas the tangential velocity components are unaffected [33]. First, we define the normal vector:

$$\hat{\mathbf{n}} = \frac{\mathbf{r}_{p,i} - \mathbf{r}_{q,i}}{|\mathbf{r}_{p,i} - \mathbf{r}_{q,i}|}. \tag{9}$$

The normal and tangential component of the particle p velocity is then given by $(\mathbf{v}_{p,i} \cdot \hat{\mathbf{n}})\hat{\mathbf{n}}$ and $\mathbf{v}_{p,i} - (\mathbf{v}_{p,i} \cdot \hat{\mathbf{n}})\hat{\mathbf{n}}$, respectively. One-dimensional elastic collision formula can be applied for the normal directions. The velocity after the collision for particle p is given by:

$$\mathbf{v}_{p,i}^{\mathrm{col,pp}} = \frac{(\mathbf{v}_{p,i} \cdot \hat{\mathbf{n}})\hat{\mathbf{n}}(m_p - m_q) + 2m_q(\mathbf{v}_{q,i} \cdot \hat{\mathbf{n}})\hat{\mathbf{n}}}{m_p + m_q} + \mathbf{v}_{p,i} - (\mathbf{v}_{p,i} \cdot \hat{\mathbf{n}})\hat{\mathbf{n}}. \tag{10}$$

It can be noted that the first term of Equation (10) is identical to the 1D elastic collision formula along the normal axis. Simplifying Equation (10) gives:

$$\mathbf{v}_{p,i}^{\mathrm{col,pp}} = \mathbf{v}_{p,i} - \frac{2m_p}{m_p + m_q}\left[(\mathbf{v}_{p,i} - \mathbf{v}_{q,i}) \cdot \hat{\mathbf{n}}\right]\hat{\mathbf{n}} \tag{11}$$

Swapping the subscripts p and q in Equation (11) gives the modified velocity equation for particle q.

In addition to particle–particle collisions, collisions with a wall/solid-surface may also occur in LOC systems (e.g., collision with bottom substrate, sidewalls of microfluidic channels, etc.). Assuming that the walls are much more massive than the particles, the velocity of a particle after collision can be calculated using a reflection operation. If particle p collides with a flat wall with surface normal $\hat{\mathbf{n}}_{\text{col,pw}}$, then the velocity is modified as:

$$\mathbf{v}_{p,i}^{\text{col,pw}} = \mathbf{v}_{p,i} - 2\left(\mathbf{v}_{p,i} \cdot \hat{\mathbf{n}}_{\text{col,pw}}\right)\hat{\mathbf{n}}_{\text{col,pw}} \tag{12}$$

Using Equations (11) and (12), most common collision mechanics occurring in LOC environments can be modeled.

5. Physics Integration

We incorporate all the aforementioned physics into the particle dynamics solver. The solver starts with randomly populated position and velocity vectors for each particle. A vector white noise term is generated from a normal distribution. The particle position depended diffusion tensor is calculated from Equation (6). Then Equation (3) and (4) is used to find predicted particle velocity and position for the next time step (i.e., $\mathbf{v}_{i+1}$ and $\mathbf{r}_{i+1}$, respectively). As the Langevin equation does not take into account the presence of any obstacles in the particle path, the calculated position vector may represent an unphysical condition (i.e., two particles overlapping, particles going through walls, etc.). To correct for this, the Euclidean distances between each particle pair and each particle-wall pair are calculated to check for collisions. If a collision is detected, then position and velocity given by the Langevin equations needs to be adjusted. Using Equation (11) or Equation (12), the velocity vector is adjusted as:

$$\mathbf{v}_{p,i+1}^{\text{adj}} = \begin{cases} \mathbf{v}_{p,i+1}^{\text{col,pp}}, & \text{if collision with particle.} \\ \mathbf{v}_{p,i+1}^{\text{col,pw}}, & \text{if collision with wall.} \end{cases} \tag{13}$$

The corresponding adjusted position vector is:

$$\mathbf{r}_{p,i+1}^{\text{adj}} = \mathbf{r}_{p,i} + \mathbf{v}_{p,i+1}^{\text{adj}}\Delta t\,. \tag{14}$$

The velocity and position vectors are replaced by the adjusted quantities (i.e., $\mathbf{v}_{p,i+1} \leftarrow \mathbf{v}_{p,i+1}^{\text{adj}}$ and $\mathbf{r}_{p,i+1} \leftarrow \mathbf{r}_{p,i+1}^{\text{adj}}$) before proceeding to the next time step. The Python code implements the entire process in parallel for all the particles (i.e., $p = 0, 1, \cdots N_p - 1$) inside a single time loop. The implemented wall collision physics is general and works for any arbitrary polygon geometry. The flowchart shown in Figure 2 shows all the calculation steps.

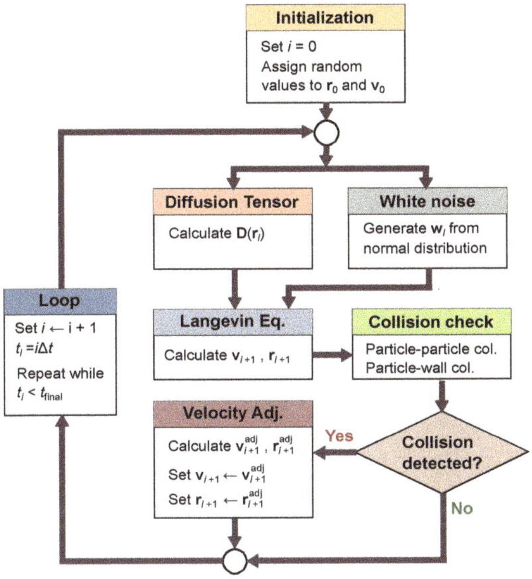

Figure 2. Flowchart of particle dynamics simulation steps.

6. Results

Using the developed algorithm, we present simulation results of two example LOC setups. The first example is a typical optical trapping setup with a Gaussian beam. The second example is a microfluidic LOC device that uses dielectrophoresis to sort or separate live and dead yeast cells.

6.1. Optical Trap

We consider an optical trapping experiment with a Gaussian beam. We implement a time dependent external force, modeling an optical beam that is turned ON and OFF at specific time instances. For simplicity, we assume a conservative gradient-force profile resulting from a Gaussian potential well, $u(\mathbf{r})$, centered around the origin. The potential is defined as:

$$u(\mathbf{r}) = \frac{4}{3}\pi r_0^3 A_d e^{-\left(\frac{r}{w}\right)^2}. \tag{15}$$

Here, $A_d = 120 k_B T\ \mu\text{m}^{-3}$ is the depth of the volume density of the potential well (i.e., depth of the potential well normalized with respect to particle volume), r_o is the radius of a particle, $w = 10\,\mu\text{m}$ is representative of the width of the well, and r is the radial distance from the trap center (assumed to be at the origin) to the particle position. The corresponding force profile, $\mathbf{F}_{\text{opt}}(\mathbf{r})$, is given by [7,34]:

$$\mathbf{F}_{\text{opt}}(\mathbf{r}) = -\nabla u(\mathbf{r}). \tag{16}$$

The force profile is shown in Figure 3. Due to the symmetry of the potential well, we only plot the radial component of the force, F_r. The x, y and z components of the force can easily be calculated from Equation (16) as well. Both the potential profile and the force profile are assumed to scale linearly with particle volume. This is a commonly used approximation for optical traps [35]. The conservative force-profile approximation is also commonly accepted for most cases. However, for some cases, especially involving near-field trapping, it is possible to have non-conservative force components [7]. Instead of Equation (16), more generalized approaches are necessary to model the force for those cases. For the current case, we define the $\mathbf{F}_{\text{ext}}$ term in the particle dynamics model as:

$$\mathbf{F}_{\text{ext}}(\mathbf{r},t) = \begin{cases} \mathbf{F}_{\text{opt}}(\mathbf{r}), & \text{if } 1\,\text{s} < t < 8\,\text{s}. \\ 0, & \text{otherwise.} \end{cases} \tag{17}$$

This represents a case when the optical beam is ON during the time interval $1\,\text{s} < t < 8\,\text{s}$. The interval boundaries are set arbitrarily. It should be noted that other possible external force terms (e.g., gravitational sedimentation, thermophoresis, etc.) can also be added here.

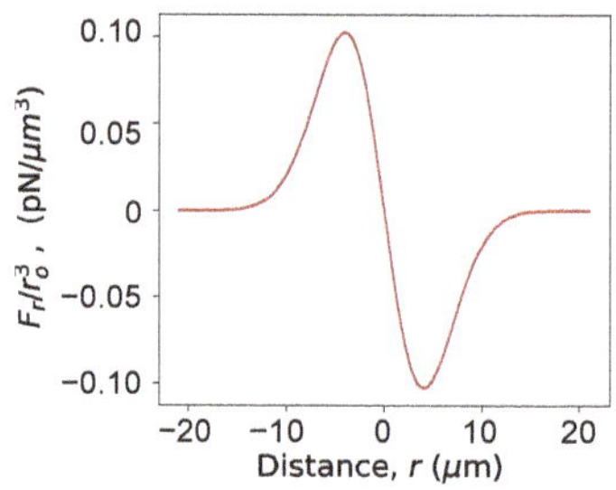

Figure 3. Force profile for the optical trap with Gaussian potential well.

We consider $N_p = 4$ polystyrene beads (density, $\rho = 1055\,\text{kg/m}^3$) suspended in water (dynamic viscosity, $\eta = 8.9 \times 10^{-4}\,\text{Pas}$, and temperature, $T = 300\,\text{K}$) with radii $r_{o,0} = 1.5\,\mu\text{m}$, $r_{o,1} = 2.5\,\mu\text{m}$, and $r_{o,2} = r_{o,3} = 2\,\mu\text{m}$. The corresponding particle masses can be calculated from the volume and density values. A solid bottom surface is assumed to be located at the $z = 0$ plane. With these conditions, the simulation was run from $t = 0$ to $t = 12\,\text{s}$ with a step size of $\Delta t = 0.01\,\text{s}$. The simulations results are shown in Figures 4 and 5. From Figure 4, we can see the particle positions at an arbitrary time instance ($t = 6.3\,\text{s}$), when they are trapped on the optical spot. The associated multimedia file 1 is an animated video that shows the same results for all time instances. Figure 5 show the positions and velocities of the particles as a function of time. The position plots show that the particles start moving towards the origin at around $t = 1\,\text{s}$, when the Gaussian optical spot (centered around the origin) is turned ON. As the particles move towards the optical spot, they start colliding with each other. This is represented by the spiky nature of the plots during the time interval from $t \approx 2\,\text{s}$ to $t \approx 8\,\text{s}$. After $t > 8\,\text{s}$, the optical spot is turned OFF and the particles are no longer trapped. During this time the Brownian motion dominates as the particles move randomly. It should be noted that a particle does not always become trapped when the optical spot is active. Trapping occurs only when a particle is within the capturing range of the trap [36,37]. In other instances of the simulation, where the random initial position of a particle is significantly away from the optical spot, the particle does not become trapped.

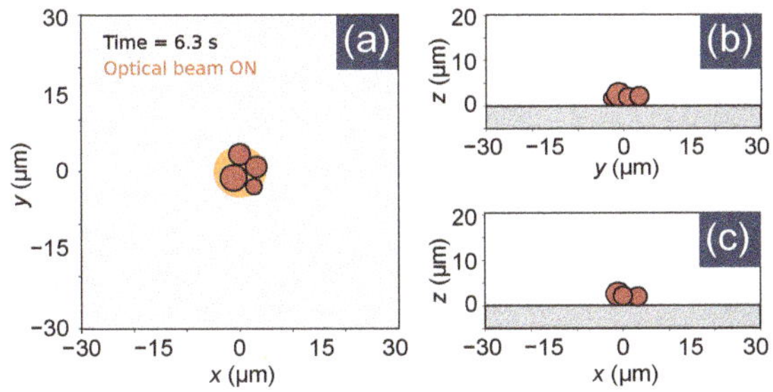

Figure 4. Position of particles in an optical trap at the time instance $t = 6.3$ s. (**a**) xy plane, (**b**) yz plane, and (**c**) xz plane view. The yellow circle in (**a**) indicates the optical spot. Multimedia file 1 is an animated video that shows the particle positions at different time instances.

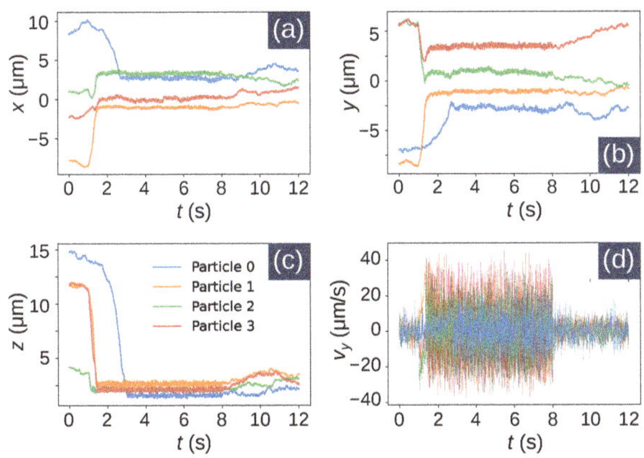

Figure 5. Position and velocities of particles in an optical trap as a function of time. (**a**) x position, (**b**) y position, (**c**) z position, and (**d**) y velocity component as functions of time. All the plots share the same legend.

6.2. Dielectrophoretic Cell Sorting or Separation

For our second example, we simulate a microfluidic device that uses DEP forces for cell sorting or separation which is a common application of LOCs [15–17]. Specifically, we focus on the separation of viable (live) and non-viable (dead) yeast cells using the device proposed by Doh et al. in their paper [15]. The device uses tapered electrodes embedded within a microfluidic chamber. The structure is shown in Figure 6a. The fluid walls are 50 μm high. The electrodes having a height of 0.12 μm can be considered planar. The electrodes have a length of 350 μm (along the x axis) and the gap between them is 16 μm. The top and bottom electrodes have a width (along y axis) of 20 μm, whereas the width of the tapered middle electrode spans 46–70 μm. The top and bottom electrodes are electrically shorted externally. An AC voltage of 4 V at 5 MHz is applied between the center electrode and the other electrodes. A strong field is created between the electrodes generating dielectrophoretic forces. As a solution of cells flows from left to right, the electrodes exert different forces on live and dead cells, modifying their trajectories. Thus, specific cells can be diverted to specific microfluidic channels. We choose this device as an example as it incorporates some of essential particle manipulation mechanics (e.g., fluid flow, dielectrophoretic force, etc.) used in many LOC devices.

To use our numerical solver to simulate this device, first we have to model the forces acting on the cells. The time averaged DEP force, $\mathbf{F}_{\text{DEP}}$, on the cells can be approximated as [3,15,17]:

$$\mathbf{F}_{\text{DEP}} = 2\pi\epsilon_0\epsilon_m r_o^3 \Re\big[\tilde{C}_M\big] \nabla |\mathbf{E}|^2 . \tag{18}$$

Here, ϵ_0 is the permittivity of free space, $\epsilon_m = 80$ is the relative permittivity of the medium, $r_o = 3\,\mu\text{m}$ [38] is the cell radius, $\tilde{C}_M$ is the Clausius–Mossotti factor, $\Re[\cdot]$ is the operator that outputs the real part of its argument, and $\mathbf{E}$ is the electric field. $\mathbf{E}$ depends on the geometry of the electrodes and the device. It can be noted that the sign of $\Re[\tilde{C}_M]$ determines whether $\mathbf{F}_{\text{DEP}}$ is attractive (positive DEP) or repulsive (negative DEP). Since $\tilde{C}_M$ depends on the material properties of the cells, they differ from live cells to dead cells [39]. We use the multi-shell model to calculate $\tilde{C}_M$ for viable and non-viable yeast cells [3]. The procedure is discussed in detail in the Supplementary Information Document. We take the geometrical and material parameters of the cells and the device dimensions from the literature [15,40,41]. Plugging in these values, we find that at $f = 5\,\text{MHz}$, the viable and non-viable cells have relative permittivities of $\epsilon_{\text{viable}} = 199.94$, and $\epsilon_{\text{nonviable}} = 18.82$, respectively, and have conductivities of $\sigma_{\text{viable}} = 0.36\,\text{S/m}$ and $\sigma_{\text{nonviable}} = 0.013\,\text{S/m}$, respectively. These values correspond to $\Re\big[\tilde{C}_{M,\text{ viable}}\big] = 0.945$ and $\Re\big[\tilde{C}_{M,\text{ nonviable}}\big] = -0.25$. Thus, the viable cells experience positive DEP, whereas the non-viable cells experience negative DEP.

The electric field term in Equation (18) can be calculated using any of the widely available numerical electromagnetic field solvers for any arbitrary device geometry. We use Comsol Multiphysics® for this purpose. The calculated electric potential and field distributions are shown in Figure 6b,c. The corresponding force values on a viable yeast cell (calculated using Equation (18)) are shown in Figure 6d. It can be seen that the force pulls the viable cells near the gap regions between the electrodes where the field intensity is maximum. For a non-viable cell, the sign of the force would be reversed and scaled by a factor of $|\Re\big[\tilde{C}_{M,\text{ nonviable}}\big]|/|\Re\big[\tilde{C}_{M,\text{ viable}}\big]| \approx 0.26$.

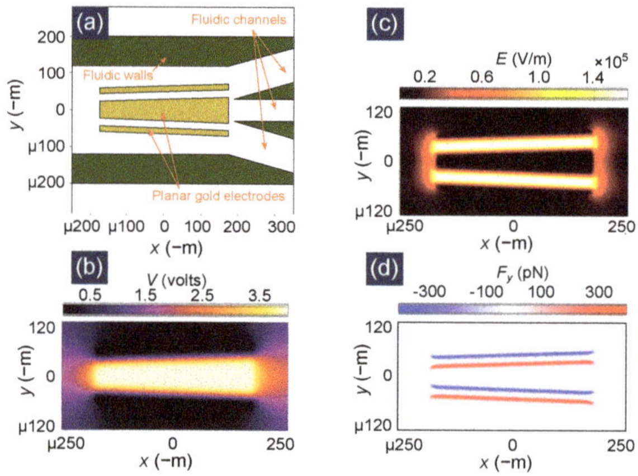

Figure 6. The geometry and electromagnetic simulation of the dielectrophoretic cell sorter device. (**a**) Top view of the device, (**b**) Electric potential distribution, (**c**) Electric field distribution, and (**d**) Force experienced by a viable yeast cell. All figures depict the xy plane view of the three-dimensional device. For simulation, the large middle electrode is excited with 4 V, 5 MHz AC voltage, whereas the other two electrodes are grounded.

Unlike the optical tweezers example, the cell sorter device works with a flowing fluid instead of a stationary pool of liquid. A left to right fluid flow with velocity of $150\,\mu\text{m/s}$ is considered. Due to the presence of the microfluidic channels, the fluid velocity does not maintain the purely left-to-right direction throughout the device. Considering the fact that the average fluid flow direction is tangential to the channel surface, we use the following simplified piece-wise model:

$$\mathbf{v}_f(\mathbf{r},t) = \begin{cases} 150\hat{\mathbf{x}} + 30\hat{\mathbf{y}}\,, & \text{if } x > 170\,\mu\text{m}\,, y > 30\,\mu\text{m}\,. \\ 150\hat{\mathbf{x}} - 30\hat{\mathbf{y}}\,, & \text{if } x > 170\,\mu\text{m}\,, y < -30\,\mu\text{m}\,. \\ 150\hat{\mathbf{x}}\,, & \text{otherwise.} \end{cases} \tag{19}$$

Here, $\mathbf{v}_f(\mathbf{r},t)$ has units of µm/s. The y component of the fluid velocity for a given x component is approximated from the slope of the channels so that the average flow is tangential to the fluid wall. A 3D moving average filter was applied to make the velocity profile smoother. It should be stated that this flow profile is an approximation. We also ignore any fluid flow that might be induced by the non-uniform electric field [42]. It is possible to plug in more accurate data from an external solver in our code. However, the piece-wise model is sufficiently accurate for the current demonstration problem. In the future, we plan to implement a numerical fluid flow solver and integrate it with our particle dynamics solver.

From Equations (18) and (19), we plug in the $\mathbf{F}_{\text{ext}} = \mathbf{F}_{\text{DEP}}$ and the fluid flow velocity in Equations (3) and (4) to solve the particle dynamics in the time range $t = 0$ to $t = 6\,\text{s}$ with a time step of $\Delta t = 5\,\text{ms}$. Like the optical trapping example, the particle–particle collision and particle–wall collision models are included in the simulation. We simultaneously simulate the trajectory of 17 viable yeast cells and 17 non-viable yeast cells with random initial positions located in the domain $x \in [-480, -200]$ µm, $y \in [-280, 280]$ µm and $z \in [6, 80]$ µm. The simulation results are shown in Figure 7. Four time instances are depicted here. The entire time evolution of the cell positions can be seen in the attached multimedia file 2. Although both live and dead cells experience the same fluid flow, they experience different DEP forces as they have different $\Re[\tilde{C}_M]$ values. We note that as the cells are carried left-to-right by the fluid flow, the live cells (depicted in red) become deflected top/bottom (near the electrode gaps) whereas the dead cells (depicted in blue) are pushed towards the center electrode. This is consistent with the force profile shown in Figure 6d. As the cells travel to the right of the device near the output fluid channels, the cells are grouped as can be seen in $t = 1.3\,\text{s}$ and $t = 2.7\,\text{s}$ frames of Figure 7. The top and bottom output fluid channels end up having a much higher concentration of live cells and the center channel contains most of the dead cells.

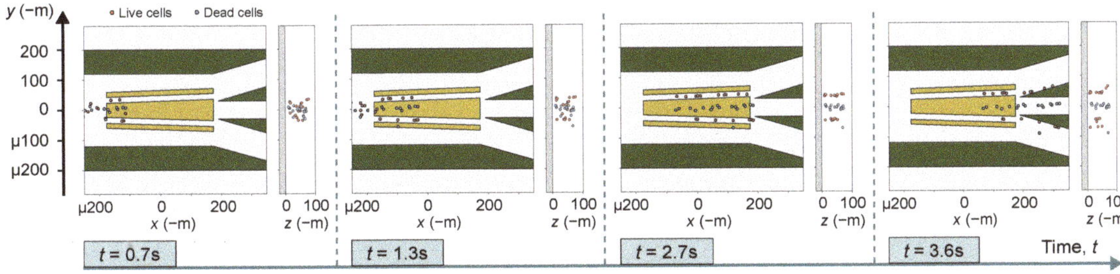

Figure 7. Position of live and dead yeast cells at four time instance. All figures share the same y axis. At each time instance, the particle position is plotted in xy and zy planes. The times instances are separated by dashed lines. The red circles represent live cells and the blue circles represent dead cells. The green regions are microfluidic walls, whereas the gold region are the planar electrodes. The gray region represents the substrate of the device. Multimedia file 2 contains an animation showing data of all simulated time instances.

The sorting mechanics can be analyzed further by plotting the trajectory of a few cells, as shown in Figure 8. During $0 < t < 0.8\,\text{s}$, the y position of the all the cells remain constant as they are not yet near the electrodes to experience the dielectrophoretic force.

At this stage, driven by the fluid flow, their motion is limited to the x direction. During $0.8\,\mathrm{s} < t < 3.5\,\mathrm{s}$, the cells are within the region of influence of the electrodes. The live cells move toward the high field region (the region between the electrodes) and dead cells move away from there. Since the $y = 0$ line is at the middle of the two high field regions (note Figure 6c), the dead cells accumulate around this line. On the other hand, the live cells accumulate near the top or bottom electrodes. At $t > 3.5\,\mathrm{s}$, each cell enters one of the three output microfluidic channels. The fluid flows in the top and bottom channels change directions creating a shift in cell trajectory. This can be noted in Figure 8 as the live cells follow a trajectory parallel to the top or bottom channels. The dead cells follow a constant y path through the center channel. This is consistent with the experimental results reported by Doh et al. in their paper [15]. Thus, the simulation accurately demonstrates the cell sorting or separation mechanics.

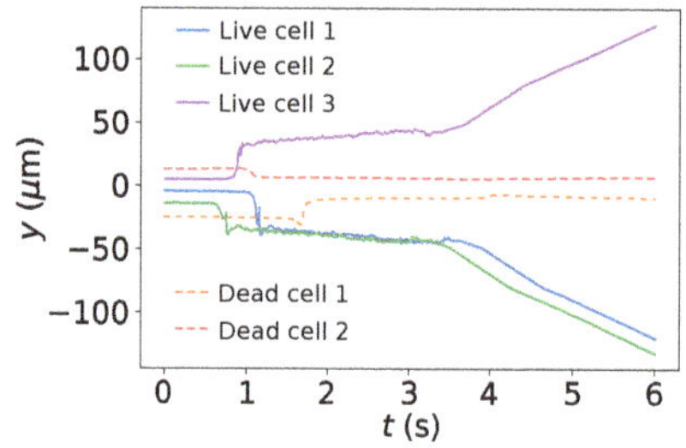

Figure 8. y trajectory data of three live cells and two dead cells.

7. Conclusions

A numerical model that can simulate behavior of particles in LOC devices was developed. Using the model, we successfully simulated the operations of an optical trap and a dielectrophoretic cell sorter. The numerical solver is generally applicable and can simulate most LOC devices once the external force terms are defined. It can be used for the development and optimization of new and existing LOC devices. The solver was coded using open-source Python libraries and can be run on freely accessible compilers. We also made our code for the solver publicly available [26]. Due to the general nature, the developed solver can be of interest to researchers working on biomicrofluidics and LOC devices.

Supplementary Materials: The following are available online at https://www.mdpi.com/article/10.3390/mi12101265/s1, The Supplementary Material contains a detailed description of Clausius–Mossotti factor and how the dielectric function of cells were modeled. Multimedia files showing animation of the simulations are also included in the supplementary materials.

Author Contributions: Conceptualization, M.A.Z. and L.H.; methodology, M.A.Z., M.W. and P.P.; validation, M.W., P.P. and M.A.J.; writing—original draft preparation, M.A.Z.; visualization, M.A.Z. and M.W.; supervision, L.H. and R.W.D. All authors have read and agreed to the published version of the manuscript.

Funding: This work was partially supported by the National Institute of Health (NIH) Grant 5R21HG009758 and R01GM138716.

Institutional Review Board Statement: Not applicable.

Informed Consent Statement: Not applicable.

Data Availability Statement: The data that support the findings of this study are available from the corresponding author upon reasonable request.

Acknowledgments: The authors thank Paul C. Hansen, Yao-Te Cheng and Max Yuen for their helpful suggestions.

Conflicts of Interest: The authors declare no conflicts of interest.

References

1. Markarian, N.; Yeksel, M.; Khusid, B.; Farmer, K.R.; Acrivos, A. Particle motions and segregation in dielectrophoretic microfluidics. *J. Appl. Phys.* **2003**, *94*, 4160–4169. [CrossRef]
2. Arefi, S.M.A.; Yang, C.W.T.; Sin, D.D.; Feng, J.J. Simulation of nanoparticle transport and adsorption in a microfluidic lung-on-a-chip device. *Biomicrofluidics* **2020**, *14*, 044117. [CrossRef]
3. Pethig, R. Review Article—Dielectrophoresis: Status of the theory, technology, and applications. *Biomicrofluidics* **2010**, *4*, 022811. [CrossRef]
4. Zaman, M.A.; Padhy, P.; Ren, W.; Wu, M.; Hesselink, L. Microparticle transport along a planar electrode array using moving dielectrophoresis. *J. Appl. Phys.* **2021**, *130*, 034902. [CrossRef]
5. Hyoung Kang, K.; Xuan, X.; Kang, Y.; Li, D. Effects of dc-dielectrophoretic force on particle trajectories in microchannels. *J. Appl. Phys.* **2006**, *99*, 064702. [CrossRef]
6. Enger, J.; Goksör, M.; Ramser, K.; Hagberg, P.; Hanstorp, D. Optical tweezers applied to a microfluidic system. *Lab Chip* **2004**, *4*, 196–200. [CrossRef]
7. Zaman, M.A.; Padhy, P.; Hesselink, L. Near-field optical trapping in a non-conservative force field. *Sci. Rep.* **2019**, *9*, 649. [CrossRef]
8. Lu, D.; Gámez, F.; Haro-González, P. Temperature Effects on Optical Trapping Stability. *Micromachines* **2021**, *12*, 954. [CrossRef]
9. Hsiao, Y.C.; Wang, C.H.; Lee, W.B.; Lee, G.B. Automatic cell fusion via optically-induced dielectrophoresis and optically-induced locally-enhanced electric field on a microfluidic chip. *Biomicrofluidics* **2018**, *12*, 034108. [CrossRef]
10. Wu, M.C. Optoelectronic tweezers. *Nat. Photonics* **2011**, *5*, 322–324. [CrossRef]
11. Zaman, M.A.; Padhy, P.; Cheng, Y.T.; Galambos, L.; Hesselink, L. Optoelectronic tweezers with a non-uniform background field. *Appl. Phys. Lett.* **2020**, *117*, 171102. [CrossRef]
12. Alshareef, M.; Metrakos, N.; Perez, E.J.; Azer, F.; Yang, F.; Yang, X.; Wang, G. Separation of tumor cells with dielectrophoresis-based microfluidic chip. *Biomicrofluidics* **2013**, *7*, 011803. [CrossRef]
13. Patel, S.; Showers, D.; Vedantam, P.; Tzeng, T.R.; Qian, S.; Xuan, X. Microfluidic separation of live and dead yeast cells using reservoir-based dielectrophoresis. *Biomicrofluidics* **2012**, *6*, 034102. [CrossRef] [PubMed]
14. Calero, V.; Garcia-Sanchez, P.; Ramos, A.; Morgan, H. Combining DC and AC electric fields with deterministic lateral displacement for micro- and nano-particle separation. *Biomicrofluidics* **2019**, *13*, 054110. [CrossRef] [PubMed]
15. Doh, I.; Cho, Y.H. A continuous cell separation chip using hydrodynamic dielectrophoresis (DEP) process. *Sens. Actuators A Phys.* **2005**, *121*, 59–65. [CrossRef]
16. Hughes, M.P. Fifty years of dielectrophoretic cell separation technology. *Biomicrofluidics* **2016**, *10*, 032801. [CrossRef]
17. Kazemi, B.; Darabi, J. Numerical simulation of dielectrophoretic particle separation using slanted electrodes. *Phys. Fluids* **2018**, *30*, 102003. [CrossRef]
18. Mathew, B.; Alazzam, A.; Abutayeh, M.; Gawanmeh, A.; Khashan, S. Modeling the trajectory of microparticles subjected to dielectrophoresis in a microfluidic device for field flow fractionation. *Chem. Eng. Sci.* **2015**, *138*, 266–280. [CrossRef]
19. Choi, S.; Lee, W.I.; Lee, G.H.; Yoo, Y.E. Analysis of the Binding of Analyte-Receptor in a Micro-Fluidic Channel for a Biosensor Based on Brownian Motion. *Micromachines* **2020**, *11*, 570. [CrossRef]
20. Volpe, G.; Volpe, G. Simulation of a Brownian particle in an optical trap. *Am. J. Phys.* **2013**, *81*, 224–230. [CrossRef]
21. Reeves, D.B.; Weaver, J.B. Simulations of magnetic nanoparticle Brownian motion. *J. Appl. Phys.* **2012**, *112*, 124311. [CrossRef]
22. Wei, Y.F.; Hsiao, P.Y. Unfolding polyelectrolytes in trivalent salt solutions using dc electric fields: A study by Langevin dynamics simulations. *Biomicrofluidics* **2009**, *3*, 022410. [CrossRef]
23. Banerjee, A.; Kihm, K.D. Experimental verification of near-wall hindered diffusion for the Brownian motion of nanoparticles using evanescent wave microscopy. *Phys. Rev. E* **2005**, *72*, 042101. [CrossRef]
24. Choi, C.; Margraves, C.; Kihm, K. Examination of near-wall hindered Brownian diffusion of nanoparticles Experimental comparison to theories by Brenner (1961) and Goldman et al. (1967). *Phys. Fluids* **2007**, *19*, 103305. [CrossRef]
25. Svoboda, K.; Block, S.M. Biological Applications of Optical Forces. *Annu. Rev. Biophys. Biomol. Struct.* **1994**, *23*, 247–285. [CrossRef] [PubMed]
26. Zaman, M.A. Brownian Dynamics in a Time-Varying Force-Field. 2021. Available online: https://github.com/zaman13/Brownian-dynamics-in-a-time-varying-force-field (accessed on 7 September 2021).
27. Ermak, D.L.; McCammon, J. Brownian dynamics with hydrodynamic interactions. *J. Chem. Phys.* **1978**, *69*, 1352–1360. [CrossRef]
28. Katayama, Y.; Terauti, R. Brownian motion of a single particle under shear flow. *Eur. J. Phys.* **1996**, *17*, 136–140. [CrossRef]
29. Drossinos, Y.; Reeks, M.W. Brownian motion of finite-inertia particles in a simple shear flow. *Phys. Rev. E* **2005**, *71*, 031113. [CrossRef]
30. Higham, D.J. An algorithmic introduction to numerical simulation of stochastic differential equations. *SIAM Rev.* **2001**, *43*, 525–546. [CrossRef]
31. Dimits, A.; Cohen, B.; Caflisch, R.; Rosin, M.; Ricketson, L. Higher-order time integration of Coulomb collisions in a plasma using Langevin equations. *J. Comput. Phys.* **2013**, *242*, 561–580. [CrossRef]
32. Cromer, A. Stable solutions using the Euler approximation. *Am. J. Phys.* **1981**, *49*, 455–459. [CrossRef]
33. Čepič, M. Elastic collisions of smooth spherical objects: Finding final velocities in four simple steps. *Am. J. Phys.* **2019**, *87*, 200–207. [CrossRef]

34. Xu, H.; Käll, M. Surface-Plasmon-Enhanced Optical Forces in Silver Nanoaggregates. *Phys. Rev. Lett.* **2002**, *89*. [CrossRef]
35. Neuman, K.C.; Block, S.M. Optical trapping. *Rev. Sci. Instrum.* **2004**, *75*, 2787–2809. [CrossRef]
36. Zaman, M.A.; Padhy, P.; Hansen, P.C.; Hesselink, L. Dielectrophoresis-assisted plasmonic trapping of dielectric nanoparticles. *Phys. Rev. A* **2017**, *95*, 023840. [CrossRef]
37. Zaman, M.A.; Padhy, P.; Hesselink, L. Capturing range of a near-field optical trap. *Phys. Rev. A* **2017**, *96*, 043825. [CrossRef]
38. Vahey, M.D.; Pesudo, L.Q.; Svensson, J.P.; Samson, L.D.; Voldman, J. Microfluidic genome-wide profiling of intrinsic electrical properties in Saccharomyces cerevisiae. *Lab Chip* **2013**, *13*, 2754. [CrossRef] [PubMed]
39. Ettehad, H.M.; Wenger, C. Characterization and Separation of Live and Dead Yeast Cells Using CMOS-Based DEP Microfluidics. *Micromachines* **2021**, *12*, 270. [CrossRef]
40. Talary, M.S.; Burt, J.P.H.; Tame, J.A.; Pethig, R. Electromanipulation and separation of cells using travelling electric fields. *J. Phys. D Appl. Phys.* **1996**, *29*, 2198–2203. [CrossRef]
41. Asami, K.; Hanai, T.; Koizumi, N. Dielectric properties of yeast cells. *J. Membr. Biol.* **1976**, *28*, 169–180. [CrossRef]
42. Green, N.G.; Ramos, A.; González, A.; Morgan, H.; Castellanos, A. Fluid flow induced by nonuniform ac electric fields in electrolytes on microelectrodes. III. Observation of streamlines and numerical simulation. *Phys. Rev. E* **2002**, *66*, 026305. [CrossRef]

micromachines

MDPI

Article

Travelling-Wave Dipolophoresis: Levitation and Electrorotation of Janus Nanoparticles

Touvia Miloh * and Jacob Nagler

School of Mechanical Engineering, University of Tel-Aviv, Tel-Aviv 69978, Israel; syankitx@gmail.com
* Correspondence: miloh@eng.tau.ac.il

Abstract: We present a theoretical study of the hydrodynamic and electrokinetic response of both metallic spherical polarized colloids as well as metallodielectic Janus particles, which are subjected to an arbitrary non-uniform ambient electric field (DC or AC forcing). The analysis is based on employing the linearized 'standard' model (Poisson–Nernst–Planck formulation) and on the assumptions of a 'weak' field and small Debye scale. In particular, we consider cases of linear and helical time-harmonic travelling-wave excitations and provide explicit expressions for the resulting dielectrophoretic and induced-charge electrophoretic forces and moments, exerted on freely suspended particles. The new analytic expressions thus derived for the linear and angular velocities of the initially uncharged polarizable particle are compared against some available solutions. We also analyze the levitation problem (including stability) of metallic and Janus particles placed in a cylindrical (insulating or conducting) pore near a powered electrode.

Keywords: electrokinetics of metallodielectric Janus particles; dipolophoresis; dielectrophoresis; induced-charge electrophoresis; travelling waves and non-uniform fields; electrorotation; levitation and stability

Citation: Miloh, T.; Nagler, J. Travelling-Wave Dipolophoresis: Levitation and Electrorotation of Janus Nanoparticles. *Micromachines* **2021**, *12*, 114. https://doi.org/10.3390/mi12020114

Academic Editor: Antonio Ramos

Received: 22 December 2020

Accepted: 14 January 2021

Published: 22 January 2021

1. Introduction

One of the preferred techniques for manipulating and controlling phoretic motions of micro/nano polarizable particles freely suspended in an electrolyte, is by means of externally applying ambient electric fields. The applied field can be steady (DC) or time-harmonic (AC), as well as uniform (constant) or spatially non-uniform. The nanoparticle (NP) itself can be conducting or dielectric, uncharged or initially charged, chemically active or inert, homogeneous or non-homogeneous (i.e., meta-material). In this study, we theoretically consider the case of a metallic (perfectly conducting) spherical NP, as well as metallodielectric (MD) Janus particle (JP) comprising of two hemispheres, one metallic and the other dielectric. Both NP and JP are subjected to an arbitrary time-harmonic non-uniform electric field and are assumed inert and initially uncharged. Nanofluidic applications of such colloids are abundant, especially those involving JP due to their inherent symmetry-breaking features and potential use as cargo carriers (see for example the recent reviews [1,2]). Nevertheless, it should be mentioned that in spite of the current interest in JP electrokinetics and the growing number of publications on the subject, theoretical studies on the dynamics (i.e., induced linear and angular velocities) of JP's under general non-uniform AC excitations, are rather scarce [3].

When exposed to a non-uniform ambient field, an initially uncharged polarized colloid experiences a dielectrophoretic (DEP) force and torque, determined by the multipole system (including the leading dipole term) within the NP and the corresponding partial derivatives of the ambient potential evaluated at these singularities [4]. Due to polarization, a distribution of induced charges (Poisson equation) is created in the electrolyte (decaying exponentially away from the NP) by the electric field. The above induced-charge density interacts with the electric field and inflicts a low-Reynolds (creeping) fluid motion around the NP. The resulting force (integration of viscous stresses over the NP) is defined as the 'induced –charge electrophoretic' force (ICEP) acting on the colloid [5,6]. The sum of these two forces (DEP +ICEP) has been coined by [7] as the dipolophoretic (DIP) force. At least for spherical polarized colloids (especially at low forcing frequencies), DEP and ICEP generally act in opposite directions. Interestingly, it has been shown by [8], that for a *constant-gradient* field (i.e., linear in Cartesian coordinates), DEP and ICEP for a spherical NP precisely cancel each other so that the total DIP is practically null! However, as demonstrated in [9,10], this result holds only under the limit of infinitely thin electric double layer (EDL) and for time-independent (DC) electric forcing.

In most cases, the applied fields are taken as uniform. They may be unidirectional or have two out of phase uniform components, acting along orthogonal directions resulting in electrorotation (ROT) [4,11]. In these cases, the DEP force is null but there is a finite DEP torque leading to particle rotation. Explicit expressions for the DIP angular velocity of, say metallic nanowire [12], sphere [13–15] or tri-axial ellipsoid [16], can be found by separately considering the DEP and ICEP contributions to the torque exerted on the NP. Alternately, the ambient field may combine uniform and 'constant-gradient' terms [8,9,17] (typical for active colloids [18]), where both DEP and ICEP act simultaneously in opposite directions. The most general electric forcing however, involves arbitrary time-harmonic (single or multiple frequencies) non-uniform ambient fields, acting on a colloid of unrestricted EDL. The underlying nonlinear electrostatic problem is first solved by linearization and employing the PNP standard model based on the 'weak' field assumption [19]. The hydrodynamic problem is next analyzed by considering the non-homogeneous Stokes equation with a Columbic forcing.

The special case of an infinitely polarizable (metallic) free spherical NP placed under AC excitations [5,19,20], is of particular interest since it is amenable to analysis and so is (for the same reason) the special case of a travelling- wave electric forcing [4,11]. One of the earlier studies on sinusoidal time-harmonic travelling-wave electrokinetics, is due to Huang et al. [21], coining the 'twDEP' terminology and demonstrating that unlike conventional DEP, twDEP is related to the imaginary rather to the real part of the NP dipole- term. The idea of pumping fluids by means of travelling-wave electro-osmosis, using an array of equally spaced planar or spiral elec-trodes, was proposed in several studies [22–25]. The corresponding twDIP axisymmetric problem of freely suspended perfectly conducting spherical NP, placed along the center of a cylindrical pore, was analyzed by Miloh and Boymelgreen [26]. Recently, a theoretical/numerical twDIP study of free metallic spherical (or cylindrical) NP's, lying in unbounded space and subject to a sinusoidal travelling wave AC electric forcing, has been also presented by Flores et al. [27]. However, the above mentioned twDIP studies, can be considered only as 'long-wave length' approximations of the full problem (by considering only the first two linear terms) and practically fail when the typical wavelength of the electric forcing, is of the same order as the size of the NP (Rayleigh's assumption). One of the main results of the present study, is obtaining a closed-form (Bessel function) results for the twDIP force and torque acting on a metallic spherical NP under a general 3D linear or helical travelling-wave electric forcing. These new results are exact in the sense that they are valid for arbitrary wavelengths and are shown to reduce (to leading-order) to the available approximate (long-o wave) solutions for metallic spherical NP's.

Unlike homogeneous NP's, DIP theoretical studies on non-homogeneous (two-face) Janus particles are scarce [28,29], due to assitional symmetry breaking complexities. Indeed, most of the recent publications (mostly experimental) on JP's electrokinetics, consider metallodielectic (including effect of coating) spherical particles under a *uniform* DC ac AC fields [3,8,18,30–38]. As stated, the growing interest in JP's dynamics is connected with their special self-propulsion features as cargo carriers, mobile electrodes and unique frequency dispersion characteristics. A spherical JP that is subjected to a uniform field will not experience a DEP force (ignoring boundary or wall effects) and any non-symmetric particle (contrary to a *spherical* NP), will acquire a finite phoretic velocity due to ICEP, as first demonstrated in the DC limit by Squires and Bazant [8]. When exposed to a non-uniform ambient field (such as linear or helical travelling wave excitations), the JP is subjected to both DEP and ICEP resulting in an induced phoretic motion (translation or rotation). Nevertheless, as demonstrated in the sequel, the DIP response of a JP is essentially different from that of an homogeneous NP, even under the same electric forcing. By taking advantage of the large disparity between the dielectric properties of the two hemispheres comprising the metallodielectric JP, we are able to further simplify the DIP analysis [28,29] and present for the first time an explicit DIP solution for a spherical JP excited by an arbitrary travelling wave. We also consider the pertinent DEP levitation problem [4,39–43] of a freely suspended JP placed axi-symmetrically in a cylindrical pore near a powered circular electrode. In addition to the practical problem of electric levitation, there is much interest in other modes of physical levitation, such as acoustic and optical techniques [44–46] for manipulating free nanoparticles under AC fields in a gaseous environment. By analyzing the stability (radial and axial) of the above DEP levitation problem, we demonstrate that a spherical JP is more responsive to levitation, since its height is 2–3 larger compared to that of an equivalent NP (same forcing). We also provide a simple expression for this equilibrium elevation in terms of the physical parameters.

In this work we present a theoretical framework for determining the dynamic response (forces and torques) of spherical metallic NP"s or metallodielectric Janus particles, freely suspended in an unbounded electrolyte that are exposed to a spatially non-homogeneous time- harmonic ambient electric field. Analytic expressions are provided for both the DEP and ICEP dynamic loads (forces and torques) acting on a single colloid, thus providing the total DIP phoretic response of the NP (i.e., linear and angular velocities), under the creeping flow (Stokes) hypothesis. In particular, we consider cases of NP or JP that are subjected to linear or helical travelling wave AC excitations and provide exact solutions for the loads in terms of the forcing wavelength and frequency. The different dynamic response between NP and JP to the same electric excitations, are also discussed in the context of the levitation problem of an isolated colloid placed in a vertical cylindrical pore above a powered electrode. Explicit expressions are provided for the equilibrium levitation height and passive stability in both radial and axial directions.

The structure of the paper is as follows: In Section 2, we provide a summary of the general methodology for solving the non-linear electrostatic problem of an initially uncharged freely suspended spherical NP or JP lying in an unbounded symmetric binary electrolyte. Linearization is then enforced by assuming 'weak' field AC excitations ('standard' model), resulting in a Robin-type boundary condition applied on the surface of the polarizable colloid. The DEP problem is first discussed in Section 3, by obtaining explicit expressions for the internal higher -order multipoles (DC and AC) in axisymmetric and non-symmetric electric excitations. In particular we demonstrate the above methodology for linear and helical travelling- wave forcing. Next, we consider in Section 4, the corresponding ICEP problem for the same electric forcing using Teubner's [47] approach under the assumption of an infinitely thin EDL. The frequent case of the DIP torque exerted on NP or JP under helical (circumferential) wave forcing, is presented in Section 5 and compared against the available electrorotation (ROT) solutions. The special shape of a metallodielectric JP, consisting of two hemispheres which have unique symmetry-breaking properties affecting both DEP and ICEP, is discussed in Section 6 including a comparison to the known DC-ICEP solution under a uniform field forcing. Finally, we present, in Section 7, detailed stability analysis of the pertinent levitation problem of both NP and JP placed symmetrically in a vertical cylindrical pore (insulating or conducting walls) above a powered circular electrode. It is demonstated that JP's are more responsive to levitation compared to NP's (by a factor 2–3) under the same forcing. We conclude with a short Section 8, including a summary and discussion.

2. General Formulation

We consider a spherical metallodielectric (MD) micro/nano Janus particle (JP), which is exposed to an ambient arbitrary non-uniform alternative current (AC) electric field (such as a travelling- wave). The free JP of radius a consists of two hemispheres with large contrast between the corresponding permittivities and is suspended in a symmetric monovalent aqueous unbounded electrolyte. Using a spherical coordinate system (R, θ, φ) centered at the JP, the general imposed non-homogenous harmonic electric field can be conveniently expressed as:

$$\overline{\chi}_{am}(R,\theta,\varphi) = -\sum_{n=1}^{\infty}\sum_{m=0}^{2n+1} A_n^m \widetilde{R}^n P_n^m(\mu)e^{im\varphi}; \ \widetilde{R} = R/a \tag{1}$$

where A_n^m are complex amplitudes, $\mu = \cos\theta$ and $P_n^m(\mu)$ denote the associate Legendre polynomials. Written in terms of a Cartesian coordinate system, having the same origin, $x_1 = R\mu$ and $x_2 + ix_3 = R\sin\theta e^{i\varphi}$, the AC electric potential χ of frequency ω can be expressed in terms of its phasor Equation (1), as $\chi = \mathrm{Re}\{\overline{\chi}e^{-i\omega t}\}$.

The partially polarized initially uncharged JP is assumed impervious to both anions and cations. Because of polarization and by virtue of Gauss's law, an induced-charge distribution [5,6] will be formed around the particle, exhibiting an exponential decay away from the polarized JP. The induced-charge density Q is governed by the Helmholtz equation and depends on the Debye scale λ_0, which for most electrolyte solutions is of the order of few tenths of a nanometer. By ignoring surface conductance (small Dukhin number) and under the assumptions of a 'weak' field (i.e., below the thermal scale) and thin ($\lambda_0 \ll a$) EDL (electric double layer), which corresponds to the traditional 'Standard model' [6–10] can be written in terms of its phasor as:

$$2\overline{\phi} = -\left(\frac{\lambda}{\lambda_0}\right)^2\overline{Q} + \overline{\chi}; \ \frac{1}{\lambda^2} = \frac{1}{\lambda_0{}^2} - \frac{i\omega}{\mathbf{D}} \tag{2}$$

where $(\overline{Q}, \overline{\phi}, \overline{\chi})$ satisfy $\lambda^2\nabla^2\overline{Q} = \overline{Q} = -2\lambda_0^2\nabla^2\overline{\phi}$ and $\nabla^2\overline{\chi} = 0$. Here λ denotes the complex Debye scale and $\mathbf{D}$ represents the diffusivity of the symmetric electrolyte. Thus, following [10], $\overline{Q}$ (Helmholtz) and $\overline{\chi}$ (Laplace) are expressed as:

$$\begin{aligned} \overline{Q} &= -2\sum_{n=1}\sum_{m=0} C_n^m K_n\left(\widetilde{R}/\lambda\right)P_n^m(\mu)e^{im\varphi}; \\ \overline{\chi} &= -2\sum\left[A_n\widetilde{R}^n + \left(\tfrac{n}{n+1}A_n + D_n\right)\widetilde{R}^{-(n+1)}\right]P_n^m(\mu)e^{im\varphi} \end{aligned} \tag{3}$$

where $\mathrm{K}_n\left(\widetilde{R}/\lambda\right) = K_{n+1/2}\left(\widetilde{R}/\lambda\right)/\widetilde{R}^{1/2}K_{n+1/2}(1/\lambda)$ and $K_{n+1/2}$ is the modified Bessel function.

Let us next assume that the polarized nano-particle (NP) is initially uncharged, implying that $C_0 = 0$. Enforcing the zero ion-flux boundary conditions $\frac{\partial}{\partial \widetilde{R}}(2\overline{\phi} + \overline{Q}) = 0$ across the NP surface [10,19], renders.

$$\left[1 - (\lambda/\lambda_0)^2\right]\frac{\partial\overline{Q}}{\partial\widetilde{R}} + \frac{\partial\overline{\chi}}{\partial\widetilde{R}} = 0 \text{ or } (n+1)D_n^m = i\Omega C_n^m; \ \Omega = \frac{\omega a\lambda_0}{D} \tag{4}$$

where Ω represents the RC frequency, since for a relatively thin EDL $\lambda \approx \lambda_0$ and $\frac{\partial}{\partial\widetilde{R}}K_n\left(\widetilde{R}/\lambda\right)\Big|_{\widetilde{R}=1} \approx -a/\lambda_0$ [9,10].

The interface between the two homogenous hemispheres of the MD JP (Figure 1) is perpendicular to the x_1 axis, such that the surface of the metallic half sphere is given by $0 \geq \mu \geq -1$, $0 \leq \varphi \leq 2\pi$ and that of the dielectric part by $1 \geq \mu \geq 0$, $0 \leq \varphi \leq 2\pi$. It is also assumed that on the dielectric (vanishing small permittivity) surface, the induced-charge density (see Equation (4a)) is practically null and thus $\frac{\partial\overline{\chi}}{\partial\widetilde{R}} = 0$. On the other hand, the polarized metallic surface (infinitely large permittivity) is considered as equipotential, which suggests following Equations (2) and (3) that on the metallic (coated) part, $\overline{Q} = \overline{\chi}$ and $\frac{\partial \overline{Q}}{\partial R} = -\overline{Q}(a/\lambda_0)$ (thin EDL).

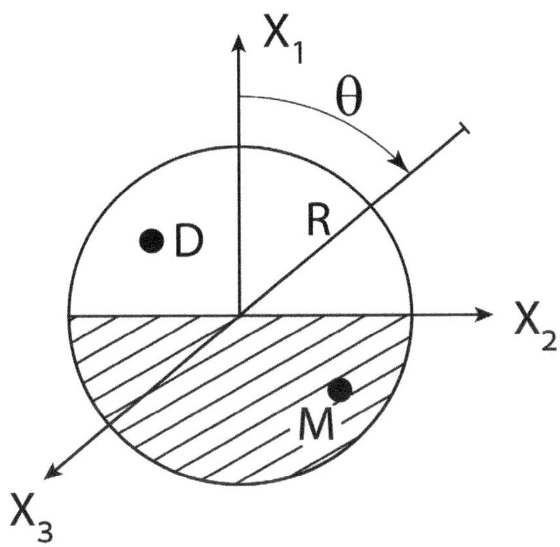

Figure 1. The interface between two homogenous hemispheres of a metallo-dielrctric (MD) Janus that is subjected to non-uniform ambient electric fields expressed in spherical coordinates (Equation (3b)).

Finally, substituting these relations back into Equation (4a) renders the Robin-type boundary conditions $\frac{\partial \overline{\chi}}{\partial R} = -i\Omega\overline{\chi}$ on the metallic surface, which is the common boundary condition enforced on an ideally polarizable particle in terms of its surface capacitance [17].

In order to solve the following mixed boundary problem:

$$\frac{\partial \overline{\chi}}{\partial R} = \begin{Bmatrix} -i\Omega\overline{\chi} \\ 0 \end{Bmatrix} \begin{Bmatrix} 0 \geq \mu \geq -1 \\ 1 \geq \mu \geq 0 \end{Bmatrix} \tag{5}$$

with $\overline{\chi}$ given in Equation (3), we apply the Fourier—Legendre methodology [45], by multiplying Equation (5) by $P_l^m(\mu)$ and integrating the resulting expression over the surface of each hemisphere. Using the Legendre orthogonal properties leads to:

$$[2(n+1) - i\Omega]D_n^m = i\Omega\frac{2n+1}{n+1}A_n^m - i\Omega\frac{(2n+1)(n-m)!}{(n+m)!}\sum_{l\neq n}\left[\left(\frac{2l+1}{l+1}\right)A_l^m + D_l^m\right]\gamma_{l,n}^m \tag{6}$$

where

$$\gamma_{l,n}^m = \int_0^1 P_l^m(\mu)P_n^m(\mu)d\mu. \tag{7}$$

Equation (6) is the sought relation for evaluating the various multipole coefficients D_n^m in (3b) depending on the prescribed amplitude coefficients A_n^m of the non-homogenous ambient field Equation (1). Once Equation (6) is solved to determine D_n^m in terms of the dimensionless RC forcing frequency Ω, the induced-charge coefficients C_n^m in (3a) can be directly found from (4b). It is important to note here that Equation (6) was obtained for a spherical MD JP that is exposed to an arbitrary AC non-uniform applied field. In the DC limit ($\Omega = 0$), we readily find from (4b) that $D_n^m = 0$, which is a consequence of imposing the Neumann boundary condition $\frac{\partial \overline{\chi}}{\partial R} = 0$ on $\overline{R} = 1$ for $1 \geq \mu \geq -1$ and of the special form selected for the exterior potential in (3b). Furthermore, it is also worth mentioning that

Equation (6) can be readily reduced to the corresponding relation, which holds for a (full) *homogenous* metallic spherical NP by letting in Equation (6):

$$\gamma_{e,n}^{m} = \frac{1}{2}\int_{-1}^{1} P_e^m(\mu)P_n^m(\mu)d\mu = \frac{(n+m)!}{(2n+1)(n-m)!}\delta_{mn} \tag{8}$$

Thus, in the DC limit, substituting Equation (8) into Equation (7) simply renders by virtue of (4a):

$$[(n+1) - i\Omega]D_n^m = i\Omega\frac{2n+1}{n+1}A_n^m;\ [(n+1) - i\Omega]C_n^m = (2n+1)A_n^m. \tag{9}$$

We conclude this section by bearing in mind that the present analysis holds both for a homogeneous ideally polarized particle (NP) as well as for a Janus spherical colloid (JP), subject to either DC or AC excitations of arbitrary non-uniform applied electric fields. In the case of a JP, the corresponding higher-order multipoles (D_n^m) can be found by inverting Equation (6), whereas for an homogeneous particle (NP), they are given explicitly by (9) in terms of the prescribed forcing amplitudes A_n^m. Next we will consider few dipolophoretic (DIP) DC and AC scenarios, involving both NP and JP under different non-uniform electric excitations. In particular, we will separately discuss cases of dielectrophresis (DEP) and induced-charge electrophoresis (ICEP), involving either NP or JP under general 3D travelling—wave harmonic forcing (linear and helical) in connection with evaluating the DIP dynamical loads (force and torque) exerted on the freely suspended particle.

3. DEP

3.1. Theoretical Background

Due to the spatial non-uniformity of the ambient electric field, the particle (regardless of its polarization) will experience a DEP force, resulting in a finite phoretic motion. The DEP force depends on the multipole system within the colloid and on the partial derivatives of the applied field evaluated at the locations of these higher-order singularities inside the particle. In particular and following the notations of [19], the general DEP force acting on the NP under non-uniform AC excitation can be expressed as:

$$\vec{F}_{DEP} = -\pi\left(\frac{\partial}{\partial x_1}, \frac{\partial}{\partial x_2}, \frac{\partial}{\partial x_3}\right)\sum_s G_{\alpha\beta\gamma}^{*}(-1)^{\alpha+\beta+\gamma}\frac{\partial^{\alpha+\beta+\gamma}}{\partial x_{1s}^{\alpha}\partial x_{2s}^{\beta}\partial x_{3s}^{\gamma}}(\overline{\chi}_{am}) \tag{10}$$

where the upper (*) stands for the complex conjugate, (α, β, γ) are positive integers such that $G_{\alpha\beta\gamma}$ represent higher-order multipoles of order (the familiar $n = \alpha + \beta + \gamma$ dipole term corresponds to denotes summation over the positions of all $\sum\limits_{S}$) and $n = 1$ internal multipoles located at (x_{1s}, x_{2s}, x_{3s}). Next, we note following Equations (11) and (3) that $\overline{\chi} = 2\overline{\chi'} + 2\overline{\chi}_{am}$, where the disturbance potential is given by

$$\overline{\chi'} = -\sum_n\sum_m \widetilde{D}_n^m R^{-(n+1)}P_n^m(\mu)e^{im\varphi} = -\sum_s G_{\alpha\beta\gamma}(-1)^{\alpha+\beta+\gamma}\frac{\partial^{\alpha+\beta+\gamma}}{\partial x_{1s}^{\alpha}\partial x_{2s}^{\beta}\partial x_{3s}^{\gamma}}\left(\frac{1}{R}\right). \tag{11}$$

Here, $R^2 = \sum\limits_{i=1}^{3}(x_i - x_{is})^2$ and $\widetilde{D}_n^m = \frac{n}{n+1}A_n^m + D_n^m$. Equation (11) can be used (as shown below) to determine the multipole outputs $G_{\alpha\beta\gamma}$ in terms of the ambient amplitudes A_n^m and the coefficients D_n^m arising from solving the electrostatic problem around the NP.

Since we use concurrently two coordinate systems (spherical & Cartesian) with common origin at the NP center, it is useful to employ the following relations between external and internal spherical harmonics expressed in the two systems [48]:

$$P_n^m(\mu)R^{-(n+1)}e^{im\varphi} = \frac{(-1)^n}{(n-m)!}\frac{\partial^{n-m}}{\partial x_1^{n-m}}\left(\frac{\partial}{\partial x_2} + i\frac{\partial}{\partial x_3}\right)^m\left(\frac{1}{R}\right) \tag{12}$$

and

$$\lim_{R\to 0}\left\{\frac{\partial^{n-m}}{\partial x_1^{n-m}}\left(\frac{\partial}{\partial x_2} + i\frac{\partial}{\partial x_3}\right)^m\left(R^n P_n^m(\mu)e^{im\varphi}\right)\right\} = \left\{\begin{array}{c} n! \\ \frac{1}{2}(n+m)! \end{array}\right\} \text{for} \left\{\begin{array}{c} m=0 \\ m\neq 0 \end{array}\right\}. \tag{13}$$

The above procedure will be first demonstrated for an axisymmetric electric forcing whereby $m = 0$.

3.2. Axisymmetric Forcing

Consider for example the JP depicted in Figure 1 which is subjected to axisymmetric AC ambient fields expressed in cylindrical coordinates (x_1, r), $e^{i(kx_1-\omega t)}I_0(kr)$ or $e^{\pm\lambda x_1 - i\omega t}J_0(\lambda r)$, where I_0 and J_0 denote the common Bessel functions. Both potentials exhibit azimuthal symmetry with respect to the axial x_1 axis, which is chosen later in the direction of gravity. The first potential describes a simple travelling- wave $e^{i(kx_1-\omega t)}$ with a wave number k, moving along the x_1 axis ($r = 0$) with a velocity ω/k_1. The second potential corresponds to the induced potential in a (vertical) cylindrical pore of radius b bounded by two circular electrodes; say one powered and other grounded. The Eigen-functions λ depend on the specific boundary conditions applied on the cylindrical wall. For the case of insulating cylindrical walls, λ are determined by $J_1(\lambda b) = 0$ and for conducting walls by $J_0(\lambda b) = 0$. As shown in the sequel, the particular form of the applied potential uniquely determines the ambient amplitudes A_n in Equation (1).

Taking advantage of axial symmetry, we note that the induced DEP force acts along the x_1 axis. Furthermore, the multipoles G_{n00} in Equation (10) can be expressed in this case according to Equations (11) and (12) (see also [19]) simply as $G_{n00} = \frac{\tilde{D}_\mu}{n!}$. Finally, substituting Equation (13) for $m = 0$ in Equation (10) and using Equation (1), renders the following rather compact expression for the axial DEP force (Re denotes the real part):

$$F_{DEP}^{(1)} = -2\pi\text{Re}\sum_{n=1}[nA_n + (n+1)D_n]A_{n+1}^* \tag{14}$$

Thus, we may conclude that Equation (14) provides the sought expression for the DEP force acting on a spherical NP exposed to a non-uniform AC forcing. In the particular DC limit ($\Omega = 0$), we can deduce following (4b) that $D_n = 0$ and thus Equation (14) is reduced to the following rather simple quadratic expression (valid to both NP and JP):

$$F_{DEP}^{(1)} = -4\pi\sum_{n=1} nA_nA_{n+1}. \tag{15}$$

It can be readily verified that the above general DC DEP solution agrees with that given in Equation (24) of [9]. As far as the AC case is considered, one can examine for example the 'constant gradient' approximation for a NP [7–9,17], where $D_1 = \frac{3}{2}\frac{i\Omega}{(2-i\Omega)}A_1$ and (A_1, A_2) are non-zero and real coefficients, yielding $F_{DEP}^{(1)} = 4\pi A_1A_2\left(\Omega^2 - 2\right)/\left(\Omega^2 + 4\right)$ in agreement with Equation (17) of [17].

Scrutiny of the above relations reveal that the DEP force (for spherical NP's), arises only due to interactions between neighboring modes, i.e., one even and the other odd. Clearly, if all modes of the ambient field Equation (1) are either even or odd, the resulting DEP force exerted on an homogenous NP is null. However, the situation for a JP is quite different as demonstrated below. Let us next discuss several DEP cases involving DC forcing and then the more intriguing cases of AC excitations.

3.2.1. Standing Wave

Expressed in a cylindrical coordinate system (x_1, r) placed at the origin of the spherical NP, the ambient electric potential of a harmonic standing wave of unit amplitude and wave length $2\pi/k$, can be written as,

$$\overline{\chi}_{am} = e^{ik(x_1-c)} I_0(kr) \; ; \; x_2^2 + x_3^2 = r^2 \tag{16}$$

where c denotes the distance (Figure 2) between the center of the spherical colloid lying on the x_1 axis ($r = 0$) to the nearest node of the standing wave, where $k(x_1 - c) = n\pi, n = 1, 2, 3, \ldots$. Clearly, for $c = 0$, the colloid is placed precisely at one of the wave nodal points.

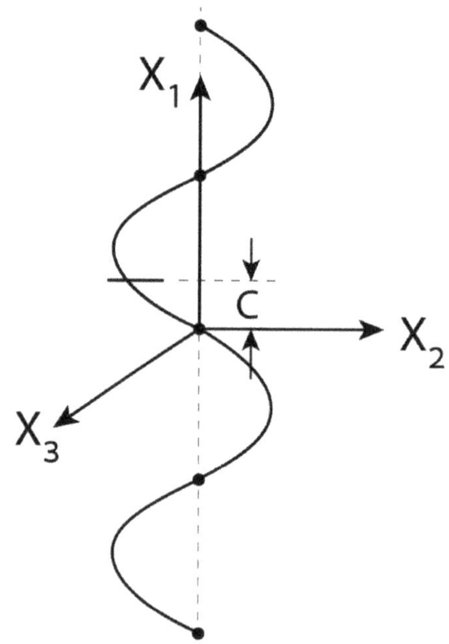

Figure 2. A free polarizable micro/nano NP or JP lying on the x_1 axis ($r = 0$) at a distance c from the nearest nodal point of a standing wave.

Expanding both Equations (1) and (16) in a Taylor series in terms of x_1, evaluated near the origin and along the axis of symmetry ($r = 0,\ \widetilde{R} = x_1/a,\ m = 0$), readily implies by analytic continuation that:

$$A_n = -e^{-ikc}\frac{(ika)^n}{n!} \tag{17}$$

since $P_n(1) = I_0(0) = 1$. Finally, substituting Equation (17) into Equation (15) yields:

$$F_{DEP}^{(1)} = 4\pi(ka)^3 \mathrm{Re}\left\{ ie^{-2ikc} \sum_{n=1}^{\infty} \frac{(-1)^n (ka)^{2(n-1)}}{(n-1)!(n+1)!} \right\} = 4\pi(ka) J_2(2ka)\sin(2kc) \tag{18}$$

Note that Equation (18) is exact and shows that when subjected to a DC standing wave excitation, the DEP acting on a spherical NP is null when $c = 0$ (i.e., colloid placed at a wave nodal point), or obviously under the long wavelength approximation (uniform field), whereby $ka \to 0$.

3.2.2. Cylindrical Pore

Here, we consider the case of a free spherical colloid (including JP) placed at a distance c above a powered circular electrode along the axis of a cylindrical pore ($r = 0$) of radius b and height H (see Figure 3). The corresponding electrical potential expressed in a cylindrical coordinate system attached to the colloid is:

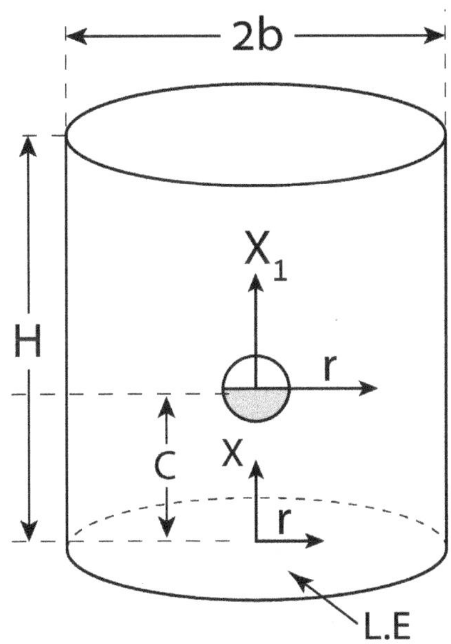

Figure 3. Freely suspended spherical colloid (NP/JP) placed at a distance c above a powered circular electrode along the axis of a cylindrical pore ($r = 0$) of radius b and height H.

$$\overline{\chi}_{am} = \frac{\sinh[\lambda(H - x_1 - c)]}{\sinh(\lambda H)} J_0(\lambda r) \tag{19}$$

such that the lower electrolyte ($x_1 = -c$) is powered with a potential $J_0(\lambda r)$, whereas the upper ($x_1 = H - c$) electrode is grounded. If the cylindrical wall of the pore is taken as insulating, then λ (the lowest eigenvalue) is determined by imposing $J_1(\lambda b) = 0$, namely $\lambda b = 3.8317$, whereas if the wall is conducting $J_0(\lambda b) = 0$, then $\lambda b = 2.408$. Expanding both Equations (1) and (19) in a Taylor series around the origin ($r = 0$, $\widetilde{R} = x_1/a$, $m = 0$), renders:

$$A_n = \frac{(-1)^{n+1}}{n!} \frac{(\lambda a)^n}{\sinh(\lambda H)} \left\{ \begin{array}{c} \cosh \lambda(H - c) \\ \sinh\lambda(H - c) \end{array} \right\}; \left\{ \begin{array}{c} n - odd \\ n - even \end{array} \right\}. \tag{20}$$

Finally, substituting Equation (20) into Equation (15) yields:

$$F^{(1)}_{DEP} = \frac{2\pi(\lambda a)^3 \sinh[2\lambda(H - c)]}{\sinh^2(\lambda H)} \sum_{n=1} \frac{(\lambda a)^{2(n-1)}}{(n-1)!(n+1)!} = 2\pi(\lambda a) I_2(2\lambda a) \frac{\sinh[2\lambda(H - c)]}{\sinh^2(\lambda H)} \tag{21}$$

Again, the right-hand side of Equation (21) is exact, showing that under an ambient non-uniform electric potential Equation (19), a metallic homogenous NP situated in a circular pore $r = 0$, $\lambda \neq 0$), will always experience a DEP force depending on the geometric parameters

(b, H) and the colloid position (c). Nevertheless, we see that for $\lambda \to 0$, Equation (19) reduces to a linear field (in x_1) for which the DEP force Equation (21) vanishes.

As far as we understand, these analytic (closed form) expressions for the DEP force experienced by a spherical colloid under DC axisymmetric non-uniform forcing, are given here for the first time. In the next section, we will demonstrate how the foregoing analysis can be extended for non-symmetric (azimuthal) electric forcing.

3.3. Non-Symmetric Forcing

As an example of an *asymmetric* ambient potential expressed in a cylindrical coordinate system (x_1, r, φ), let us consider the case of a general *helical* (spiral) wave- field given by $e^{i(kx-\omega t+m\varphi)}I_m(kr)$, where k (as before) represents the wave number and m is taken here as a positive integer. The particular choice of $m = 0$ clearly corresponds to the previously discussed case of an axisymmetric (rectilinear) sinusoidal-wave along the x_1 axis. The special choice of $m = 1$ is also of special interest, since it describes a non-uniform rotating wave field (ROT), which is further analyzed in the sequel. It is worth mentioning that the above (Laplace) electric field, can be considered as a limiting case of the following optical (Helmholtz) Bessel beam of TM polarization $e^{i(k_x x-\omega t+m\varphi)}J_m(k_r r)$ whereby $k_x^2 + k_r^2 = k_0^2$ in the limit where the wavelength k_0 is vanishingly small, implying that $k_r = \pm i k_x$. Again, the case of $m = 0$ (corresponds optical analogy) to linear polarization whereas that of $m = 1$ to a circular polarization [49].

A general wave field contains an infinite number of modes $A_n^m(1)$, where n is a positive integer and $m = 0, 1, 2, \ldots, n$. In the case of axisymmetric forcing ($m = 0$), it has been shown that the DEP force results from interactions between neighboring (i.e., even and odd) modes, namely $A_n A_{n+1}$. By using symmetry arguments, one finds that this force acts only along the axis of symmetry. However, for $m \neq 0$, the resulting DEP force has also finite components along the transverse axes (x_2, x_3). These terms arise from a $A_n^m A_{n+1}^{m\pm1}$-type interaction [50].

In order to keep the analysis tractable and for illustrating the general approach, let us select $m = 1$ and arbitrary n. Such a case is of particular interest, since it generalizes the common [4,11] ROT (electro-rotation) AC forcing (under the long-wave approximation), which will be discussed in the sequel. It should be noted however that for a single m-mode, the DEP force along the transverse (x_2, x_3) axes is null, but still there is a finite DEP force acting in the axial (x_1) direction. Thus, we refer to the ambient asymmetric wave field in the form of Equation (1):

$$\overline{\chi}_{am}(R, \theta, \varphi) = -\sum_{n=1}^{\infty} A_n^1 R^n P_n^1(\mu) e^{i\varphi} \tag{22}$$

where A_n^1 are prescribed complex amplitudes. Following Equations (11) and (13), we can express the disturbance potential as:

$$\overline{\chi}' = -\sum_{n=1}^{\infty} \widetilde{D}_n^1 P_n^1(\mu) R^{-(n+1)} e^{i\varphi} = -\sum_{n=1}^{\infty} \frac{\widetilde{D}_n^1}{(n-1)!} \frac{\partial^{n-1}}{\partial x_1^{n-1}} \left(\frac{\partial}{\partial x_2} + i \frac{\partial}{\partial x_3} \right) \left(\frac{1}{R} \right) \tag{23}$$

which implies that the corresponding multipoles in Equations (10) and (11) are simply given by $G_{n-1,1,0} = G_{n-1,0,1} = \widetilde{D}_n^1/(n-1)!$. Using the same relations and Equation (13), one also gets:

$$\lim_{R \to 0} \left\{ \frac{\partial^{n-1}}{\partial x_1^{n-1}} \left(\frac{\partial}{\partial x_2} + i \frac{\partial}{\partial x_3} \right) \left(R^n P_n^1(\mu) e^{i\varphi} \right) \right\} = \frac{1}{2}(n+1)! \tag{24}$$

Finally, substituting Equation (24) into Equation (10), renders the following expression for the axial DEP force exerted on a spherical NP under *asymmetric* ($m = 1$) AC excitation:

$$F_{DEP}^{(1)} = -\pi \mathrm{Re} \sum_{n=1} n(n+2)\left[nA_n^1 + (n+1)D_n^1\right] \mathrm{A}_{n+1}^{*1} \tag{25}$$

Clearly, under DC forcing ($\omega = 0$), the coefficients D_n^1 are null by virtue of (4b) and thus Equation (25) is simply reduced to:

$$F_{DEP}^{(1)} = -2\pi \sum_{n=1} n^2(n+2) A_n^1 \mathrm{A}_{n+1}^1 \tag{26}$$

Below we analyze few cases of non-uniform asymmetric wave forcing.

3.4. Helical Wave

In a similar manner to Equation (16), let us consider a helical (circumferential) wave ($m = 1$) propagating along the x_1 axis with a linear velocity ω/k. For the particular case, where the position of the colloid on the axis of symmetry is displaced at a distance c from the nearest wave mode, one can express the ambient potential in a body-attached coordinate system as:

$$\overline{\chi}_{am} = e^{i[k(x_1-c)+\varphi-\omega t]} I_1(kr) = -\sum_{n=1}^{\infty} A_n^1 \widetilde{R}^n P_n^1(\mu) e^{i\varphi} \tag{27}$$

Expanding Equation (27) in a Taylor series near the origin (colloid center) in powers of x_1 ($\widetilde{R} = x_1/a$)) for $r = 0$ ($\mu = 1$), by recalling that $\mathrm{I}_1(kr) \to \frac{1}{2}\sqrt{1-\mu^2}kx_1$, $P_n^1(1) \to \frac{1}{2}\sqrt{1-\mu^2}n(n+1)$ and $e^{ikx} = \sum\limits_{n=0} \frac{(ikx)^n}{n!}$, yields the following explicit expression:

$$A_n^1 = -\frac{i^{n-1}e^{-ikc}}{(n+1)!}(ka)^n, \quad n \geq 1. \tag{28}$$

where the frequency dependent term has been suppressed. Finally, substituting Equation (28) into Equation (26) renders for the steady DEP force ($\omega = 0$):

$$\begin{aligned} F_{DEP}^{(1)} &= -2\pi(ka)^3 \mathrm{Re}\left\{ ie^{-2ikc} \sum_{n=1}^{\infty} (-1)^n \left(\tfrac{n}{(n+1)!}\right)^2 (ka)^{2(n-1)} \right\} \\ &= \tfrac{\pi}{2} \sin(2kc) \zeta^2 \tfrac{d}{d\zeta} \left\{ \zeta \tfrac{d}{d\zeta} \left[\tfrac{J_0(2k\zeta)-1}{\zeta^2} \right] \right\}; \end{aligned} \tag{29}$$

where $\zeta = ka$. It is again worth mentioning that if the particle lies at one of the nodal points of the spiral wave, namely when $2kc = n\pi$, the axial DEP force Equation (29) vanishes and reaches a maximum value for $2kc = (n+1/2)kc$.

Extending the analysis for AC forcing ($\omega \neq 0$), is straightforward when using the additional multipole term related to D_n^1 in Equation (26). Note however that under AC excitations and for a homogenous NP, D_n^m are given explicitly in terms of the amplitudes A_n^m of the ambient field (9a) and the RC frequency Ω. The same expression also applies for a JP where D_n^m is determined by solving Equation (7).

Before concluding this section, which provides a general methodology for evaluating the DEP forces and presenting next the corresponding ICEP analysis, it is appropriate to discuss the recent DEP results on traveling- wave of metallic NP's reported in [27]. Flores-Mena et al. [27], considered a special axisymmetric two-mode travelling wave excitation ($c = 0$) given by $-e^{i(kx-\omega t)} I_0(kr)$, for which case Equation (17) renders $A_1 = ika$, $A_2 = -(ka)^2/2$. We also observe from Equation (9a) that $D_1 = \frac{3}{2}i\Omega/(2-i\Omega)$ and thus, since according to Equation (14) $F_{DEP}^{(1)} = -2\pi \mathrm{Re}\{(A_1 + 2D_1)A_2^*\}$, we finally obtain $F_{DEP}^{(1)} = -6\pi(ka)\Omega/\left(4+\Omega^2\right)$, which coincides with Equation (25) of [27]. Nevertheless, aside from being valid only for the particular case where $c = 0$, this expression should be considered only as the leading-order 'weakly' non-uniform field (long- wavelength) approximation (i.e., $ka << 1$) of the exact expression found by substituting Equations (17) and (9a) into Equation (14), namely:

$$F_{DEP}^{(1)} = -2\pi\text{Re}\left\{\sum_{n=1}^{\infty}(n+1)D_n A_{n+1}^*\right\} = -2\pi(ka)^3\Omega\sum_{n=1}^{\infty}\frac{(2n+1)(ka)^{2(n-1)}}{(n!)^2\left[(n+1)^2+\Omega^2\right]} \tag{30}$$

The above *exact* expression can be also easily extended for the more general 'displaced' case, namely $c \neq 0$. The DEP result of [27] is indeed seen to be the leading term $(n = 1)$ of the infinite summation Equation (30).

4. ICEP

4.1. General Formulation

In addition to the DEP force, which arises because of the non-uniformity of the ambient electric field, the free colloid also experiences an ICEP force due to the flow field driven by the induced-charge electro-osmosis in the surrounding solute. The total force (DEP + ICEP) exerted on the NP, is referred to as dipolophoresis (DIP), where in general DEP and ICEP act (especially at low frequencies) in different directions [8,9,17]. Following Teubner's [47] formulation and using Lorentz reciprocal relation, the ICEP force acting on a freely suspended polarizable particle subject to an AC field, can be written [3] as:

$$F_{ICEP}^{(i)} = -\frac{1}{8\lambda_0^2}\int_{\forall}\overline{Q}^*\left(u_j^{(i)}-\delta_{ij}\right)\frac{\partial\overline{\chi}}{\partial x_j}d\forall \tag{31}$$

where λ_0 is the Debye length scale (EDL), $\overline{Q}^*$ (conjugate) is the induced charge density Equation (3a) and $\overline{\chi}$ (harmonic) is the electric potential Equation (3). In addition, $\forall$ represents the semi-unbounded fluid volume and $u_j^{(i)}$ is a generic solution of the homogenous Stokes equation [51]. In particular, $u_j^{(i)}$ denotes the i-*th* components of the Stokesian velocity induced in the fluid due to unit velocity of a rigid particle moving along the j axis. For example, for a spherical colloid, one has [51]:

$$u_j^{(i)} - \delta_{ij} = \left(\frac{3}{4\widetilde{R}} + \frac{1}{4\widetilde{R}^3} - 1\right) + \frac{3}{4}\frac{x_i x_j}{\widetilde{R}^2}\left(\frac{1}{\widetilde{R}} - \frac{1}{\widetilde{R}^3}\right) \tag{32}$$

Making use of the fact that $\lambda_0/a \to 0$ (thin EDL) and $u_j^{(i)} = \delta_{ij}$ (δ_{ij} being the Kronecker delta tensor) on the colloid surface S, Equation (32) together with Equation (33) can be reduced to the following surface integral, since $\overline{Q}$ decays exponentially away from the colloid as $e^{-(R-a)/\lambda_0}$:

$$F_{ICEP}^{(i)} = -\frac{1}{8}\int_S \overline{Q}^*\frac{\partial\overline{\chi}}{\partial x_j}\frac{\partial}{\partial R}\left(u_j^{(i)}-\delta_{ij}\right)\Big|_{\widetilde{R}=1} dS = \frac{3}{16a}\int_S \overline{Q}^*\left(\frac{\partial\overline{\chi}}{\partial x_i} - n_i\frac{\partial\overline{\chi}}{\partial R}\right)dS + O(\lambda_0/a) \tag{33}$$

where $n_i = x_i/R$ denotes the external normal (unit) vector to S.

4.2. Axisymmetric Traveling Wave

Let us first consider the axial ICEP force exerted on a spherical colloid under a non-uniform axisymmetric AC loading, by noting that $\frac{\partial\overline{\chi}}{\partial x_1} - n_1\frac{\partial\overline{\chi}}{\partial R}\Big|_{\widetilde{R}=1} = \frac{1}{a}\left(1-\mu^2\right)\frac{\partial\overline{\chi}}{\partial\mu}$. Next, employing this relation together with Equation (3) and the following identity [48]:

$$(2n+1)\left(1-\mu^2\right)\frac{dP_n(\mu)}{d\mu} = n(n+1)[P_{n-1}(\mu) - P_{n+1}(\mu)] \tag{34}$$

in the second surface integral in Equation (33) and using orthogonality, finally yields:

$$F_{ICEP}^{(1)} = 3\pi \mathrm{Re}\left\{ \sum_{n=1} \frac{n(n+1)}{2n+1} \left[\frac{C_{n-1}^*}{2n-1} - \frac{C_{n+1}^*}{2n+3} \right] \left(A_n + \widetilde{D}_n \right) \right\};\ \widetilde{D}_n = \frac{n}{n+1} A_n + D_n. \tag{35}$$

The above is the sought expression for the axial ICEP force exerted on a freely suspended spherical particles (including JP) that is exposed to an arbitrary non-uniform axisymmetric AC electric forcing. Considering for example, the case of a perfectly polarized (metallic) NP. Imposing the equipotential boundary condition $\overline{\phi} = 0$ on $\overline{R} = 1$ and noting that according to Equations (2) and (3) $C_n = A_n + \widetilde{D}_n$ (thin EDL), Equation (35) is simply reduced to:

$$F_{ICEP}^{(1)} = 6\pi \mathrm{Re}\left\{ \sum_{n=1} \frac{(n+1)}{(2n+1)(2n+3)} C_n C_{n+1}^* \right\}. \tag{36}$$

Clearly, in the DC limit, Equation (36) renders [9]:

$$F_{ICEP}^{(1)} = 12\pi \sum_{n=1} \frac{(n+1)}{(2n+1)(2n+3)} C_n C_{n+1} = 12\pi \sum_{n=1} \frac{A_n A_{n+1}}{n+2} \tag{37}$$

since under DC Equation (3), $D_n = 0$ and $C_n = \frac{2n+1}{n+1} A_n$.

Next, combining Equation (37) with Equation (15), readily yields the final expression for the total DIP (DEP + ICEP) force acting on a metallic spherical particle under a steady (DC) non-uniform axisymmetric forcing:

$$F_{DIP}^{(1)} = -4\pi \sum_{n=1}^{\infty} \frac{(n-1)(n+3)}{n+2} A_n A_{n+1} \tag{38}$$

in agreement with [9]. It is rather interesting to observe [6–8] that by considering only the first two modes of the ambient field ('constant gradient' approximation), namely keeping only A_1 and $A_2 (n = 1)$ in Equation (38), the corresponding DIP force for a spherical NP vanishes, since the DEP and ICEO components are equal but act in opposite directions. However, as demonstrated in [9,10] and in the sequel, this intriguing result is true only under the limit of an infinitely thin EDL and DC forcing.

In the case of an AC ambient axisymmetric non-uniform fields, the coefficients C_n can be readily found for a metallic NP by realizing that in addition to the surface S being an equipotential ($\phi = 0$), one has to satisfy following (4a) $(n+1)D_n = i\Omega C_n$ and $C_n = A_n + \widetilde{D}_n$, so that:

$$C_n = \frac{(2n+1)A_n}{(n+1) - i\Omega},\quad D_n = \frac{(2n+1)i\Omega A_n}{(n+1)[(n+1) - i\Omega]}. \tag{39}$$

Substituting Equation (39) in Equation (37) lastly renders:

$$F_{ICEP}^{(1)} = 6\pi \mathrm{Re}\left\{ \sum_{n=1} \frac{(n+1) A_n A_{n+1}^*}{[(n+1) - i\Omega][(n+2) + i\Omega]} \right\}. \tag{40}$$

which reduces to (37b) for $\Omega = 0$.

For the purpose of illustration, let us consider the phasor of the previously discussed axisymmetric travelling wave excitation ($c = 0$), namely $-e^{ikx_1} I_0(kr)$, for which case we have $A_n = \frac{(ika)^n}{n!}$. Substituting these amplitudes back into Equation (40) leads to:

$$F_{ICEP}^{(1)} = 6\pi\Omega(ka)^3 \sum_{n=1} \frac{(ka)^{2(n-1)}}{[(n+1)^2 + \Omega^2][(n+2)^2 + \Omega^2](n!)^2}. \tag{41}$$

Note that if we consider only the first (leading) term in Equation (40), namely $n = 1$, the infinite sum in Equation (41) simply reduces to $\Omega/\left[4+\Omega^2\right]\left[9+\Omega^2\right]$, which coincides with Equation (27) of [27]. Similarly, for the 'constant gradient' type forcing [7–9,17], where the only two (real) surviving coefficients in Equation (40) are (A_1, A_2) one gets $F^{(1)}_{ICEP} = 12\pi(6+\Omega^2)A_1A_2/[(4+\Omega^2)(9+\Omega^2)]$, which again agrees with Equation (16) of [17].

4.3. Asymmetric Helical Wave

Here, we wish to determine the ICEP force acting on a spherical NP placed along the axis of a helical (spiral) wave $\overline{\chi}_{am} = -\frac{2}{ka}e^{i(kx-\varphi-\omega t)}I_1(kr)$. Following Equation (27), the corresponding amplitudes A^1_n of such a circumferential wave-field are given for $c = 0$ by $A^1_n = \frac{2(ika)^{n-1}}{(n+1)!}$, $(n > 1)$. A homogenous NP or JP with its two-phase interface lying in the $x_1 = 0$ plane (Figure 1) will experience no ICEP force along the transverse axes (x_1, x_2) since this helical field $(m = 1)$ is uniform in the transverse directions. Nevertheless, there is still a finite ICEO force acting in the axial (x_1) direction, given (for thin EDL) by Equation (35), where $\overline{Q}$ and $\overline{\chi}$ are furnished by Equation (3) with $m = 1$.

In order to evaluate the resulting ICEP surface integral in Equation (33), we make use of the following relation involving the corresponding Legendre polynomials [48]:

$$(2n+1)\left(1-\mu^2\right)\frac{dP^1_n(\mu)}{d\mu} = (n+1)^2P^1_{n-1}(\mu) - n^2P^1_{n+1}(\mu) \tag{42}$$

Substituting Equation (42) into Equation (35) and using orthogonally, finally yields:

$$F^{(1)}_{ICEP} = \frac{3\pi}{2}\mathrm{Re}\left\{\sum_{n=1}\frac{n(n+1)}{2n+1}\left[\frac{(n-1)(n+1)}{2n-1}C^{*1}_{n-1} - \frac{n(n+2)}{2n+3}C^{*1}_{n+1}\right]\left(A^1_n+\widetilde{D}^1_n\right)\right\}. \tag{43}$$

It is worth noting that Equation (43) holds for both metallic NP and JP. In the case of a perfectly conducting spherical (metallic) NP, one gets $A^1_n + \widetilde{D}^1_n = C^1_n$ (since S is equipotential). Thus, for a freely suspended ideally polarized NP, placed under any AC axisymmetric $(m = 1)$ forcing, Equation (43) reduces to:

$$F^{(1)}_{ICEP} = 3\pi\mathrm{Re}\left\{\sum_{n=1}\frac{n(n+1)(n+2)}{(2n+1)(2n+3)}C^1_nC^{*1}_{n+1}\right\} \tag{44}$$

The transverse DIP force is given by the sum of the DEP Equation (25) and ICEP Equation (44). In the DC limit, we readily get $F^{(1)}_{DIP} = -2\pi\sum\limits_{n=1} n(n-1)(n+3)A^1_nA^1_{n+1}$. Thus, in the 'constant gradient' (linear) approximation, where only the first two terms (A^1_1, A^1_2) of the ambient field are kept, the transverse DIP (similar to the axial component Equation (38)) also vanishes regardless of the magnitude of the corresponding amplitudes.

Under the present helical wave forcing Equation (39), $C^1_n = \frac{2(2n+1)(ika)^{n-1}}{(n+1)![n+1-i\Omega]}$ and the axial (x_1) ICEP component Equation (44) is given by,

$$F^{(1)}_{ICEP} = 6\pi(ka)\Omega\sum_{n=1}^{\infty}\frac{(ka)^{2(n-1)}}{(n-1)!(n+1)!\left[(n+1)^2+\Omega^2\right]\left[(n+2)^2+\Omega^2\right]} \tag{45}$$

which, as expected (compare for example to the corresponding DEP Equation (30)), is null both for a DC forcing $(\Omega = 0)$ as well as under the long-wavelength approximation $(ka \to 0)$, where the ambient field is uniform (i.e., $\overline{\chi}_{am} \simeq -x_2 + ix_3$). This limiting case corresponds to the common rotating electric field (ROT) excitation [4], where $A^1_n = \delta(n-1)$ and $n \geq 1$. It is also worth noting, that for a homogenous spherical NP forced by ROT, both DEP and ICEP force components vanish and the polarized particle is only subjected to a finite DIP torque causing the NP to rotate around the x_1 axis.

5. DIP Torque

In order to complete the prevalent DIP formulation, we consider below the corresponding expressions for the ICEP and DEP torque components exerted on a spherical NP, which is exposed to a general helical wave field excitation, where $A_n^1 = 2\frac{(ika)^{n-1}}{(n+1)!}$. The electric potential and the induced charge density in the surrounding liquid phase, are given in Equation (3) as a summation over n for $m = 1$. The ICEP torque, following Teubner's [47] formulation (assuming thin EDL), can then be expressed, in a similar manner to Equation (34) by ($\overline{r}$ denotes the radius vector):

$$\vec{T}_{ICEP} = \frac{3}{8}\int_S \overline{Q}^*\left(\vec{r} \times \overline{\chi}\right)d\forall + O(\lambda_0/a) \tag{46}$$

Considering the real part of Equation (46), we deduce that the only torque component for the above helical wave field acts in the axial x_1 direction and is given by:

$$T_{ICEP}^{(1)} = 6\pi \text{Im}\left\{\sum_{n=1} \frac{n(n+1)}{2n+1} C_n^{*1}\left(A_n^1 + \widetilde{D}_n^1\right)\right\} \tag{47}$$

where Im denotes here the imaginary part.

Note however, that for a perfectly polarized NP, imposing the equipotential boundary conditions on S to Equation (2), implies that for an infinitely thin EDL $A_n^1 + \widetilde{D}_n^1 = C_n^1$ and thus the ICEP torque Equation (47) is null, in agreement with the ROT result reported for ($n = 1$) in [15]. It is also important to note, that Equation (47) is not restricted only for helical-waves and instead holds for any transverse excitation ($m = 1$), providing $\lambda_0/a \to 0$. Thus, our conclusion that under the thin EDL limit, the ICEP torque vanishes for metallic spherical NP's is quite general. Nevertheless, one should bear in mind that this is not necessarily true for a non-homogeneous spherical JP, as demonstrated in the sequel even under the thin EDL limit. Before considering this interesting JP case, we provide below for reasons of completeness, the corresponding expression for the DEP torque acting on a spherical NP under *asymmetric* (transverse) forcing.

Using the notation in Section 3.1, the DEP torque exerted on a free NP is given by,

$$\vec{T}_{DEP} = -\pi \text{Re}\{\sum_S G_{\alpha\beta\gamma}^*(-1)^{\alpha+\beta+\gamma}\frac{\partial^{\alpha+\beta+\gamma}}{\partial x_{1s}^{\alpha}\partial x_{2s}^{\beta}\partial x_{3s}^{\gamma}}(\overline{r} \times \overline{\chi}_{am})\} \tag{48}$$

For a spherical particle all multipoles are located at the origin and thus Equation (48) has to be evaluated at $\overline{r} = 0$. Furthermore, making use of the following relation [48,50]

$$\frac{P_n^1(\mu)e^{-i\varphi}}{R^{n+1}} = \frac{(-1)^n}{(n-1)!}\frac{\partial^{n-1}}{\partial x_1^{n-1}}\left(\frac{\partial}{\partial x_2} - i\frac{\partial}{\partial x_3}\right)\left(\frac{1}{R}\right) \tag{49}$$

implies that the multipoles in Equation (48) are given by $G_{n-1,1,0} = G_{n-1,0,1} = \frac{(-1)^n}{(n-1)!}\widetilde{D}_n^1$ for $n \geq 1$, where $\widetilde{D}_n^1 = \frac{n}{n+1}A_n^1 + D_n^1$ (see Equation (3)).

In the case of a helical wave field ($m = 1$) acting on a spherical NP, symmetry arguments suggest that the resulting DEP torque has only one component around the x_1 axis. Thus, substituting Equation (49) into Equation (48) yields for $\overline{r} = 0$,

$$T_{DEP}^{(1)} = \pi \text{Im}\{\sum_S \frac{(-1)^n}{(n+1)!}\widetilde{D}_n^{*1}\frac{\partial^{n-1}}{\partial x_1^{n-1}}\left(\frac{\partial}{\partial x_2} + i\frac{\partial}{\partial x_3}\right)\overline{\chi}_{am}\} \tag{50}$$

Finally, recalling Equation (1) that $\overline{\chi}_{am} = -2\sum A_m^1 R^m P_m^1(\mu)e^{-i\varphi}$ and using Equation (42), implies that

$$\lim_{R\to 0}\frac{\partial^{n-1}}{\partial x_1^{n-1}}\left(\frac{\partial}{\partial x_2}+i\frac{\partial}{\partial x_3}\right)\left(R^m P_m^1(\mu)e^{-i\varphi}\right)\to 2\delta(m-n) \tag{51}$$

Finally, substituting Equation (51) in Equation (50), provides the sought expression of the axial DEP torque:

$$\overrightarrow{T}^{(1)}_{DEP} = -4\pi \mathrm{Im}\left\{\sum_{n=1}\frac{(-1)^n}{(n+1)!}A_n^1\widetilde{D}_n^{*1}\right\} \tag{52}$$

According to Equations (47) and (52) a metallic (perfectly symmetric) spherical NP, can experience a finite DIP torque only under AC excitations, since for $\Omega = 0$ the quadratic terms in these summations are real!

Consider for example a free homogenous NP, which is subjected to the same helical field ($c = 0$) discussed in Section 4.3, where (28)$A_n^1 = \frac{2(ika)^{n-1}}{(n+1)!}$. Equation (52) together with Equation (9) then render the following analytic expression:

$$T^{(1)}_{DEP} = -16\pi\Omega\sum_{n=1}^{\infty}\frac{(-1)^n(2n+1)}{[(n+1)!]^3}\frac{(ka)^{2(n-1)}}{\left[(n+1)^2+\Omega^2\right]} \tag{53}$$

Recalling next that under the long wavelength limit $ka \to 0$, the helical wave field simply reduces to a *uniformly* rotating field (ROT) with a phasor $\overline{\chi}_{am} \simeq -x_2 + ix_3$, for which case $A_n^1 = (1/2)\delta(n-1)$. Substituting this value together with Equation (9) in Equation (52) readily renders the well-known ROT expression:

$$T^{(1)}_{DEP} = -\frac{6\pi\Omega}{4+\Omega^2} + O(ka)^2 \tag{54}$$

previously reported in [13–15,17]. Thus, the counter-field ROT/DEP torque has a Lorentzian ('bell') shape which vanishes both for zero and infinitely large frequencies with a maximum spectrum amplitude at $\Omega_{\max} = 2$. Note that Equation (54) can be also considered as the leading-order long wave approximation of Equation (53), namely keeping only the first term in the summation. Since it has been shown that the ICEP torque for a helical wave forcing is null for a vanishingly small EDL, the resulting dipolophoretic (DIP) torque for a spherical NP consists of only a DEP torque given in Equation (53) in terms of the dimensionless wave number *ka*.

6. Janus Particle

The general expressions obtained so far for the DEP and ICEP forces and torques exerted on spherical particles suspended in a non-uniform (DC or AC) electric fields, are valid both for homogenous metallic NP's, as well as for MD JP's. The only difference between the two cases is in the corresponding expressions for the multipoles D_n^m Equation (3) in terms of the amplitudes A_n^m of the ambient field, namely Equation (6) for a JP and Equation (9) for a spherical NP.

Let us first consider the DC case of a JP, which is subjected to an arbitrary axisymmetric non—uniform ambient field Equation (1) with $m = 0$. Since for $\Omega = 0$ Equation (6) renders $D_n = 0$, the DC DEP force is given for both JP and NP by (15) as the sum of sequential amplitude terms. Note that this force is null for a uniform field ($A_n = 0$ for $n \neq 1$). Nevertheless, for the general case involving non-uniform ambient fields, the DC ICEP force according to Equation (35) is given by:

$$F^{(1)}_{ICEP} = 6\pi \mathrm{Re}\left\{\sum_{n=1} nA_n\left[\frac{C_{n-1}}{2n-1}-\frac{C_{n+1}}{2n+3}\right]\right\}. \tag{55}$$

The corresponding expression for the AC ICEP force is given by Equation (35), where the coefficients D_n for a JP are found by inverting Equation (6). It is also worth noting that under a uniform ambient field ($A_n = 0$ for $n > 1$), Equation (55) vanishes for metallic NP's ($C_1 \neq 0$). Yet, for a JP Equation (55) yields $-\frac{6\pi}{5}\mathrm{Re}\{A_1C_2\}$ for DC and $-\frac{3\pi}{5}\mathrm{Re}\{A_1C_2^*\}$ for AC excitations respectively, since following Equation (6) both $(C_n, D_n) \neq 0$ for any n (even integer under a uniform electric forcing).

Letting, for example, $A_1 = 1$ in Equation (6) and truncating the infinite series after two terms, renders the following set for $\gamma_{1,2} = 1/8$ Equation (7):

$$\begin{aligned} (4-i\Omega)D_1 + \tfrac{3}{8}i\Omega D_2 &= \tfrac{3}{2}i\Omega \\ \tfrac{5}{8}i\Omega D_1 + (6-i\Omega)D_2 &= -\tfrac{15}{16}i\Omega \end{aligned} \tag{56}$$

which yields (4b)

$$D_1 = \frac{i\Omega C_1}{2} = \frac{3\Omega(3i + \frac{49}{128}\Omega)}{24 - 10i\Omega^2 - \frac{49}{64}\Omega^2};\ D_2 = \frac{i\Omega C_2}{3} = -\frac{\frac{15}{4}i\Omega}{24 - 10i\Omega^2 - \frac{49}{64}\Omega^2} \tag{57}$$

Note that (57b) renders $C_2 = -15/32$ for $\Omega = 0$. Finally, substituting this value into Equation (55) and recalling (charge conservation) that $C_0 = 0$, simply yields $F^{(1)}_{ICEP} = -\frac{9\pi}{16}$ or (dividing by 6π) $U_{ICEP} = -\frac{3}{32}$, in agreement with Equation (3.16) of Squires & Bazant [8]. The resulting JP ICEP velocity is directed along the x_1 axis, namely from the metallic toward the dielectric hemisphere. Thus, it is shown that due to symmetry- breaking, a JP (unlike common NP) will experience a finite ICEP force/velocity even under a uniform DC electric excitation.

A similar procedure can be readily applied to evaluate the corresponding forces acting on a JP which is subject to a uniform DC or AC ambient field applied in the transverse directions (x_2, x_3). This case can be envisaged by letting $m = 1$ in Equation (1) and considering for example $A^1_n = \delta(n-1)$. Since according to Equation (9), $(C^1_n, D^1_n) = 0$ for $n \neq 1$, the ICEP force Equation (35) for a metallic NP is again null as expected. Nevertheless, the corresponding ICEP force Equation (44) for a JP is generally non-zero (even for a uniform ambient field) and is given by

$$F^{(1)}_{ICEP} = 3\pi \mathrm{Re}\left\{\sum_{n=1} n(n+2)\left[\frac{n+1}{2n+1}A^1_{n+1}C^{*1}_n - \frac{n}{2n+3}A^1_n C^{*1}_{n+1}\right]\right\} \tag{58}$$

which for $A^1_1 = 1$ simply yields $F^{(1)}_{ICEP} = -(9\pi/5)\mathrm{Re}\{A^1_1 C^{*1}_2\}$. Recalling next that $\gamma^1_{1,2} = 3/4$ Equation (7), substituting this value into Equation (6) and using again only a two-term expansion, yields for (D_1, D_2):

$$\begin{aligned} (4-i\Omega)D^1_1 + \tfrac{9}{8}i\Omega D^1_2 &= \tfrac{3}{2}i\Omega \\ \tfrac{5}{8}i\Omega D^1_1 + (6-i\Omega)D^1_2 &= -\tfrac{15}{16}i\Omega \end{aligned} \tag{59}$$

Solving the above set for $\Omega = 0$ renders $C^1_2 = D^1_2/(3i\Omega) \to -\frac{15}{32}$. Thus, the corresponding ICEP force is $-\frac{27\pi}{32}$ or $U^{(1)}_{ICEP} = -\frac{9}{64}$, which again coincides with Equations (3.16) of [8]. Finally, we note that such a transverse forcing (parallel to JP interface), will also induce an ICEP velocity whereby the JP tends to move in the axial direction towards its dielectric part.

As for a JP that is subjected to an arbitrary (non-uniform) AC field, the preceding analysis holds as well, providing the coefficients D^m_n are determined by inverting Equation (6) in terms of the prescribed forcing amplitudes A^m_n. The expressions obtained in the previous sections for travelling-waves (i.e., linear $m = 0$ or circumferential $m = 1$), may be also applied for a JP. It is worth mentioning that the common case of a *uniform* field (axial or transverse), either under DC or AC forcing, can be directly obtained from the solutions found for linear or helical travelling waves under the limit of the 'long-wave' approximation, i.e., $ka \to 0$.

We demonstrate below how the familiar ROT spectra for a freely suspended spherical NP, can be easily extended for the case of a MD JP. For this purpose, we make use of the general expression found for the DEP torque Equation (52), by noting that for a metallic NP it renders Equation (54), where $D_1^1 = \frac{3}{2}i\Omega/(2 - i\Omega)$. The same expression Equation (52) still holds for a JP providing the D_n^1 coefficients are found from Equation (6) or Equation (59). In particular, if we approximate D_1^1 by solving Equation (59) (i.e., including only the two leading terms), then Equation (52) simply yields for $n = 1$ and $A_1 = 1$:

$$T_{DEP}^{(1)} = -6\pi \mathrm{Im}\{\frac{3i\Omega + \frac{19}{128}\Omega^2}{24 - 10i\Omega - \frac{19}{64}\Omega^2}\}. \tag{60}$$

On the other hand, if we consider only 'one-term ' approximation in Equation (59), one gets $D_1^1 = \frac{3}{2}i\Omega/(4 - i\Omega)$ and the corresponding JP torque is again of a Lorentzian type, given by $T_{DEP}^{(1)} = -12\pi\Omega/(16 + \Omega^2)$. These two approximate (i.e., 'one' and 'two" term) solutions for a spherical JP, can be compared against the prevalent ROT solution Equation (54) for a spherical (metallic) NP.

Finally, it is important to note that unlike perfectly conducting spherical colloids, for which the ROT spectrum is known Equation (54), the corresponding spectrum for a metallo-dielectric Janus particle has not been obtained. The common practice to estimate ROT spectra for MD JP's, is by taking the average between the effective CM (Clausius—Mossotti) coefficients of the dielectric and metallic phases [52]. This procedure, when applied to a MD JP, reduces the spectrum amplitude approximately by a factor of 2 (since the dielectric permittivity is ignored with respect to the metallic), but still its peak remains at $\Omega_{\max}^{(NP)} = 2$. Measurements of ROT spectrum [52], for a Pt-silica JP (normalized with respect to the volume of the metallic phase) in DI, indicates that the JP spectrum is indeed shifted (compared to NP) to higher frequencies (see Figure 4a in [52]), in accordance with the above simplified 'one-term' approximation resulting in $\Omega_{\max}^{(JP)} = 4$.

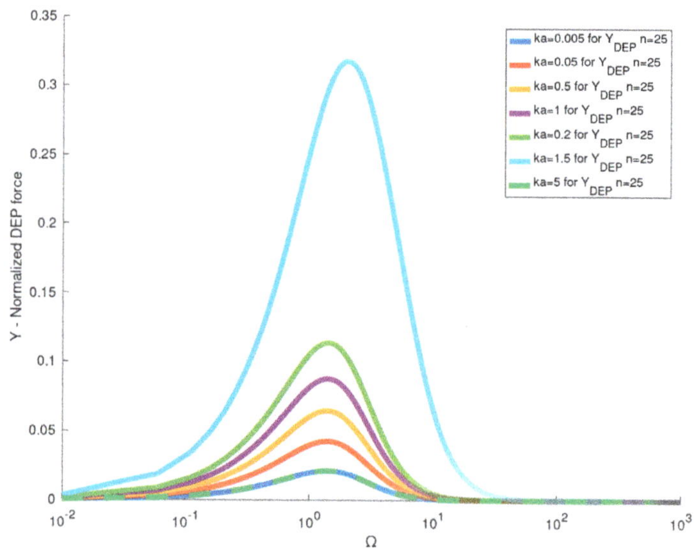

Figure 4. Comparison of the exact expression for the twDEP force (Equation (41)) for different *ka* values against the 'long wave' approximation ($n = 1$).

7. Levitation: JP in a Pore

In order to demonstrate the preceding methodology for a Janus particle, let us consider the levitation problem of a MD JP of radius a that is freely suspended in a solute within a vertical cylindrical pore of height H and bounded by two circular electrodes (Figure 3). If both electrodes are grounded, the heavy JP will rest due to gravity on the bottom of the container with its metallic hemisphere facing downwards. We use a cylindrical coordinate system (x, r) attached to the lower electrode (LE) at $x = 0$, which is assumed to be powered to a potential $-V_0 J_0(\lambda r)$, where the upper electrode ($x = H$) remains grounded. Here V_0 represents the maximum potential at the center ($r = 0$) of LE ($b \geq r \geq 0$) and λ is an arbitrary parameter. The electric potential (Laplace) induced in the cylindrical domain is clearly given by $-V_0 \frac{\sinh[\lambda(H-x)]}{\sinh(\lambda H)} J_0(\lambda r)$. The reader is reminded that $\lambda = 38317/b$ if the cylindrical ($r = b$) walls are insulating ($J_1(\lambda b) = 0$) and $\lambda = 2.408/b$ if the walls are conducting ($J_0(\lambda b) = 0$). The polarized JP will then experience both DEP and ICEP axial forces which will either push (levitate) the JP away from the LE or pull it toward the bottom depending on the ambient forcing. We will next discuss the levitation dynamics of a JP under such non-uniform excitations.

Assuming that under the combined DIP (DEP+ICEP) forces, both NP and JP, are levitated (positive DEP) to a distance $x = c$ (Figure 3) above the LE. Following Section 3.2.2 and when expressed in a body-attached coordinate system (x_1, r) such that $x_1 = x - c$, the ambient potential is given by Equation (19) (multiplied by V_0). Let us first consider a spherical NP, where the DEP (pointing upward) force is given explicitly by Equation (21). The corresponding ICEP and DIP expressions, are given by Equations (37) and (38) respectively in terms of the prescribed amplitudes Equation (20). Substituting Equation (20) into Equation (38), renders the following dimensional expression for the vertical DIP force acting on a NP under such exponentially decaying field:

$$F^{(1)}_{DIP(NP)} = 2\pi\varepsilon V_0^2 \frac{\sinh[2\lambda(H-c)]}{\sinh^2(\lambda H)} \Theta(\lambda a); \ \Theta(z) = zI_2(2z) - \frac{3I_2(2z)}{z} + \frac{3z}{2} \tag{61}$$

where ε is the solute dielectric constant $I_2(z)$ and denotes the Bessel function. Note that Equation (61) is *exact* and that the DIP force in this case Equation (19) is pointing upward (opposite to gravity) since Θ is positively definite. As a result, a metallic NP of effective mass m_{eff} (including buoyancy) will be always lifted (levitated) from the bottom to an equilibrium distance $\bar{c}$ from the LE. Assuming next that $H/b >> 1$ (or $\lambda H >> 1$) as well as $2\pi\varepsilon V_0^2 \Theta(\lambda a) > m_{eff} = (4/3)\pi a^3 g(\rho_M - \rho_F)$ and recalling that under static equilibrium $F^{(1)}_{DIP} = m_{eff} g$ results in:

$$\bar{c}_{NP} = \frac{1}{2\lambda} \ln\left(\frac{3\varepsilon V_0^2 \Theta(\lambda a)}{\tilde{\rho}_{NP} g a^3}\right); \ \tilde{\rho}_{NP} = \rho_M - \rho_F \tag{62}$$

Here, ρ_M denotes the density of the metallic NP, ρ_F represents the density of the fluid and g is the gravity acceleration. Further simplifications of Equation (62) are possible for $\lambda a << 1$ (or $b \gg a$), by noting following Equation (61b) that $\lim_{z\to 0} \Theta(z) = \frac{5}{16} z^5 + O(z^7)$ where the term in the parenthesis in Equation (62) can be replaced by $15\lambda^5 \varepsilon V_0^2 a^2 / 16\tilde{\rho}_{NP} g$.

It is important to note that this levitation equilibrium point is *stable* to disturbances in the radial direction, since the maximum potential (positive CM coefficient), is located on the axis ($r = 0$). Thus, any small radial displacement of the NP from the axis will result in a finite restoring DEP force directed toward the axis (similar to an optical tweezer [53]). As far as the stability in the axial direction is concerned, we recall following Equation (21) that $F^{(1)}_{DIP}(x_1) = F^{(1)}_{DIP}(0)e^{-2\lambda x_1}$, where x_1 denotes a small axial displacement from the equilibrium point, i.e., $\bar{x} = \bar{c} + x_1$ and $F^{(1)}_{DIP}(0)$ is given in Equation (61a). Linearizing the dynamic equation, which governs the axial motion of a metallic NP around the equilibrium point ($x_1 = 0$), finally yields [4,40]:

$$\frac{4\pi}{3} g\rho_M a^3 (1+q)\ddot{x}_1 + 6\pi\eta a\dot{x}_1 + 2\lambda F^{(1)}_{DIP}(0)x_1 = 0 \tag{63}$$

where $q = \rho_F/(2\rho_M)$ (added—mass correction) and η denotes the dynamic viscosity of the ambient fluid. The first term in Equation (63) represents NP inertia and the second is the (Stokes) damping force (ignoring wall effects). Since both $F^{(1)}_{DIP}(0)$ and λ are positive, stability in the axial direction is also assured. Thus, the NP exhibits a *passive* stable levitation behavior [40,41], by executing small damped oscillations about the equilibrium point $x = \bar{c}$.

The corresponding levitation problem for a JP (in contrast to a NP) under the same electric excitation, is somewhat more intricate since the expression for the DIP force has to be modified in order to account for symmetry- breaking effects related to the large disparity between the dielectric constants of the two hemispheres comprising the MD JP. The DEP force acting on a JP, is still given by Equation (21) (exact), namely $F^{(1)}_{DEP} \sim 2\pi(\lambda a)^3 e^{-2\lambda\bar{c}} + O(\lambda a)^5$ for $\lambda a \to 0$ and $\lambda H >> 1$. Nevertheless, the JP ICEP force, can be found from Equation (35) by noting that the multipole coefficients D_n are given by Equation (6) in terms of the ambient amplitudes A_n. In particular, using Equation (4b) together with Equation (6) implies for example that in the DC limit ($\Omega = 0$), one gets the following explicit expression:

$$\frac{2}{2n+1}C_n = \frac{A_n}{n+1} - \sum_{l\neq n} \frac{2l+1}{l+1}\gamma_{e,n}A_l \tag{64}$$

and thus following Equation (35);

$$F^{(1)}_{ICEP} = 6\pi\mathrm{Re}\left\{\sum_{n=1}\left(\frac{n+1}{2n+1}C_n A_{n+1} - \frac{n}{2n+3}C_{n+1}A_n\right)\right\}. \tag{65}$$

Recalling next that for $\lambda H >> 1$, Equation (20) implies that $A_n = \frac{(-1)^{n+1}}{n!}(\lambda a)^n e^{-\lambda\bar{c}}$ and thus the coefficients A_n are asymptotically small for $\lambda a \to 0$. Therefore, the leading- order ICEP force for a JP, can be written following Equations (64) and (65) ($\gamma_{1,2} = 1/8$) simply as $6\pi A_1[\frac{3}{32}A_1 + \frac{1}{3}A_2] + (\lambda a)^4$. Finally, combining ICEP with the leading 'two- term' DEP expression Equation (15), namely $-4\pi A_1 A_2$ Equation (15), we obtain the corresponding DIP expression for a JP (assuming $\lambda a \ll 1$):

$$F^{(1)}_{DIP(JP)} = 6\pi A_1\left[\frac{3}{32}A_1 - \frac{1}{3}A_2\right] + O(\lambda a)^4 = 3\pi(\lambda a)^2 e^{-2\lambda\bar{c}}\left[\frac{3}{16} + \frac{\lambda a}{3}\right] + O(\lambda a)^4. \tag{66}$$

Consider for example the case of an ambient DC electric field with a 'constant gradient' [8,9,17], namely where only A_1 and A_2 are nonzero. We recall following Equation (38) that the DIP force acting on a NP is null regardless of the values of these two amplitudes, whereas the corresponding DIP force for a JP Equation (66) is finite. It is also worth mentioning that according to Equation (66) the JP DIP force vanishes to $O(\lambda a)^4$ providing $A_2 = \frac{9}{32}A_1$.

As far as the corresponding levitation problem of a JP is concerned, one can repeat the analysis leading to Equation (62) and show that a JP of same radius and forcing as that of a NP, is levitated instead to an equilibrium height $\bar{c}_{JP} >> \bar{c}_{NP}$, where

$$\bar{c}_{JP} = \frac{1}{2}\ln\left(\frac{27\varepsilon V_0^2(\lambda a)^2}{32\tilde{\rho}_{JP} g a^3}\right); \quad 2\tilde{\rho}_{JP} = \rho_M + \rho_D - \rho_F \tag{67}$$

Here, ρ_D denotes the density of the dielectric and ρ_M the metallic parts of the JP hemispheres respectively. Finally, comparing Equations (62) and (67) we find that ($\lambda a \ll 1$) levitation effects are more pronounced (enhanced) for a JP compared to a NP ($\sim$ by a factor of 5/2). It is also worth mentioning that dividing Equation (66) by the Stokes drag coefficient 6π, indicates that when exposed to a uniform ambient field (i.e., $A_n = 0$ for $n \neq 1$), a MD JP acquires a dimensionless ICEP velocity (directed towards its dielectric part) given by 3/32 in agreement with Equation 3.16 of [8].

8. Summary and Discussions

In this study we present a general framework for calculating the DEP and ICEP dynamic loads (forces and moments) acting on initially uncharged perfectly conducting (metallic) spherical nano/micro and Janus particles exposed to an arbitrary (DC or AC) non-uniform ambient electric fields. The analysis is carried under the assumption of 'weak' field (ignoring convection & surface conductance) and infinitely small Debye scale. The above procedure enables us to solve the coupled linearized PNP system and obtain a closed form solution for both electrostatic and hydrodynamic problems. Special attention is payed to metallo-dielectric JP and to inhomogeneous travelling -wave type electric forcing with a prescribed wavelength and frequency. Analytic expressions can be thus found for twDEP and twICEP, which allows us to check the accuracy of the available approximations against the exact value for different wavelengths.

The imposed non-uniform ambient field is expanded in general spherical harmonics in terms of prescribed (complex) amplitudes A_n^m, where n and $m \leq 2n+1$ are positive integers, denoting the order and mode of the harmonic forcing respectively. Expressed in a Cartesian coordinate system (x_1, x_2, x_3) such that x_1 denotes the axis of symmetry, implies that $m = 0$ corresponds to the case of an axisymmetric loading and n represents the order of the polynomial in the Cartesian coordinates. It is first demonstrated that under general inhomogeneous electric forcing, the NP experiences DEP and ICEP axial loadings (in x_1 direction) due to $\mathrm{Re}\{A_n^m A_{n+1}^m\}$ type interactions (namely between odd and even amplitudes of same order). In a similar manner, transverse DEP and ICEP forces (along x_2 or x_3 directions), arise from $\mathrm{Re}\left\{A_n^m A_{n+1}^{m\pm1}\right\}$ type interactions. On the other hand, a spherical NP is subjected to a torque acting in the axial direction as a result of $\mathrm{Im}\{A_n^m A_{n+1}^m\}$ type interaction and a similar one in the transverse direction, where Re and Im denote the real and imaginary parts respectively. Corresponding analytic expressions for the dynamic reactions on a MD JP (exhibiting material symmetry breaking) are also provided in terms of the coefficients D_n^m, which are related to A_n^m through Equation (6). Using the above formulation, one can demonstrate that under the 'constant gradient' or 'linear' approximation, where only the first two terms corresponding to $n = 1$ and $n = 2$ are considered, the DEP and ICEP (at least for low frequencies) act in opposite directions. Moreover, in the DC limit, it is verified that the DIP (sum of DEP and ICEP), indeed vanishes for spherical NP. However, as shown this interesting property does not hold under AC forcing (and even in DC for finite EDL's).

In order to demonstrate the above methodology, we consider in particular electric forcing of a freely suspended NP by both standing and travelling waves in a cylindrical container. Exact expression is found for the DEP force exerted on a NP placed a standing wave Equation (16) in terms of its wavelength, size of NP and a parameter c (Figure 2) representing its position with respect to the nearest wave nodal point Equation (18). The DEP vanishes if c is null. A similar analytic form Equation (21) is found for an exponentially decaying (converging) field (Figure 3) due to powered and grounded circular electrodes Equation (19). These axisymmetric solutions are then extended for helical wave excitations Equation (27) again resulting in rather simple exact expressions Equation (29). A similar procedure can be used to calculate ICEP for travelling waves in terms of the characteristic (RC) frequency. For example, closed form solution can be obtained for both DEP and ICEP forces acting on a perfectly polarized spherical NP subject to a general travelling wave of wavelength k and frequency ω. Such solutions are given in Equations (30) and (41) in terms of the dimensionless wavenumber ka and frequency Ω and thus can be finally compared against the available approximate solutions.

Figure 4 is a plot of the exact ICEP expression Equation (41) versus ka and Ω against the 'constant-gradient' ($n = 1$) approximation recently provided by Equation (27) in [27]. It shows that the two-term 'constant-gradient' approximation can be used only for extremely small dimensionless wavelength ka of the order of 10^{-2}! Similar exact expressions can be obtained for the DEP torque acting on a NP under a helical (circumferential) travelling wave Equation (27) (note that the ICEP torque vanishes in this case). The exact solution for the DEP torque Equation (53) is compared in Figure 5 against the common leading-order Equation (54) ROT solution [13,17]. Again, one finds that Equation (54) can be considered as the small wavelength ($ka \to 0$) limit (Rayleigh) of the exact solution which can be used for $ka \leq 0.05$. The DEP force (Figure 4) and torque (Figure 5) spectra are of Lorentzian (bell) shape, vanishing for small and large frequencies with a distinct maximum at $\Omega \sim 2$. The same approach can also be used to evaluate the gradient force (optical tweezer) exerted on a NP which is subjected to a non-uniform BB (Bessel beam) under the Rayleigh approximation.

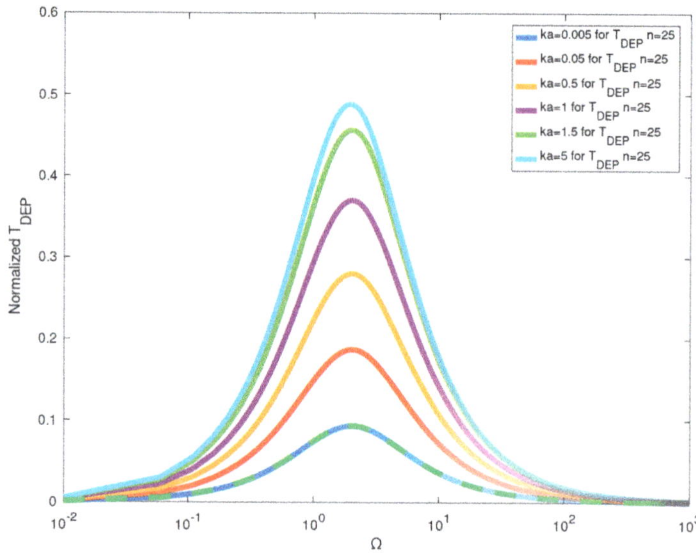

Figure 5. A comparison between the exact (Equation (53)) twDEP torque spectrum (normalized by 16π) for various ka values, against the corresponding 'long-wave' approximation (Equation (54)).

Unlike homogeneous NP, spherical JP's generally experience ICEP even under *uniform* ambient fields Equation (58). In particular we find that for $\Omega = 0$ (DC limit), the dimensionless ICEP velocities in the direction normal to the interface (toward the dielectric hemisphere) are given by $9/64$ and $3/32$ for a field directed parallel or normal to JP interface respectively, in full agreement with [8]. An analytical expression for the ICEP loading is also provided Equation (58) for arbitrary AC and non-uniform electric forcing in terms of the complex amplitudes of the ambient field. The DIP dynamic loads acting on both NP and JP exposed to a non- homogeneous ambient field, depend on the interactions between the ambient amplitudes A_n^m and the multipole term D_n^m Equation (3). Note that for a metallic NP, the coefficients D_n^m are given explicitly in terms of A_n^m Equation (9a), whereas for JP they are found by solving a linear system Equation (6).

The case of a JP, which is subjected to ROT ambient field consisting of two orthogonal out-of- phase components parallel to the JP interface, is also of special interest. Letting for example $A_1^1 = 1$, Equation (52) implies that the DEP torque depends on the imaginary part of the coefficient $\widetilde{D}_1^1$ Equation (35b), found by inverting the linear system Equation (6). A one-term solution of Equation (6) is $D_1^1 = (3i/2)/(4 - i\Omega)$ resulting Equation (54) in $T_{DEP}^{(1)} = -12\pi\Omega/(16+\Omega^2)$. Equation (57a) represents the corresponding 'two-term' solution of Equation (6) leading to the JP torque spectrum given by Equation (60). Comparing these two approximate spectra in Figure 6 against the equivalent one for a NP Equation (54), indicates that the JP spectra are generally lower (depending on coating thickness) and are shifted towards higher frequencies in qualitative agreement with the available measurements [53].

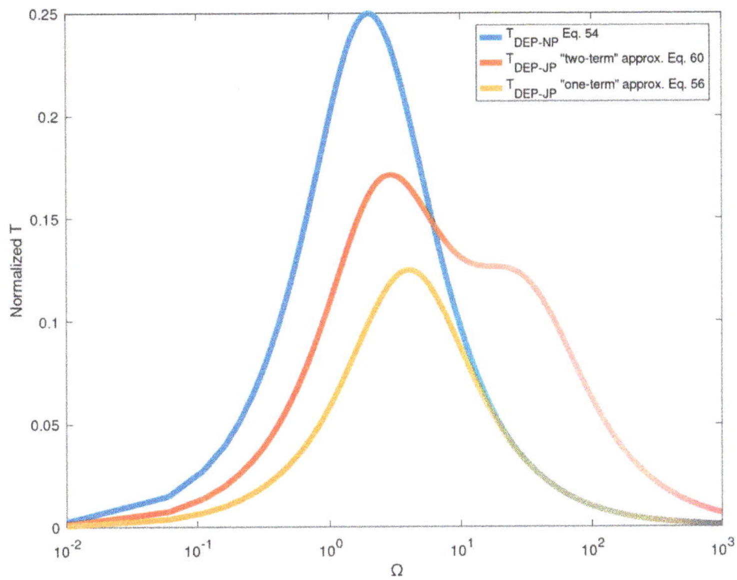

Figure 6. Comparison between the 'two-term' (Equation (60)) approximate ROT spectra for the JP torque against the corresponding spectrum for a NP (Equation (54)). Also shown is the 'one-term' JP approximation.

The weak 'double-peak' frequency response curve depicted in Figure 6, may be a unique feature of JP (compared to the Lorentzian-type NP dispersion). Nevertheless, recall that the above JP spectrum Equation (60) has been obtained as a 'two-term' approximation of Equation (6), whereas the corresponding 'one-term' approximation is again of a 'bell' shape. Obtaining more accurate JP frequency spectra are underway. It is also important to note that so far, no attempt has been made to analytically calculate the JP spectra for AC excitation and the common practice is to express it in terms of the average between the CM coefficients of the metallic and dielectric phases. Such an approximation clearly fails to predict the reported physical JP frequency ROT shift with respect to a NP [53] as depicted in Figure 6. The theoretical model also indicates that the characteristic frequency Ω_{max} Equation (4c) is inversely proportional to the JP radius and EDL thickness (through the $x^{-1/2}$ dependence on solute conductivity), in agreement with experimental findings.

Levitation and stability issues of both NPs and JPs under general non-uniform electric excitations and in particular under travelling wave and converging (decaying) fields have been also discussed. It is demonstrated that freely suspended spherical polarizable NP's and JP's placed near the lower powered electrode along the axis of a cylindrical (insulating or conducting walls) pore, are levitated due to positive DEP (overcoming gravity) to an equilibrium distance $\bar{c}$ from the bottom. As shown, this equilibrium point is stable with respect to both radial and radial disturbances. It is also interesting to note that JP's, as compared to common NP's, are more amenable to DEP levitation (by a factor 2–3), exhibiting yet another remarkable property of asymmetric Janus particles in response to ambient non-uniform electric fields.

Author Contributions: T.M. is responsible for methodology and theoretical aspects and J.N. for numerical computations and draft preparation. All authors have read and agreed to the published version of the manuscript.

Funding: The partial support of BSF Grant 2018, 168 is acknowledged.

Conflicts of Interest: The authors declare no conflict of interest.

References

1. Agrawal, G.; Agrawal, R. Janus nanoparticles: Recent advances in their interfacial and biomedical applications. *Appl. Nano Mater.* **2019**, *2*, 1738–1757. [CrossRef]
2. Feng, H.; Chang, H.; Zhong, X.; Wong, T.N. Recent advancement in induced-charge electrokinetic phenomena and their micro- and nano-fluidic applications. *Adv. Colloid Interface Sci.* **2020**, *280*. [CrossRef] [PubMed]
3. Boymelgreen, A.; Balli, T.; Miloh, T. Active colloids as mobile microelectrodes for unified label-free selective cargo transport. *Nat. Commun.* **2018**, *9*, 760. [CrossRef] [PubMed]
4. Jones, T.B. *Electromechanics of Particles*; Cambridge University Press: England, UK, 2005.
5. Murtsovkin, V. Nonlinear flows near polarized disperse particles. *Colloid J.* **1996**, *58*, 341.
6. Squires, T.M.; Bazant, M.Z. Induced- charge electro-osmosis. *J. Fluid Mech.* **2004**, *509*, 217–252. [CrossRef]
7. Shilov, V.N.; Estrela, L.V.R. *Surface Forces in Thin Films and Disperse Systems*; Nauka: Moscow, Russia, 1979; p. 35.
8. Squires, T.M.; Bazant, M.Z. Breaking symmetries in induced-charge electro-osmosis and electrophoresis. *J. Fluid Mech.* **2006**, *560*, 65–101. [CrossRef]
9. Miloh, T. Dipolophoresis of nanoparticles. *Phys. Fluids* **2008**, *20*. [CrossRef]
10. Miloh, T. Nonlinear alternating electric field dipolophoresis of spherical nanoparticles. *Phys. Fluids* **2009**, *21*. [CrossRef]
11. Jones, T.B. Basic theory of dielectrophoresis and electrorotation. *IEEE Eng. Med. Biol. Mag.* **2003**, *22*, 33–42. [CrossRef]
12. Arcenegui, J.J.; García, S.P.; Morgan, H.; Ramos, A. Electro-orientation and electrorotation of metal nanowires. *Phys. Rev. E* **2013**, *88*. [CrossRef]
13. Ren, Y.K.; Morganti, D.; Jiang, H.R.; Ramos, A.; Morgan, H. Electrorotation of metallc microspheres. *Langmuir* **2011**, *27*, 2128. [CrossRef] [PubMed]
14. Garcia, S.P.; Ren, Y.K.; Arceengui, J.J.; Morgan, H.; Ramos, A. Alternating current electrokinetic properties of gold-coated microspheres. *Langmuir* **2012**, *28*, 39.
15. García, S.P.; Ramos, A. Electrorotation of a metal sphere immersed in an electrolyte of finite Debye length. *Phys. Rev. E* **2015**, *92*. [CrossRef]
16. Miloh, T.; Goldstein-Weis, B. 3D controlled electrorotation of conducting in-axial ellipsoidal nanoparticles. *Phys. Fluids* **2017**, *29*. [CrossRef]
17. Ramos, A.; García, S.P.; Morgan, H. AC electrokinetics of conducting microparticles: A review. *Curr. Opin. Colloid Interface Sci.* **2016**, *24*, 79–90. [CrossRef]
18. Liebchen, B.; Löwen, H. Which interactions dominate in active colloids? *J. Chem. Phys.* **2019**, *150*. [CrossRef] [PubMed]
19. Miloh, T. A unified theory of dipolophoresis for nanoparticles. *Phys. Fluids* **2008**, *20*. [CrossRef]
20. Miloh, T. Dipolophoresis of interacting conducting nano-particles of finite electric double layer thickness. *Phys. Fluids* **2011**, *23*. [CrossRef]
21. Huang, Y.; Wang, X.B.; Tame, J.A.; Pethig, R. Electrokinetic behavior of colloidal particles in travelling electric fields: Studies using yeast cells. *J. Phys. D Appl. Phys.* **1993**, *26*, 1528–1535. [CrossRef]
22. Cahill, B.P.; Heyderman, L.J.; Gobrecht, J.; Stemmer, A. Electro-osmotic streaming on application of traveling-wave electric fields. *Phys. Rev. E* **2004**, *70*. [CrossRef]
23. Ramos, A.; Morgan, H.; Green, N.G.; González, A.; Castellanos, A. Pumping of liquids with traveling-wave electroosmosis. *J. Appl. Phys.* **2005**, *97*. [CrossRef]
24. Ramos, A.; González, A.; García, S.P.; Castellanos, A. A linear analysis of the effect of Faradaic currents on traveling-wave electroosmosis. *J. Colloid Interface Sci.* **2007**, *309*, 323–331. [CrossRef] [PubMed]
25. García, S.P.; Ramos, A.; González, A.; Green, N.G.; Morgan, H. Flow Reversal in Traveling-Wave Electrokinetics: An Analysis of Forces Due to Ionic Concentration Gradients. *Langmuir* **2009**, *25*, 4988–4997.
26. Miloh, T.; Boymelgreen, A. Travelling wave dipolophoresis of ideally polarizable nano-particles with overlapping electric double layers in cylindrical pores. *Phys. Fluids* **2014**, *26*. [CrossRef]
27. Flores, M.J.E.; García, S.P.; Ramos, A. Dipolophoresis and Travelling-Wave Dipolophoresis of Metal Microparticles. *Micromachines* **2020**, *11*, 259. [CrossRef]
28. Miloh, T.; Boymelgreen, A.M. Induced-charge electrophoresis of uncharged dielectric spherical Janus particles. *Electrophoresis* **2012**, *33*, 870–879.

29. Boymelgreen, A.M.; Miloh, T. Alternating current induced-charge electrophoresis of leaky dielectric janus particles. *Phys. Fluids* **2012**, *24*. [CrossRef]
30. Gangwal, S.; Cayre, O.J.; Bazant, M.Z.; Velev, O.D. Induced-Charge Electrophoresis of Metallodielectric Particles. *Phys. Rev. Lett.* **2008**, *100*. [CrossRef]
31. Gangwal, S.; Cayre, O.J.; Velev, O.D. Dielectrophoretic Assembly of Metallodielectric Janus Particles in AC Electric Fields. *Langmuir* **2008**, *24*, 13312–13320. [CrossRef]
32. Boymelgreen, A.; Yossifon, G.; Park, S.; Miloh, T. Spinning Janus doublets driven in uniform ac electric fields. *Phys. Rev. E* **2014**, *89*. [CrossRef]
33. Boymelgreen, A.; Yossifon, G.; Miloh, T. Propulsion of Active Colloids by Self-Induced Field Gradients. *Langmuir* **2016**, *32*, 9540–9547. [CrossRef] [PubMed]
34. Chen, Y.L.; Jiang, H.R. Particle concentrating and sorting under a rotating electric field by direct optical-liquid heating in a microfluidics chip. *Biomicrofluidics* **2017**, *11*. [CrossRef] [PubMed]
35. Mano, T.; Delfau, J.B.; Iwasawa, J.; Sano, M. Optimal run-and-tumble–based transportation of a Janus particle with active steering. *Proc. Natl. Acad. Sci. USA* **2017**, *114*, E2580–E2589. [CrossRef] [PubMed]
36. Bayati, P.; Najafi, A. Electrophoresis of active Janus particles. *J. Chem. Phys.* **2019**, *150*. [CrossRef] [PubMed]
37. Liu, W.; Ren, Y.; Tao, Y.; Yan, H.; Xiao, C.; Wu, Q. Buoyancy-Free Janus Microcylinders as Mobile Microelectrode Arrays for Continuous Microfluidic Biomolecule Collection within a Wide Frequency Range: A Numerical Simulation Study. *Micromachines* **2020**, *11*, 289. [CrossRef] [PubMed]
38. Shen, C.; Jiang, Z.; Li, L.; Gilchrist, J.F.; Ou, Y.H.D. Frequency Response of Induced-Charge Electrophoretic Metallic Janus Particles. *Micromachines* **2020**, *11*, 334. [CrossRef] [PubMed]
39. Holmes, L.M. Stability of magnetic levitation. *J. Appl. Phys.* **1978**, *49*, 3102. [CrossRef]
40. Jones, T.B. Cusped electrostatic fields for dielectrophoretic levitation. *J. Electrost.* **1981**, *11*, 85–95. [CrossRef]
41. Fuhr, G.; Arnold, W.M.; Hagedorn, R.; Muller, T.; Benecke, W.; Wagner, B.; Zimmermann, U. Levitation, holding and rotation of cells within traps made of high-frequency fields. *Biochem. Biophys. Acta* **1992**, *1108*, 215–223. [CrossRef]
42. Jie, Z.; Chuande, Z.; Fuzhong, Z.; Shuhua, L.; Miao, F.; Yike, T. Experimental and numerical modeling of particle levitation and movement behavior on traveling-wave electric curtain for particle removal. *Part. Sci. Technol.* **2019**, *37*, 741–749. [CrossRef]
43. Bukosky, S.C.; Hashemi, A.S.M.H.; Rader, S.P.; Mora, J.; Miller, G.H.; Ristenpart, W.D. Extreme Levitation of Colloidal Particles in Response to Oscillatory Electric Fields. *Langmuir* **2019**, *35*, 6971–6980. [CrossRef]
44. Brandt, E.H. Levitation in Physics. *Science* **1989**, *243*, 4889. [CrossRef]
45. Andrade, M.A.B.; Marzo, A.; Adamowski, J.C. Acoustic levitation in mid-air: Recent advances, challenges, and future perspective. *App. Phys. Lett.* **2020**, *116*, 250501. [CrossRef]
46. Marmolejo, J.T.; Isaksson, O.; Cabrera, T.R.; Giesselmann, N.C.; Hanstorp, D. A fully manipulable damped driven harmonic oscillator using optical levitation. *Am. J. Phys.* **2020**, *88*, 490–498. [CrossRef]
47. Teubner, M. The motion of charged colloidal particles in electric fields. *J. Chem. Phys.* **1982**, *76*, 5564. [CrossRef]
48. Hobson, E.W. *The Theory of Spherical and Ellipsoidal Harmonics*; Chelsea Pub. Co.: New York, NY, USA, 1965.
49. Turunen, J.; Friberg, A.T. Propagation –invariant optical fields. *Prog. Opt.* **2010**, *54*, 1–88.
50. Miloh, T. Pressure forces on deformable bodies in non-uniform inviscid flows. *Q. J. Mech. Appl. Math.* **1994**, *47*, 635–661. [CrossRef]
51. Happel, J.; Brenner, H. *Low Reynolds Hydrodynamics*; Springer: The Hauge, The Netherlands; Martinus Nijhoff: The Hauge, The Netherlands, 1983.
52. Chen, L.Y.; Jiang, H.R. Electrorotation of a metallic coated Janus particle under AC fields. *Appl. Phys. Lett.* **2016**, *109*. [CrossRef]
53. Ashkin, A. Optical Trapping and Manipulations of Neutral Particles Using Laser. *Proc. Natl. Acad. Sci. USA* **1997**, *94*, 4583–4860. [CrossRef]

micromachines

MDPI

Article

Particle-Induced Electrostatic Repulsion within an Electric Curtain Operating below the Paschen Limit

Stuart J. Williams [1,*], Joseph D. Schneider [1], Benjamin C. King [1] and Nicolas G. Green [2]

1 Department of Mechanical Engineering, University of Louisville, Louisville, KY 40292, USA; schneider.devin89@gmail.com (J.D.S.); bcking21@gmail.com (B.C.K.)
2 Electronics and Computer Science, University of Southampton, Southampton SO17 1BJ, UK; ng2@ecs.soton.ac.uk
* Correspondence: stuart.williams@louisville.edu; Tel.: +1-502-852-6340

Abstract: The electric curtain is a platform developed to lift and transport charged particles in air. Its premise is the manipulation of charged particles; however, fewer investigations isolate dielectric forces that are observed at lower voltages (i.e., less than the Paschen limit). This work focuses on observations of simultaneous dielectrophoretic and electrostatic forces. The electric curtain was a printed circuit board with interdigitated electrodes (0.020 inch width and spacing) coated with a layer of polypropylene, where a standing wave or travelling wave AC signal was applied (50 Hz) to produce an electric field below the Paschen limit. Soda lime glass beads (180–212 µm) demonstrated oscillatory rolling via dielectrophoretic forces. In addition, several particles simultaneously experienced rapid projectile repulsion, a behavior consistent with electrostatic phenomena. This second result is discussed as a particle-induced local increase in the electric field, with simulations demonstrating that a particle in close proximity to the curtain's surface produces a local field enhancement of over 2.5 times. The significance of this is that individual particles themselves can trigger electrostatic repulsion in an otherwise dielectric system. These results could be used for advanced applications where particles themselves provided triggered responses, perhaps for selective sorting of micrometer particles in air.

Keywords: electric curtain; electrostatics; dielectrophoresis

Citation: Williams, S.J.; Schneider, J.D.; King, B.C.; Green, N.G. Particle-Induced Electrostatic Repulsion within an Electric Curtain Operating below the Paschen Limit. *Micromachines* **2022**, *13*, 288. https://doi.org/10.3390/mi13020288

Academic Editor: Rodrigo Martinez-Duarte

Received: 6 January 2022
Accepted: 8 February 2022
Published: 11 February 2022

1. Introduction

The electric curtain is a platform first developed by Masuda [1,2] to lift and transport charged particles, typically in air. The platform itself consists of a series of parallel coplanar electrodes that generate a travelling wave (TW) AC electric field to simultaneously lift particles from the platform surface and translate them away. This multiphysical system is rich in electrokinetic and mechanical physics; particle motion is governed by particle electrokinetic properties (charge and/or induced charge), particle mass (i.e., inertia), field properties (voltage, frequency), platform construction (electrode geometry, dielectric coating), and medium properties (gas composition, pressure, humidity) [3–7]. Figure 1 provides an illustration of a four-phase electric curtain as well as experimental results (using the platform herein) demonstrating the repulsion of dust in ambient conditions using a signal of 600 V and 50 Hz.

Various studies have investigated the electric curtain experimentally [3,8–17] and with numerical modeling of individual particles' trajectories [4,6,7,18–22]. The traditional premise of the electric curtain is the manipulation of charged particles via Coulombic forces; it is typically assumed that these particles are either inherently charged or they obtain a net charge from absorption of gaseous ions from corona [3]. The frequency of the applied field coupled with particle inertia [19,20] account for observed oscillatory particle motions. However, seldom are investigations conducted at lower voltages (i.e., in the absence of corona, less than the Paschen limit) where strong repulsion is not typically observed. At these lower voltages dielectrophoretic forces are present but do not contribute towards repulsion, as dielectrophoresis (DEP) is always attractive in air [4,9,22].

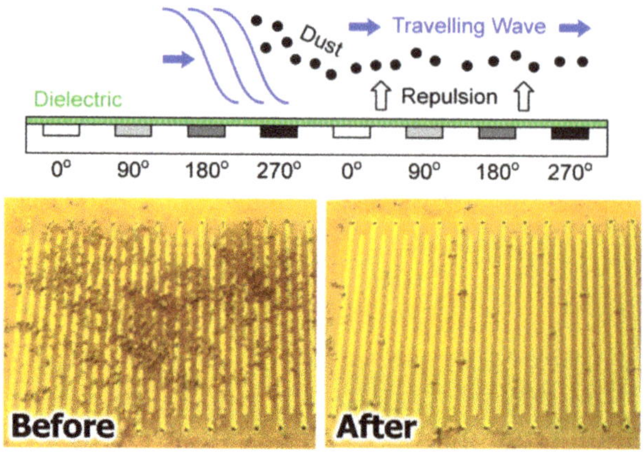

Figure 1. (**top**) Illustration of the electric curtain using four sequentially phase shifted signals at the same frequency applied to successive electrodes in an array. (**bottom**) Experimental results showing the initial and final dust distribution after the application of a 600 V, 50 Hz signal.

Results herein demonstrate that, in the absence of plasma, the particle itself can trigger localized Coulombic repulsion in an otherwise dielectric system. The following provides a brief overview of the theoretical operation of an electric curtain and an introduction to DEP.

1.1. Electric Curtain Background

The net force acting on a particle within an electric curtain is typically a combination of Coulombic forces, viscous forces, and gravitational forces. The Coulombic force governs particle translation and is dependent on the net charge of the particle. Standing wave (SW) and TW fields each affect particle translation differently: both repel from the electrode array, but the latter provide lateral translation along the array. Particles can have an inherent charge, or it can be acquired. There are several methods by which a particle can acquire a charge, including tribocharging and corona charging [23]; the latter is associated with traditional electric curtains. Corona-based electrostatic precipitators use macroscale electrodes to generate ionized gas. Similarly, though at the millimeter scale, work by Atten et al. [3] induced particle charging from dielectric barrier discharges where the gaseous ions and electrons are transferred to the particles in the immediate vicinity of the dielectric layer that coats the electrodes. Here the applied voltage needs to be greater than the Paschen voltage, the voltage necessary to start a discharge or electric arc between two electrodes and subsequently ionize surrounding gas.

Inherently, analyzing particle movement within an electric curtain provides a unique challenge, primarily due to the coincidence of the system's characteristic time and velocity parameters. In most liquid-based electrokinetic microsystems the AC frequency is high (>1 kHz) such that oscillatory electrokinetic effects can be neglected [24] and manipulated colloids reach terminal velocity within milliseconds [25]. However, these characteristics are not shared for low-frequency (<1 kHz) air-based systems, as the applied electrokinetic forces will impart a time-dependent impulsive force leading to a sudden change in particle acceleration. As such, particle mass (i.e., inertia) will be significant [19,20]. Interparticle interactions also occur [4,20], but are outside the scope of this work.

There are three forms of particle translation that have been observed: electric curtain mode [26,27], surfing mode [21], and hopping mode [6]. Electric curtain mode is characterized by high voltages (up to 30 kV) and air ionization. Particles have continuous levitation (balanced with gravity) and translate slower than the wave velocity. Surfing mode occurs with inherently charged particles in an electric field at lower voltages (no corona). Particles move synchronously with the travelling wave in sliding contact with the surface. Hopping mode is produced when charged particles are propelled forward and come to rest on the surface until the wave catches up and initiates the next jump. Interestingly, our results herein do not exhibit these typical translation modes, this is in part due to our operation mode (i.e., no corona). Instead, our findings suggest that particles exhibiting dielectric behavior can subsequently trigger electrostatic repulsion. As such, the following section provides an introduction to relevant dielectric mechanisms.

1.2. Dielectric Electrokinetics

A dielectric is a material that can be polarized (i.e., form distinct poles of charge) by an application of an electric field. When a dielectric particle is subjected to a nonuniform electric field DEP occurs. DEP is a well-known particle manipulation technique that has been able to capture, sort, concentrate, and characterize a variety of particles and biological entities as small as a few nanometers [25,28,29]. Unlike electrophoresis, particles do not need to carry a net charge in DEP. The polarization of a particle with DEP is based, in part, on the interfacial polarization at the interface of two dissimilar materials. Particles are either attracted or repelled from regions of greater field strength based on the dielectric properties (permittivity, conductivity) of the particle and the medium it exists in, as well as the applied AC frequency.

For a dipole moment $\mathbf{p} = q\mathbf{d}$, with opposite charges, q, separated by a distance, $\mathbf{d}$, subjected to a nonuniform field, $\mathbf{E}$, the resultant dielectrophoretic force (neglecting higher order terms) is [25]

$$\mathbf{F} = (\mathbf{p}\cdot\nabla)\mathbf{E}. \tag{1}$$

The dipole moment is given as

$$\mathbf{p} = \forall\alpha^{*}\mathbf{E} = 4\pi\varepsilon_m(f_{CM})a^3\mathbf{E}, \tag{2}$$

where $\forall$ is particle volume, α^* is particle polarizability, a is the radius of the particle, ε is permittivity, the subscript m refers to the medium. The term f_{CM} refers to the Clausius-Mossotti factor, defined as

$$f_{CM} = \left(\varepsilon_p^* - \varepsilon_m^*\right)/\left(\varepsilon_p^* + 2\varepsilon_m^*\right), \tag{3}$$

where the subscript p refers to particle properties. The complex permittivity ε^* is given by

$$\varepsilon^* = \varepsilon - j\sigma/(\omega), \tag{4}$$

where σ is conductivity, $\omega = 2\pi f, f$ is the AC frequency, and j^2 is -1. For a homogeneous spherical particle the dielectrophoretic force, assuming a polarized dipole, is [25,30]

$$F_{DEP} = 2\pi\varepsilon_m a^3 Re[f_{CM}]\nabla E_{rms}^2 + 4\pi\varepsilon_m a^3 Im[f_{CM}]\left(E_{rms,x}^2\nabla\varphi_x + E_{rms,y}^2\nabla\varphi_y + E_{rms,z}^2\nabla\varphi_z\right). \tag{5}$$

The first term is associated with a standing field, whereas the second term is associated with the phase-dependent portion of the field (φ). In other words, the first component is the traditional dielectrophoretic force and the second is the TW dielectrophoresis (twDEP) term which propels the particle laterally with a translating wave. The magnitude of these forces is greatest at the electrode edge (i.e., location of greatest field strength) and decrease significantly with distance.

The DEP force in Equation (5) assumes that the particle is spherical and homogeneous. Modifications are necessary for multi-shelled and non-spherical particles [31,32]. In addition, as the size of the particle approaches that of the electrode features the dipole approximation needs to be modified to include higher order multipoles (quadropole, etc.) [31,33,34]. Multipoles can increase the dielectrophoretic trapping force significantly [35].

For particles in air ($\sigma \approx 0$, $\varepsilon_m < \varepsilon_p$), $Re[f_{CM}]$ is positive and particles are attracted to greater fields; $Im[f_{CM}]$ is negative and this force is opposing the direction of the travelling wave. Numerical simulations (Figure 2) show the direction of these DEP force components. Please note that, in air, particles are attracted towards electrode edges in SW fields and would translate against the TW when on the surface.

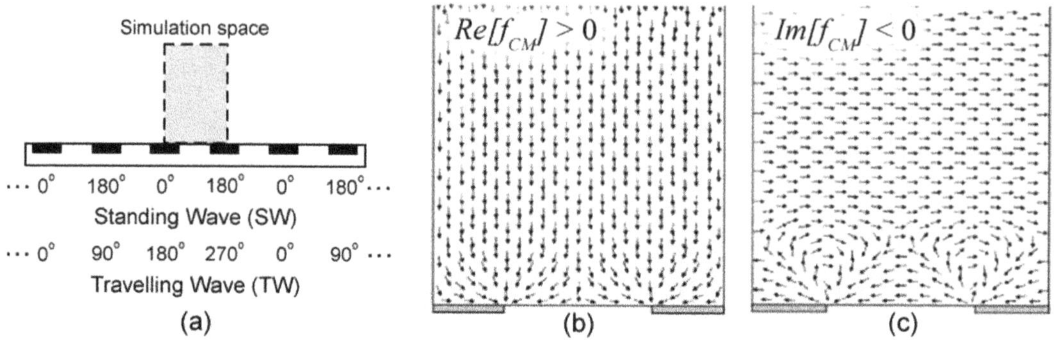

Figure 2. (**a**) Illustration depicting electrode configurations for SW and TW fields. (**b**) Direction of positive DEP force ($Re[f_{CM}] > 0$). (**c**) Direction of negative twDEP force ($Im[f_{CM}] < 0$). Both (**b**,**c**) are modified from [25].

2. Materials and Methods

This manuscript describes several investigations to study the electrokinetics of our electric curtain. (1) First, we could observe dielectrophoretic forces exerted by the curtain using a suspended AFM tip. The AFM would be polarized and actuated due to the applied AC field and dielectric phenomena could be observed. (2) Next, the electrokinetic behavior of Martian dust and spherical particles were captured with a high-speed camera. (3) Last, numerical simulations determined the influence the particle itself had on the local electric field. Details of these investigations follow.

The electric curtain platform used in all experiments consisted of a printed circuit board (ExpressPCB, Santa Barbara, CA, USA) with four sets of repeating interdigitated electrodes (0.020″ width and spacing) where a SW (0°–180°–0°–180°) or TW (0°–90°–180°–270°) AC signal was applied. The electrodes were covered with a dielectric of polypropylene tape (ε_r = 2.2–2.36, 0.0016″ polypropylene layer, 0.0009″ rubber adhesive layer, 76255A21, McMaster-Carr). A four-channel waveform generator (Model 280, Fluke Corp., Everett, WA, USA) sent SW or TW signals to a custom four-channel step-up amplifier which provided potentials up to 800 V_{rms} with frequencies up to approximately 10 kHz. Unless otherwise stated, the applied AC frequency for these tests was 50 Hz.

2.1. AFM Experiments

The electrokinetic-induced deflection of a suspended AFM cantilever was measured using an Asylum MFP-3D AFM. The cantilever chip was made of Pyrex and supported two triangle levers. The short lever was 100 μm long × 13.5 μm wide with an approximate stiffness of 0.32 N/m. In each trial, exact cantilever stiffness was calibrated from a force curve on hard silicon surface. The short lever was functionalized with a 20 ± 2 μm diameter borosilicate glass sphere. Before each test, the AFM z-piezo was used to move the lever 34 ± 2 μm away from the surface of the electric curtain. The arm of the cantilever was aligned parallel with the electrodes. Deflection of the cantilever was recorded for two seconds at 10 kHz, while a SW AC signal (50 Hz) was applied to the electric curtain.

2.2. Particle Interactions with the Electric Curtain

For initial confirmation of electric curtain behavior, Martian dust simulant (JSC MARS-1A, Orbital Technologies Corp., Madison, WI, USA) was used with a particle diameter of 1 mm and smaller with over 50% of the particles (by weight) had diameters greater than 0.25 mm. The primary chemical composition, by weight, was 34.5% SiO_2, 18.5% Al_2O_3, 9.3% Fe_2O_3, and no more than 5% of other individual compounds. A sample of dust was manually applied to the curtain before field activation

Later, experiments were conducted with a second particle type using spherical non-conductive polarizable particles which provided the advantages of consistent particle shape, homogeneity, and size over previous dust samples. Tracking of individual spheres provided additional insight into governing electrokinetic forces. A sparse sample of solid soda lime glass beads (180–212 μm, SLGMS-2.5, Cospheric, Santa Barbara, CA, USA) coated the surface of the electric curtain prior to the activation of a SW or TW field.

A high-speed camera (HiSpec II, 1000 fps) with a zoom lens (Macro 7000, Navitar) acquired videos from both "top view" and "side view" perspectives, showing the interactions and positions of particles relative to the electrodes.

2.3. Numerical Simulations

Electrostatics simulations were performed in COMSOL Multiphysics (model details in Supplementary Material) to investigate the influence of the particle position on the local electric field. In brief, the 3D electrostatics system was governed by

$$\varepsilon^* = \varepsilon - j\sigma/(\omega), \quad (6)$$

where V is electric potential and ρ is volume charge density. For simplicity, we assumed negligible volume charge ($\rho = 0$) and used $\varepsilon_r = 1$ for air, $\varepsilon_r = 7$ for the glass particle, and $\varepsilon_r = 2.28$ for the insulative layer. Here, $\varepsilon = \varepsilon_r\varepsilon_o$ with ε_r as the relative permittivity and ε_o the permittivity of free space. A spherical particle was placed over an electrode edge at various heights above the insulator layer (0.5 μm to 300 μm) and the resulting local electric field was modeled.

3. Results and Discussion

3.1. AFM Experiments

Figure 3a shows the oscillation of the cantilever deflection for a 50 Hz, 300 V signal for a 40 ms period. The cantilever position was approximately centered between two electrode strips. The oscillation period was 10 ms (i.e., 100 Hz), twice that of the applied AC frequency—this characteristic is evidence of induced dielectric forces. However, electrostatic forces were not negligible. The amplitude of every other wave was measurably greater than the preceding peak, suggesting a non-neutral charge on the suspended AFM. Figure 3b shows the downward attractive deflection of the cantilever as a function of applied voltage (100 V–300 V, 50 Hz). The results qualitatively agree with the expected SW DEP force being greatest near the electrode edges and increase with applied voltage (Figure 2b). Additional tests without a tip (cantilever only) also demonstrated similar deflection behavior suggesting that the cantilever itself was also polarized (results not shown); therefore, the magnitude of the resultant measured attractive force was not due solely by its attached glass sphere. Qualitatively, these AFM results demonstrated that dielectric forces exist for objects not in direct contact with the surface of the electric curtain. Furthermore, this setup is not too far removed from Pohl and Pethig's apparatus [36] that demonstrated positive dielectrophoresis of a suspended object in air. In the future, the use of optical tweezers [37] could be used to decouple any mechanical attachments to the suspended particle under investigation.

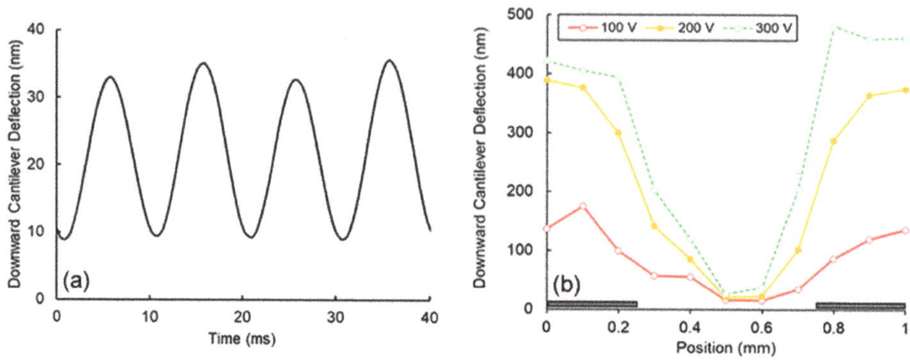

Figure 3. (**a**) A portion of the time trial demonstrating oscillatory deflection of an AFM for a 50 Hz field. (**b**) AFM deflection as a function of voltage and position. Electrodes are illustrated as dark rectangles.

3.2. Particle Interactions with the Electric Curtain

Dust was successfully repelled from the surface of an electric curtain platform using an applied voltage of 600 V at 50 Hz for both SW and TW fields (Figure 1, bottom), though the latter demonstrated repulsion in the direction of the translating wave. Dust repulsion was frequency dependent; at AC frequencies greater than 1 kHz, there was no observable particle movement. More specifically, for frequencies between 230 Hz and 1 kHz, limited particle 'agitation' was observed, consistent with previous observations [3], though no bulk repulsion occurred. However, once the AC frequency decreased to approximately 210 Hz there was sudden and immediate particle repulsion; repulsion occurred for all tested frequencies below this limit (10 Hz to 210 Hz). Several larger particles whose size was similar to the electrode gap were trapped in the dielectric-coated regions (Figure 1, bottom-right). This further suggests dipole (and likely multipole [33,34]) generation, subsequent electro-orientation (the alignment of the particle's long axis in the direction of the field [31]), and demonstrates single particle dielectrophoretic trapping.

High-speed videos (1000 fps) were acquired from a "side view" in order to observe bulk dust repulsion behavior. The video showed that dust would be repelled from the surface in a pulse-like manner whose period (10 ms) corresponded to twice the applied frequency (Supplemental Material, Video S1). These results were consistent with the AFM experiments in which the applied forces occur at twice the applied AC frequency, suggesting dielectric behavior. However, particles experiencing DEP in should be attractive, as demonstrated by the AFM results. This observed discrepancy of repulsive behavior motivated the proceeding study using homogeneous spherical glass particles of consistent size.

At 50 Hz and 600 V, spherical glass particles were repelled for both SW and TW fields. For both scenarios particles experienced low velocity oscillatory "rolling" along the surface and/or rapid repulsive projectile motion. For an SW field, rolling particles would typically oscillate about an electrode strip and about its nearest electrode neighbors (Figure 4a, blue double arrow). This behavior suggests particles experienced positive DEP and, when including inertial effects, the oscillatory rolling particle motion (blue arrows) makes sense. However, repulsed projectile particles (Figure 4a, red arrow) were observed at velocities much greater than the oscillatory velocity. These particles initiated their "launch" in close proximity to electrode edges (where the field is strongest) and projection occurred in approximately 10 ms intervals (twice the applied AC frequency).

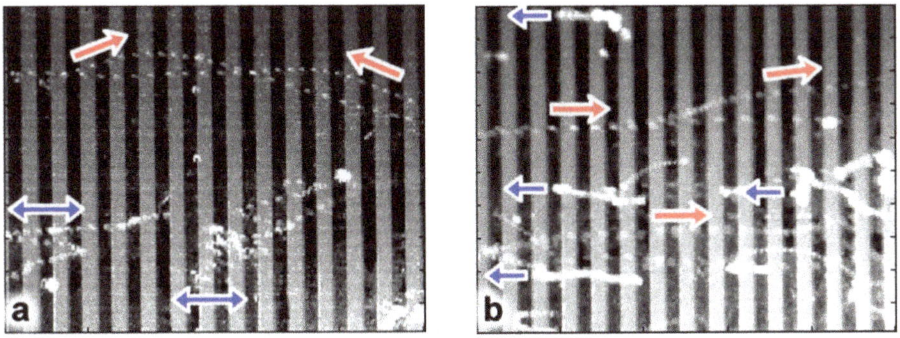

Figure 4. Consecutive overlaid images over a 0.2 s interval (2 ms between images) for both SW and TW fields. (**a**) SW manipulation of glass spheres where both oscillatory rolling (blue arrows) and high velocity repulsion (red arrows) were observed. (**b**) TW manipulation where particles would roll slowly against the field (blue arrows) or be repelled rapidly in the direction of the applied field (red arrows).

TW results also demonstrated particle rolling and projectile motion though with specific, and opposite, directionality. The particles rolled against the TW field direction relatively slowly (Figure 4b, blue arrows), while others were projected at high velocity in the direction of the wave (Figure 4b, red arrows). Due to the negative $Im[f_{CM}]$ at 50 Hz and the direction of the twDEP force near the electrode surface (Figure 2c), dielectrophoretic forces are responsible for each particle's roll direction that is opposite of the TW. Particle repulsion in the direction of the TW field implies electrostatic repulsion and is consistent with typical electric curtain performance. High speed videos of both "top view" and "side view" SW and TW experiments are available in Supplementary Materials (SW: Videos S2 and S3; TW: Videos S4 and S5).

For both SW and TW studies, the projectile motion of particles was initiated near the electrode edge where the electric field is greatest. However, in the absence of particles, arching nor plasma was not observed, even when the voltage increased to the upper limit of our equipment (800 V). From these results it was hypothesized that the particles themselves triggered local electrostatic repulsion. As DEP attracted particles towards regions of greater field strength, the particle itself would locally distort the field further, intensifying it to the point of triggering electrostatic repulsion. Thus, the following numerical simulations were conducted to determine the significance of field amplification due to the presence of the particle itself.

3.3. Numerical Simulations

Figure 5a depicts the cross-section view of the 3D electrostatics model and the simulated space (more details in Supplementary Material). The sphere was placed centered with the electrode edge and its gap above the insulator was varied. The electric field magnitude at the point directly above the electrode edge on the insulator layer was measured (Figure 5a,b). In the absence of a particle, the electric field at this location was modeled to be 12.5 kV/cm and it did not significantly increase until the gap between the particle and insulator layer was less than one particle diameter. Figure 5c, Figure 5d and Figure 5e show a zoomed-in portion of the simulation for gaps of 100 μm, 10 μm, and 1 μm, respectively. The space between the particle and the insulative layer experiences a significant increase in electric field, such that the modeled 0.5 μm gap was 34.4 kV/cm (2.8 times greater than the modeled field without the particle).

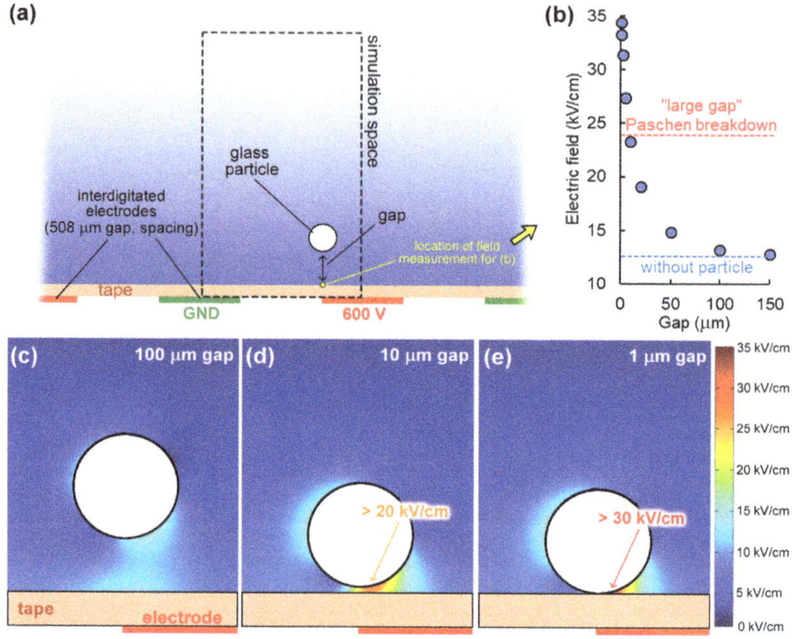

Figure 5. (**a**) Depiction of the simulation space (more details in Supplementary Material). (**b**) Magnitude of electric field at the indicated measurement point. As the gap decreases the localized electric field increases. (**c**) Numerical simulation result of the electric field magnitude near the particle for a gap of 100 μm; similarly for (**d**) 10 μm and (**e**) 1 μm gaps.

Numerical simulations definitively show that particles themselves significantly increase the electric field locally. Next, these modeled values were compared to Paschen's law, which is the breakdown voltage where a discharge would occur between two electrodes as a function of gap length [38]. For a uniform field and "very large gaps", the limiting value of the field under normal temperature and pressure is approximately 24 kV/cm (labeled in Figure 5b). It is unlikely that the field limit for our electric curtain system occurs at 24 kV/cm, as our electrodes are insulated and planar, as the breakdown voltage would differ for non-uniform fields [39]. Although we do not know the breakdown voltage for our electric curtain, the experimental and numerical evidence herein supports that particle-induced electrostatic repulsion is a reasonable phenomenological explanation of the observed behavior.

4. Concluding Remarks

Positive dielectrophoretic forces were expected [36] for our experimental conditions (i.e., without corona); however, simultaneous Coulombic repulsion was unexpected until we accounted for local particle-induced field distortions. These experiments demonstrate an interesting electrokinetic behavior where particles exhibit dipolar (DEP, "pulsed" repulsion at twice the applied frequency) whose attractive forces distort the electric field and subsequently trigger monopolar (Coulombic repulsion) forces. These findings serve as a guide for future work in investigating the interaction between particles and the electric field in electric curtains.

By no means is this experimental study comprehensive, as charge transfer mechanisms are complex. For example, work by J. Lowell demonstrated that charge transfer between two insulators can occur without tribocharging [40]. Furthermore, Lowell showed that polarization significantly influenced charge transfer energy states [41], regardless of whether the states are donors or accepters. Furthermore, particle and insulative layer polarizability is frequency dependent, and this characteristic has been explored previously where directionality of particle motion in a TW curtain was AC frequency dependent [20]. Generally, particles would translate with the wave at low frequencies but would move in the opposite direction at higher frequencies; at even greater frequencies no net particle movement would occur. Thus, understanding frequency-dependent polarizability within a dielectric curtain should be further explored in applying this technology towards particle sorting.

Supplementary Materials: The following supporting information can be downloaded at https://www.mdpi.com/article/10.3390/mi13020288/s1. A document describing the COMSOL Multiphysics numerical simulations and a table of results that was plotted in Figure 5b (Figures S1 and S2, and Table S1). A high-speed video of pulsed repulsion of dust is available (Video S1). High speed "top view" and "side view" videos of soda lime glass repulsion are available for SW configuration (Videos S2 and S3) and TW configuration (Videos S4 and S5).

Author Contributions: Conceptualization, S.J.W.; Methodology, J.D.S. and B.C.K.; Formal Analysis, S.J.W., J.D.S., B.C.K. and N.G.G.; Investigation, J.D.S. and B.C.K.; Writing—Original Draft Preparation, S.J.W.; Writing—Review & Editing, S.J.W. and N.G.G.; Supervision, S.J.W.; Project Administration, S.J.W.; Funding Acquisition, S.J.W. All authors have read and agreed to the published version of the manuscript.

Funding: This research was funded by NASA Kentucky Space Grant Consortium through grants RIDG-14-003 and UF-16-006.

Data Availability Statement: Videos are available in Supplementary Material. AFM data are available upon request from the corresponding author, S.J.W.

Acknowledgments: S.J.W. acknowledges experimental work conducted by the co-authors (J.D.S. and B.C.K.) during 2015 and 2016. It was not until recently that the authors S.J.W. and N.G.G. explored the revised hypothesis herein with the inclusion of numerical simulations.

Conflicts of Interest: The authors declare no conflict of interest.

References

1. Masuda, S.; Fujibayashi, K.; Ishida, K.; Inada, H. Confinement and transportation of charged aerosol clouds via electric curtain. *Electr. Eng. Jpn.* **1972**, *92*, 43–52. [CrossRef]
2. Masuda, S.; Washizu, M.; Iwadare, M. Separation of small particles suspended in liquid by nonuniform traveling field. *IEEE Trans. Ind. Appl.* **1987**, *23*, 474–480. [CrossRef]
3. Atten, P.; Pang, H.L.; Reboud, J.-L. Study of dust removal by standing-wave electric curtain for application to solar cells on mars. *IEEE Trans. Ind. Appl.* **2009**, *45*, 75–86. [CrossRef]
4. Chesnutt, J.K.W.; Marshall, J.S. Simulation of particle separation on an inclined electric curtain. *IEEE Trans. Ind. Appl.* **2013**, *49*, 1104–1112. [CrossRef]
5. Hemstreet, J.M. Velocity distribution on the Masuda panel. *J. Electrost.* **1985**, *17*, 245–254. [CrossRef]
6. Melcher, J.R.; Warren, E.P.; Kotwal, R.H. Theory for finite-phase traveling-wave boundary-guided transport of triboelectrified particles. *IEEE Trans. Ind. Appl.* **1989**, *25*, 949–955. [CrossRef]
7. Myers, O.D.; Wu, J.; Marshall, J.S. Nonlinear dynamics of particles excited by an electric curtain. *J. Appl. Phys.* **2013**, *114*, 154907. [CrossRef]
8. Calle, C.I.; Buhler, C.R.; Johansen, M.R.; Hogue, M.D.; Snyder, S.J. Active dust control and mitigation technology for lunar and Martian exploration. *Acta Astronaut.* **2011**, *69*, 1082–1088. [CrossRef]
9. Calle, C.I.; Buhler, C.R.; McFall, J.L.; Snyder, S.J. Particle removal by electrostatic and dielectrophoretic forces for dust control during lunar exploration missions. *J. Electrost.* **2009**, *67*, 89–92. [CrossRef]
10. Calle, C.I.; Mackey, P.J.; Hogue, M.D.; Johansen, M.R.; Yim, H.; Delaune, P.B.; Clements, J.S. Electrodynamic dust shields on the international space station: Exposure to the space environment. *J. Electrost.* **2013**, *71*, 257–259. [CrossRef]
11. Desai, A.; Lee, S.W.; Tai, Y.C. A MEMS electrostatic particle transportation system. *Sens. Actuators A Phys.* **1999**, *73*, 37–44. [CrossRef]
12. Hemstreet, J.M. Three-phase velocity distribution of lycopodium particles on the Masuda panel. *J. Electrost.* **1992**, *27*, 237–247. [CrossRef]
13. Kawamoto, H. Some techniques on electrostatic separation of particle size utilizing electrostatic traveling-wave field. *J. Electrost.* **2008**, *66*, 220–228. [CrossRef]

14. Moesner, F.M.; Higuchi, T. Traveling electric field conveyor for contactless manipulation of microparts. In Proceedings of the Industry Applications Conference, Thirty-Second IAS Annual Meeting, IAS'97, Conference Record of the 1997 IEEE, New Orleans, LA, USA, 5–9 October 1997; Volume 2003, pp. 2004–2011.
15. Schmidlin, F.W. A new nonlevitated mode of traveling-wave toner transport. *IEEE Trans. Ind. Appl.* **1991**, *27*, 480–487. [CrossRef]
16. Xing, X.; Yang, J.M.; Tai, Y.C.; Ho, C.M. Micromachined membrane particle filters. *Sens. Actuators A Phys.* **1999**, *73*, 184–191.
17. Zhao, Y.; Chung, S.K.; Yi, U.-C.; Cho, S.K. Droplet manipulation and microparticle sampling on perforated microfilter membranes. *J. Micromech. Microeng.* **2008**, *18*, 025030. [CrossRef]
18. Gartstein, Y.N.; Shaw, J.G. Many-particle effects in travelling electrostatic wave transport. *J. Phys. D Appl. Phys.* **1999**, *32*, 2176–2180. [CrossRef]
19. Liu, G.; Marshall, J.S. Particle transport by standing waves on an electric curtain. *J. Electrost.* **2010**, *68*, 289–298. [CrossRef]
20. Liu, G.Q.; Marshall, J.S. Effect of particle adhesion and interactions on motion by traveling waves on an electric curtain. *J. Electrost.* **2010**, *68*, 179–189. [CrossRef]
21. Schmidlin, F.W. Modes of traveling-wave particle-transport and their applications. *J. Electrost.* **1995**, *34*, 225–244. [CrossRef]
22. Sun, Q.; Yang, N.; Cai, X.; Hu, G. Mechanism of dust removal by a standing wave electric curtain. *Sci. China Phys. Mech. Astron.* **2012**, *55*, 1018–1025. [CrossRef]
23. Lindley, K.S.; Rowson, N.A. Charging mechanisms for particles prior to electrostatic separation. *Magn. Electr. Sep.* **1997**, *8*, 101–113. [CrossRef]
24. Castellanos, A.; Ramos, A.; Gonzalez, A.; Green, N.G.; Morgan, H. Electrohydrodynamics and dielectrophoresis in microsystems: Scaling laws. *J. Phys. D Appl. Phys.* **2003**, *36*, 2584–2597. [CrossRef]
25. Morgan, H.; Green, N.G. *AC Electrokinetics: Colloids and Nanoparticles*; Research Studies Press: Philadelphia, PA, USA, 2003.
26. Weiss, L.C. Electrodynamic behavior of textile fibers. *Text. Res. J.* **1982**, *52*, 59–65. [CrossRef]
27. Masuda, S.; Matsumoto, Y. Theoretical characteristics of standing-wave electric curtains. *Electr. Eng. Jpn.* **1973**, *93*, 71–77. [CrossRef]
28. Pethig, R. Dielectrophoresis: Status of the theory, technology, and applications. *Biomicrofluidics* **2010**, *4*, 022811. [CrossRef]
29. Wood, N.R.; Wolsiefer, A.I.; Cohn, R.W.; Williams, S.J. Dielectrophoretic trapping of nanoparticles with an electrokinetic nanoprobe. *Electrophoresis* **2013**, *34*, 1922–1930. [CrossRef]
30. Khoshmanesh, K.; Nahavandi, S.; Baratchi, S.; Mitchell, A.; Kalantar-zadeh, K. Dielectrophoretic platforms for bio-microfluidic systems. *Biosens. Bioelectron.* **2011**, *26*, 1800–1814. [CrossRef]
31. Jones, T.B. *Electromechanics of Particles*; Cambridge University Press: Cambridge, UK; New York, NY, USA, 1995.
32. Gimsa, J. A comprehensive approach to electro-orientation, electrodeformation, dielectrophoresis, and electrorotation of ellipsoidal particles and biological cells. *Bioelectrochemistry* **2001**, *54*, 23–31. [CrossRef]
33. Washizu, M.; Jones, T.B. Multipolar dielectrophoretic force calculation. *J. Electrost.* **1994**, *33*, 187–198. [CrossRef]
34. Washizu, M.; Jones, T.B.; Kaler, K. Higher-order dielectrophoretic effects—Levitation at a field null. *Biochim. Biophys. Acta* **1993**, *1158*, 40–46. [CrossRef]
35. Voldman, J.; Braff, R.A.; Toner, M.; Gray, M.L.; Schmidt, M.A. Holding forces of single-particle dielectrophoretic traps. *Biophys. J.* **2001**, *80*, 531–541. [CrossRef]
36. Pohl, H.A.; Pethig, R. Dielectric measurements using non-uniform electric field (dielectrophoretic) effects. *J. Phys. E Sci. Instrum.* **1977**, *10*, 190–193. [CrossRef]
37. Park, H.; LeBrun, T.W. Optical trap loading of dielectric microparticles in air. *J. Vis. Exp.* **2017**, *120*, 54862. [CrossRef]
38. Moscg, W.; Arora, R. *High Voltage and Electrical Insulation Engineering*; Wiley-Blackwell: Hoboken, NJ, USA, 2011.
39. Osmokrovic, P.; Krivokapic, I.; Krstic, S. Mechanism of electrical breakdown left of Paschen minimum. *IEEE Trans. Dielectr. Electr. Insul.* **1994**, *1*, 77–81. [CrossRef]
40. Lowell, J. Constraints on Contact Charging of Insulators. 1. Spatial Localization of Insulator States. *J. Phys. D Appl. Phys.* **1986**, *19*, 95–104. [CrossRef]
41. Lowell, J. Constraints on Contact Charging of Insulators. 2. Energy Constraints. *J. Phys. D Appl. Phys.* **1986**, *19*, 105–113. [CrossRef]

micromachines

Article

Dielectrophoretic Micro-Organization of Chondrocytes to Regenerate Mechanically Anisotropic Cartilaginous Tissue

Yoshitaka Takeuchi [1] and Shogo Miyata [2,*]

[1] Graduate School of Science and Technology, Keio University, 3-14-1 Hiyoshi, Kohoku-ku, Yokohama 223-8522, Japan; dai_bamboo77@yahoo.co.jp
[2] Faculty of Science & Technology, Keio University, 3-14-1 Hiyoshi, Kohoku-ku, Yokohama 223-8522, Japan
* Correspondence: miyata@mech.keio.ac.jp

Abstract: Recently, many studies have focused on the repair and regeneration of damaged articular cartilage using tissue engineering. In tissue engineering therapy, cells are cultured in vitro to create a three-dimensional (3-D) tissue designed to replace the damaged cartilage. Although tissue engineering is a useful approach to regenerating cartilage, mechanical anisotropy has not been reconstructed from a cellular organization level. This study aims to create mechanically anisotropic cartilaginous tissue using dielectrophoretic cell patterning and gel-sheet lamination. Bovine chondrocytes were patterned in a hydrogel to form line-array cell clusters via negative dielectrophoresis (DEP). The results indicate that the embedded chondrocytes remained viable and reconstructed cartilaginous tissue along the patterned cell array. Moreover, the agarose gel, in which chondrocytes were patterned, demonstrated mechanical anisotropy. In summary, our DEP cell patterning and gel-sheet lamination techniques would be useful for reconstructing mechanically anisotropic cartilage tissues.

Keywords: dielectrophoresis; chondrocyte; mechanical anisotropy; tissue engineering; cell patterning

Citation: Takeuchi, Y.; Miyata, S. Dielectrophoretic Micro-Organization of Chondrocytes to Regenerate Mechanically Anisotropic Cartilaginous Tissue. *Micromachines* **2021**, *12*, 1098. https://doi.org/10.3390/mi12091098

Academic Editor: Rodrigo Martinez-Duarte

Received: 26 August 2021
Accepted: 9 September 2021
Published: 11 September 2021

1. Introduction

Articular cartilage covers bone ends in diarthrodial joints and is an inhomogeneous, anisotropic, and viscoelastic connective tissues that serve as a low-friction load-bearing material [1–3]. Cartilage plays an important role in mammalian skeletal movements. Due to the avascular nature, low cell density, and low proliferative activity of chondrocytes, hyaline cartilage cannot regenerate after injury; wear and tear; or degeneration through common diseases, such as osteoarthritis [4,5]. Therefore, severe cartilage damage often requires surgical treatment. Current clinical approaches to generating new cartilaginous tissue include debridement, microfracture, autologous chondrocyte transplantation, and mosaicplasty. However, it is difficult to regenerate hyaline cartilage using these therapeutic approaches.

Tissue engineering approaches have been developed to restore articular cartilage damage, which involve culturing autologous chondrocytes in vitro to create a three-dimensional (3-D) tissue designed to replace the damaged tissue [6–9]. In clinical studies, chondrocyte sheets or chondrocyte cultures in atelocollagen gel were transplanted to treat osteochondral defects [6,8]. In native articular cartilage, chondrocyte organization varies with the depth of the articulating surface [10]. Collagen fibers, which are one of the main components of cartilage, align in response to the direction of cyclic deformations induced by daily walking or other activities. Moreover, this alignment is related to cell biosynthesis and tissue anisotropy. However, these inhomogeneous and anisotropic structures of articular cartilage cannot be regenerated using a basic tissue engineering approach. To address this problem, some researchers have reported that culturing cells on scaffolds with anisotropic structures would induce anisotropic remodeling of cartilaginous tissue [11,12]. Mechanical stimulation is another approach to regenerating anisotropic tissue. Lee et al. cultured scaffold-free engineered tissue with tensile stimulation to regenerate anisotropic human neocartilage [13]. Anisotropic mechanical properties were also observed in the fibroblast-seeded gels subjected to mechanical stretching [14]. Our research group also reported that cyclic compression improved the stiffness and mechanical anisotropy of the regenerated cartilaginous tissue [15]. From previous studies, anisotropic cellular organization was required to simulate "native" biological tissue. In this study, we focused on the direct control of cellular organization in hydrogels to regenerate 3-D anisotropic cartilaginous tissue. The random encapsulation of chondrocytes within hydrogels has been widely used for the in vitro culture of articular chondrocytes [16–20]. However, the control of cellular organization in hydrogels remains difficult. While articular cartilage has been one of the first tissues to be successfully treated with tissue engineering therapy, regenerating the anisotropic architecture and biomechanical properties of articular cartilage remains a challenging problem.

Recently, a novel micro-particle patterning technique was developed that utilizes DEP forces to manipulate living cells [21–27]. Albrecht et al. reported the encapsulation of living cell arrays using dielectrophoresis and photo-cross-linked hydrogels [23–25]. In our previous studies, we reported cell manipulation and separation systems [28–30] and a cell-patterning technology in hydrogels using dielectrophoresis [31]. However, there have not been any studies on remodeling the mechanical anisotropic tissue based on controlled cellular organization by dielectrophoresis.

This study aimed to perform DEP cell patterning in a hydrogel to modulate cellular organization, proliferation, and extracellular matrix synthesis to regenerate mechanically anisotropic cartilaginous tissue. First, the dielectrophoretic properties of primary articular chondrocytes were evaluated to determine the optimal conditions for manipulating the cells in agarose gel. Based on the measured dielectrophoretic properties, the chondrocytes were patterned in an anisotropic alignment inside the agarose gel using negative-DEP forces. Mechanical tensile tests were performed to characterize the mechanical anisotropy of the chondrocyte-aligned hydrogel constructs.

2. Materials and Methods

2.1. Chondrocyte Isolation for Dielectrophoresis

Articular cartilage was harvested from articular joints of 3–6-week-old calves from a local abattoir, and chondrocytes were isolated from cartilage explants by enzymatic digestion [15]. Briefly, cartilage explants were excised from the humeral head and minced into small pieces, which were then shaken gently in 0.2% collagenase type II digested with Dulbecco's modified Eagle's medium/Ham's F12 (DMEM/F12) supplemented with 5% fetal bovine serum (FBS) and antibiotics–antimycotics for 8 h at 37 °C. The cartilage-digested solution was filtered through a 70 μm nylon mesh filter to remove debris. The cells were isolated from the digest by centrifugation for 5 min and rinsed twice with phosphate-buffered saline (PBS) containing antibiotics–antimycotics. Finally, after centrifugation for 5 min, the cells were resuspended in a low-conductivity osmotically balanced buffer (LC buffer: 10 mM HEPES, 0.1 mM $CaCl_2$, and 59 mM D-glucose in sucrose solution [28–31]) for dielectrophoresis experiments. The total cell number was counted using a hemocytometer to adjust the cell concentration prior to the experiments.

2.2. Dielectrophoresis Chamber for Living Cells

A dielectrophoresis chamber for cellular analysis and accumulation was developed to establish a nonuniform electric field in a rectangular volume. The chamber was constructed by sandwiching a 500 μm silicon rubber gasket between two glass slides coated with a conductive material, indium tin oxide (ITO) (Figure 1a). The bottom-side glass was partially coated with a 1 μm film of SU-8 photoresist to insulate specific areas of the conductive surface and fabricate a parallel line electrode array with a width of 20 μm; the lines were spaced 80 μm apart (Figure 1b). A sine-wave AC voltage was applied between the upper and bottom ITO-coated glass slides to generate a non-uniform electric field. DEP is a phenomenon that occurs under an applied non-uniform electric field, inducing dipoles within a polarized cell in a buffer solution. The cell, in a non-uniform electric field, can be moved by DEP forces toward high or low electric field regions, depending on the relative electric property of cells, which is related to the cell type and function [21,28]. A high electric field is localized at the line-shaped electrodes, and cells are localized to the high or low electric field region due to positive- or negative-dielectric forces depending on their electric properties (Figure 2). The AC voltage was applied between the parallel-line electrode-fabricated glass slide and upper ITO-coated glass slide using a function generator (WF1944B, NF Corp., Yokohama, Japan) and an amplifier (BA4850, NF Corp., Yokohama, Japan). The applied voltage was monitored using an oscilloscope (TDS1001B, Tektronix, Beaverton, OR, USA) connected in parallel. The movement of cells within the DEP chamber was observed using a phase-contrast microscope (CKX41, Olympus, Tokyo, Japan) with a digital video camera. The dielectrophoretic chamber was sterilized with 70% ethanol followed by rinses with a low-conductivity (LC) buffer containing 1% Pluronic F108 (Sigma-Aldrich, St. Louis, MO, USA) prior to the dielectrophoresis experiments. Pluronic 108 was used to avoid cell adhesion to the glass slides after dielectrophoresis experiments. In our previous study [31], it was confirmed that cell viability and function would be maintained under the dielectrophoretic conditions of this study.

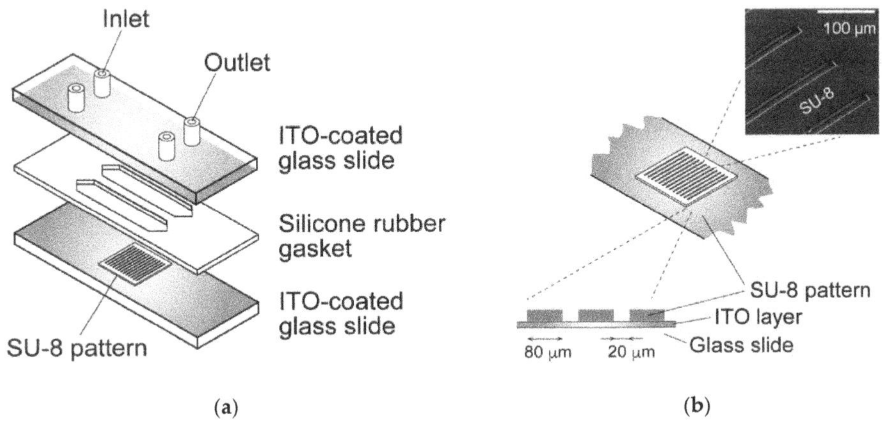

Figure 1. Schematic of the dielectrophoretic chamber. (**a**) The chamber was constructed by sandwiching a silicone rubber gasket between two glass slides coated with indium tin oxide (ITO). (**b**) The bottom slide was partially insulated by an SU-8 pattern to fabricate a parallel-line electrode array.

2.3. Dielectrophoretic Characterization of Chondrocytes

For the DEP characterization, chondrocyte suspension in LC buffer was injected into the DEP chamber and subjected to an AC voltage for 180 s after injection. The magnitude of the imposed AC voltage was 10 $V_{p\text{-}p}$ with a frequency varying from 10 kHz to 1 MHz. The behavior of chondrocytes was observed using a digital camera on the microscope, and microphotographs were captured 180 s after each AC voltage frequency was imposed. The number of cells on the electrodes (positive-DEP) and between the electrodes (negative-DEP) were counted in each captured image. The ratios of cells indicating positive- and negative-DEPs in the chamber were calculated to evaluate the crossover frequency. The dielectrophoretic property of a cell (indicating p-DEP or n-DEP) was assessed based on the region where the cell moved (Figure 3). The frequency dependency of the DEP property was evaluated by the positive-DEP ratio R_p and the negative-DEP ratio R_n, calculated as follows:

$$R_p = N_P/(N_P + N_n),\ R_n = N_n/(N_P + N_n), \tag{1}$$

where N_P and N_n are the numbers of cells indicating positive-DEP and negative-DEP, respectively. To evaluate the dielectrophoretic property of cells, the dielectrophoresis parameter was defined as R_p–R_n.

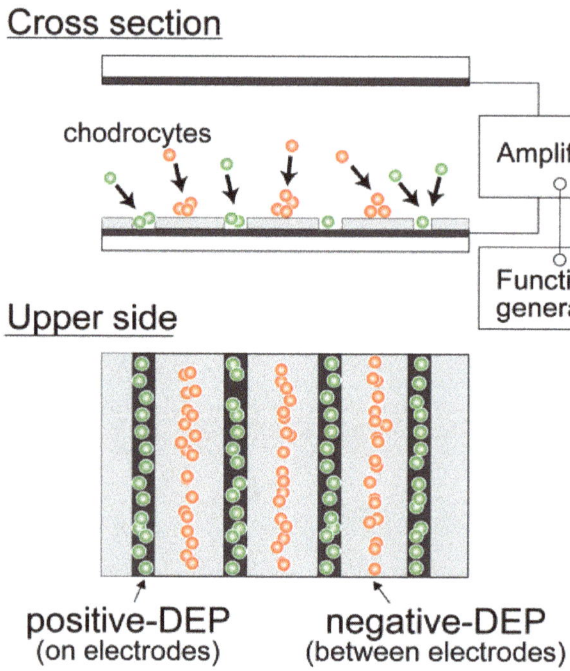

Figure 2. Dielectrophoresis of living cells. Cells were moved to a region with a low electric field (between the electrodes) or high electric field (on the electrodes).

2.4. Dielectrophoretic Cell Accumulation in Agarose Gel to Fabricate Cell-Aligned Three-Dimensional Cultures

Based on dielectrophoretic characterization, chondrocytes were patterned in agarose gel via the negative-DEP force. Chondrocyte-suspended agarose solutions were prepared as previously described [15,32–35]. The isolated chondrocytes were suspended in the LC buffer and mixed in a 1:1 ratio with 3% low-melting agarose (A2576; Sigma-Aldrich, St. Louis, MO, USA) in the LC buffer to make a 1.5% agarose solution with a cell density of 1.0×10^7 cells/mL. For cell accumulation, the chondrocyte/agarose solution was injected into the chamber. A sine-wave AC voltage of 15 $V_{p\text{-}p}$ and 10 kHz was applied for 15 min to cause rapid cell localization in the region of low electric field (negative-DEP). After chondrocyte accumulation, the chondrocyte/agarose solution in the chamber was cooled at 4 °C for 20 min to obtain a gelation agarose solution.

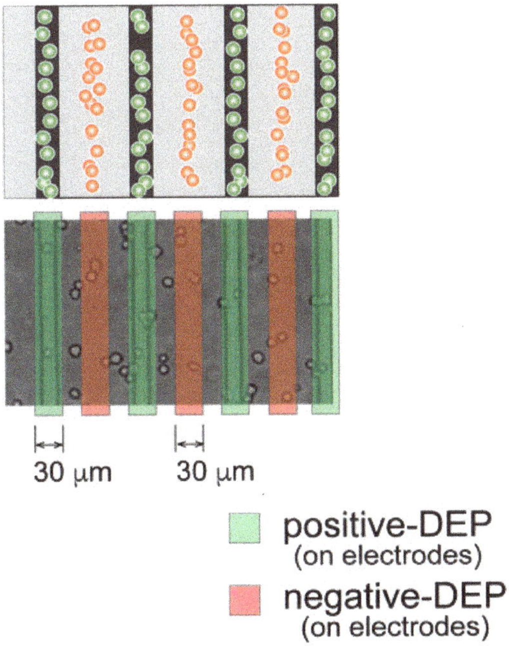

Figure 3. Evaluation of the dielectrophoretic property of chondrocytes. After dielectrophoresis, cells in red regions demonstrated a negative-DEP response, whereas those in green regions demonstrated positive-DEP response.

Three agarose gel sheets containing patterned-chondrocyte arrays were stacked to form a 3-D culture sample for mechanical anisotropy evaluation. The gel sheets in the DEP chamber were transferred to a Petri dish using a custom-made gel sheet holder using negative pressure suction. The gel sheet holder was constructed from a nylon mesh sheet and stainless-steel parts. The gel sheets were held by adsorbing the sheet onto the nylon mesh through aspiration from a syringe (Figure 4). After the first sheet was transferred to the dish, the second and third gel sheets were collected and stacked onto the already transferred gel sheet using the same procedure. Then, a 25 μL 1.5% agarose solution was dropped between the sheets as adhesive glue. The position of each gel sheet was aligned manually under a stereomicroscope. The stacked gel-sheet construct was shaped into a thin plate 5 × 15 mm^2 with a thickness of 1.5 mm (500 μm × 3 sheets) for cell culture experiments. To evaluate mechanical anisotropy, three types of specimens were fabricated; the cells were patterned parallel or perpendicular to the longitudinal direction of the gel sheet and were homogeneously dispersed to prepare three experimental groups (i.e., parallel, perpendicular, and homogeneously dispersed groups).

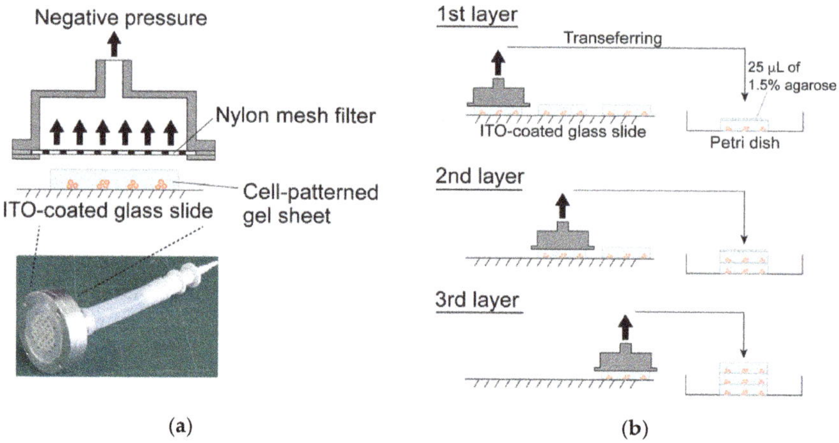

Figure 4. Fabrication of the layered gel-sheet construct. (**a**) Gel sheet transferring device using negative air pressure and (**b**) layering process of the cell-patterned agarose gel sheets.

The chondrocyte-accumulated agarose gel constructs were cultured for up to 21 days in DMEM/F12 containing 20% FBS and 50 μg/mL ascorbic acid. The cultures were maintained in a humidified tissue culture incubator at 37 °C and 5% CO_2.

2.5. Biomechanical Characterization of the Regenerated Tissue

After 7, 14, and 21 days of culturing, the mechanical properties of the cultured construct were evaluated using a custom-made material testing device. The material testing device consisted of a load cell (Kyowa Electronic Instruments, Tokyo, Japan), stepping motor driven stage (Sigma-Koki, Saitama, Japan), and stainless-steel grips for the gel sheet (Figure 5). The stacked-gel-sheet construct was gripped to expose a 5 × 5 mm^2 area, and tensile deformation was applied. The specimen was stretched at a strain rate of 0.01/s up to a strain of 0.4. The stress and strain were calculated using the measured displacement and load values. The elastic modulus, rupture stress, and rupture strain were measured from the stress–strain curve. Regarding the analysis of the hydrogel strain–stress curve, the stress at the point of maximum stress was defined as the rupture stress, and the strain at that point was defined as the rupture strain (Figure 6).

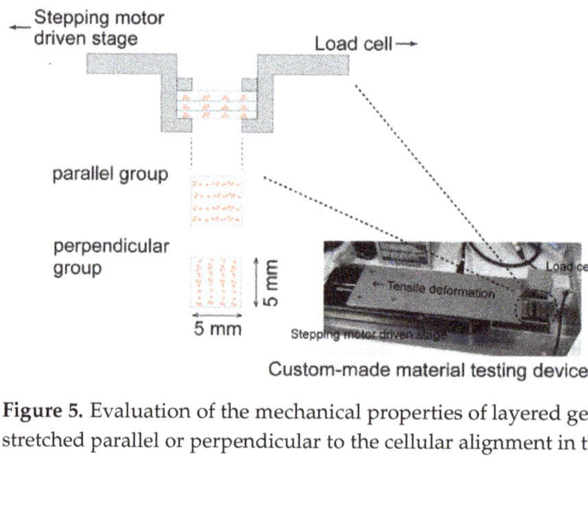

Figure 5. Evaluation of the mechanical properties of layered gel-sheet constructs. Specimens were stretched parallel or perpendicular to the cellular alignment in the gel-sheet construct.

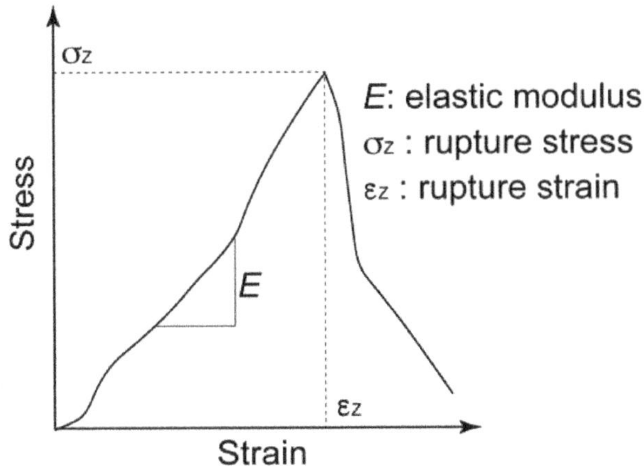

Figure 6. Definition of elastic modulus (E), rupture stress (σ_Z), and rupture strain for the stress–strain curve of the gel-sheet laminated construct.

2.6. Cell Proliferation, Viability, and Biochemical Composition of the Regenerated Tissue

The biochemical properties of the cultured constructs were evaluated after 7, 14, and 21 days of culturing. For histological analysis, the cultured samples were fixed in a 4% paraformaldehyde solution at 4 °C overnight, followed by staining with Safranin-O to evaluate the sGAG distribution. For biochemical characterization, the samples were digested in papain (125 μg/mL in PBS) at 60 °C for 6 h. Each digested lysate was independently assayed for GAG content using a dimethylmethylene blue assay [34,36] and for DNA content using a fluorometric DNA assay with Hoechst 33258 [37]. The cell number in the cultured construct was calculated from the total DNA content divided by the cellular DNA content (7.7 pg). The viability of the chondrocytes in the agarose gel was assessed using live/dead staining. The cultured constructs were incubated in 1 mL of DMEM containing 1 μg calcein-AM (Dojindo, Japan) and 2 μg propidium iodide (PI; Dojindo, Kumamoto, Japan) at 37 °C for 30 min. Fluorescent images were captured using a fluorescence microscope (CKX41, Olympus, Tokyo, Japan). Dead cells appeared red, while viable cells were green.

2.7. Statistical Analysis

Most of the data were representative of three individual experiments with similar results. The statistical significance of the experimental data was evaluated using the Tukey–Kramer method. Statistical significance was set to $p < 0.05$.

3. Results

3.1. Dielectrophoretic Property of Primary Chondrocytes

The primary chondrocytes changed the DEP responses depending on the frequency of the AC voltage. Figure 7 shows photomicrographs of chondrocytes under DEP at different frequencies. The chondrocytes showed a negative-DEP at 10 kHz, whereas a positive-DEP was observed at 500 kHz. Moreover, the chondrocytes showed both negative- and positive-DEPs at frequencies of approximately 100 kHz.

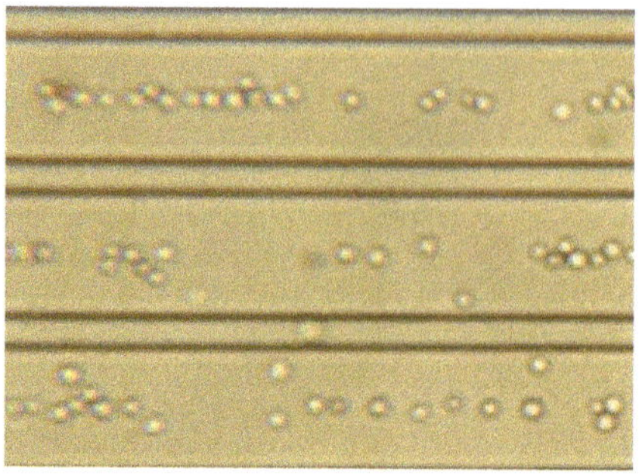

(**a**)

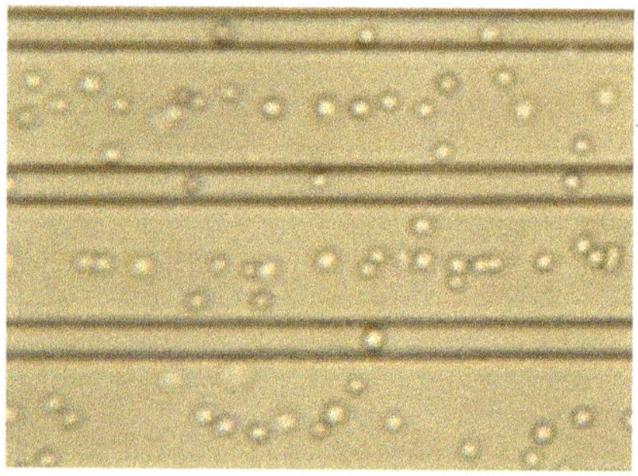

(**b**)

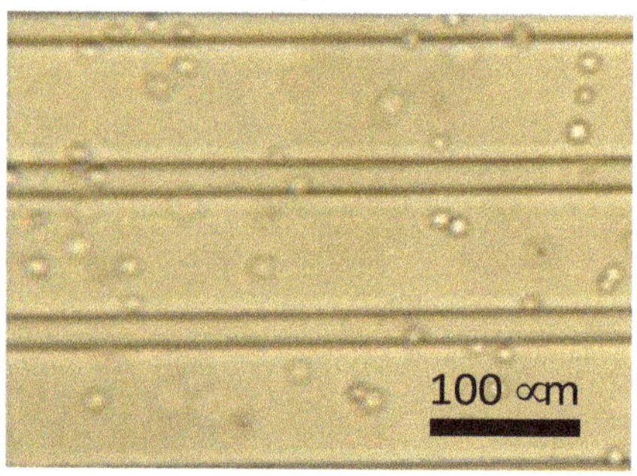

(**c**)

Figure 7. Dielectrophoresis of chondrocytes using parallel line-electrode array. (**a**) 10 kHz, (**b**) 100 kHz, and (**c**) 500 kHz of AC voltage.

To evaluate the DEP frequency dependency of the cells, we defined R_p–R_n as the frequency-dependent parameter. Figure 8 shows the ratio of chondrocytes expressing positive- and negative-DEP responses (R_p–R_n). The chondrocytes switched from positive- to negative-DEPs between 90 and 110 kHz. The result indicates that the crossover frequency of the primary chondrocyte DEPs was approximately 100 kHz (Figure 8).

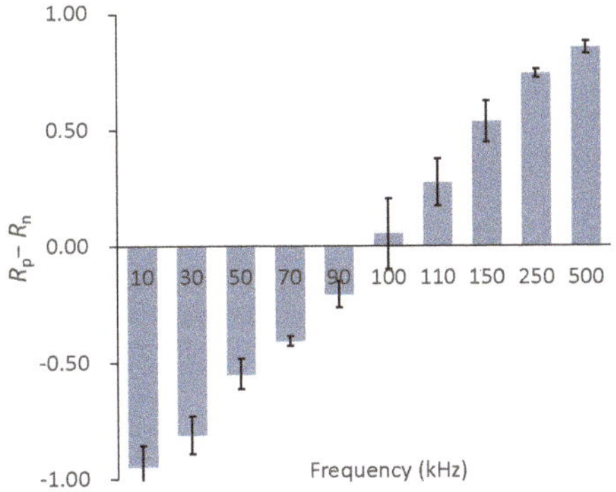

Figure 8. DEP frequency dependency parameter, $R_p - R_n$, of chondrocytes. Mean ± S.D., $n = 5$.

3.2. Tissue Reconstruction in Chondrocyte-Organized Agarose Gel Constructs

The cells were patterned in an agarose gel sheet based on the measured DEP properties of chondrocytes. Following the cell patterning in gel sheets and lamination of the cell-patterned gel sheets, the cell arrays were retained to form anisotropic tissue during the culturing period. From the safranin-O staining of the gel-sheet layered construct, cartilaginous tissue containing sGAG was reconstructed along the patterned cell array, whereas the chondrocytes seeded homogeneously in agarose gels secreted and reconstructed homogeneous tissue (Figure 9). The cartilaginous tissue in the cultured agarose gel expanded with increasing culture time. Each reconstructed tissue along the cell array was connected to adjacent tissues over a seven day culturing period. Cell viability was maintained in the gel-sheet layered constructs, and in the homogeneously chondrocyte-seeded agarose gel (Figure 10).

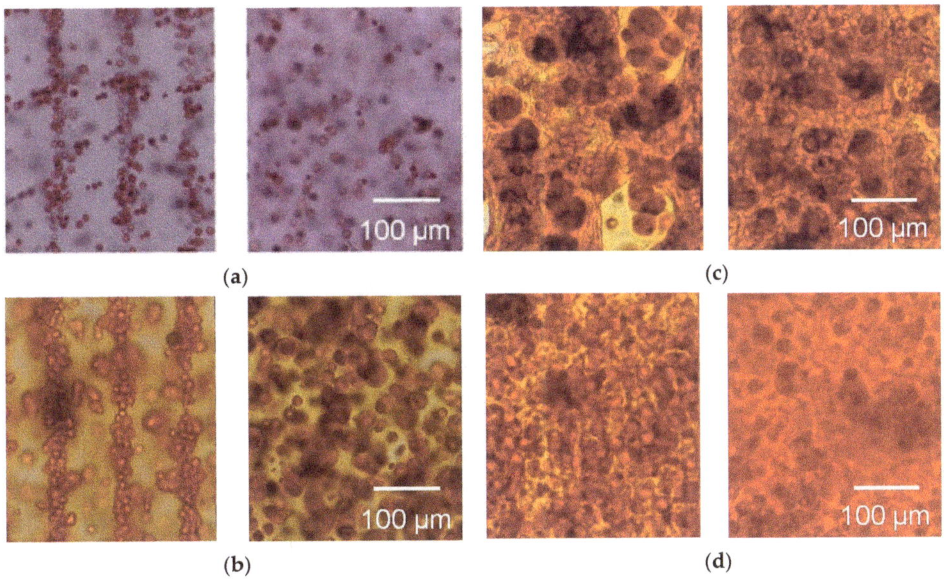

Figure 9. Safranin-O staining of the agarose gel-sheet constructs cultured for (**a**) 0, (**b**) 7, (**c**) 14, and (**d**) 21 days. Left: cell-patterned specimens; right: homogeneously cell-dispersed specimens.

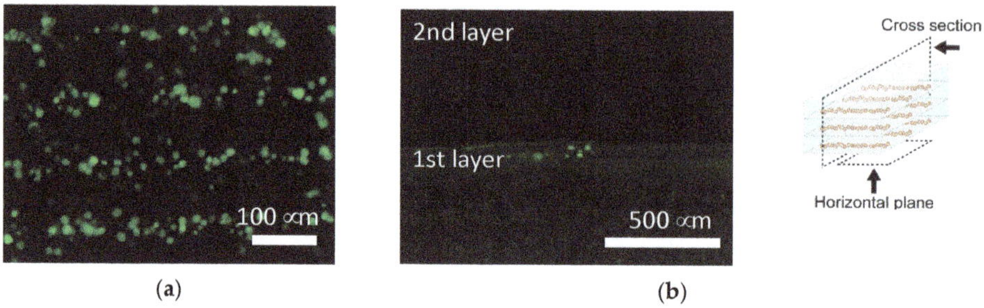

Figure 10. Calcein-AM staining of the agarose gel-sheet constructs cultured for 7 days. Fluorescence images captured from (**a**) the horizontal plane and (**b**) the cross section.

There were no significant differences in the sGAG contents of the specimens from the three experimental groups (Figure 11). There were also no significant differences in the number of cells among the three experimental groups (Figure 12). The sGAG contents and cell numbers of specimens in all experimental groups increased with increasing culturing time.

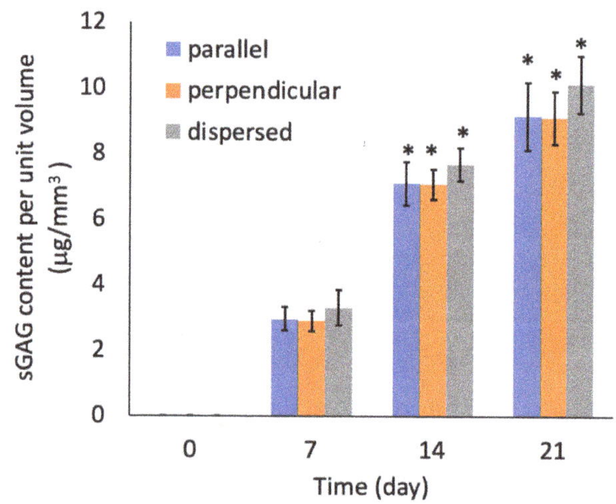

Figure 11. Sulfated GAG content of the agarose gel-sheet constructs cultured for 21 days. Mean +/− S.D., $n = 5$. * indicates a significant difference in each value compared to day 7, $p < 0.05$.

3.3. Mechanical Anisotropy of the Regenerated Tissue

At day 0, the tensile mechanical tests could not be performed because the stiffness of specimens was not sufficient to handle them. The tensile test for perpendicular specimens at day seven also could not be performed. Figure 13 shows a typical stress–strain diagram for each sample group on day 21. The results of tensile tests show that the specimens were fractured at tensile strains smaller than 0.4 for all specimen groups. In this study, the elastic modulus was calculated by linear approximation in the region of strain 0.05–0.1. The elastic modulus increased with increasing culturing time for all sample groups (Figure 14a). On days 14 and 21, the elastic modulus of the parallel group, in which chondrocytes were patterned parallel to the tensile direction, was significantly larger than that of the perpendicular group. The parallel group also showed significantly larger values of rupture stress and strain compared with the perpendicular group (Figure 14b,c).

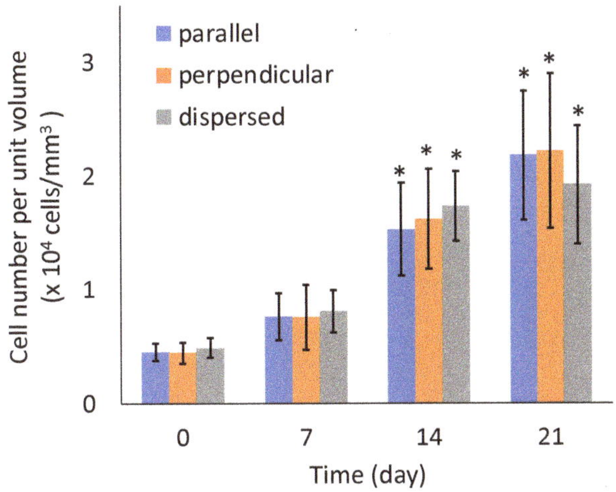

Figure 12. Cell number of the agarose gel-sheet constructs cultured for 21 days. Mean +/− S.D., $n = 5$. * indicates a significant difference in each value compared to day 0, $p < 0.05$.

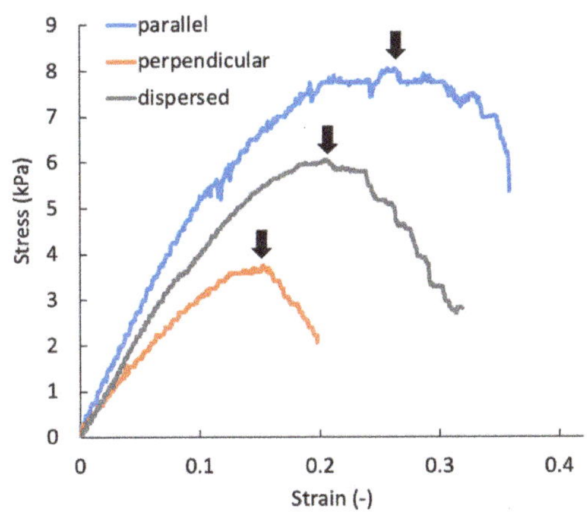

Figure 13. Representative stress–strain diagram of the agarose gel sheet-laminated construct containing chondrocytes cultured for 21 days. Black arrows indicate the rupture point of each specimen.

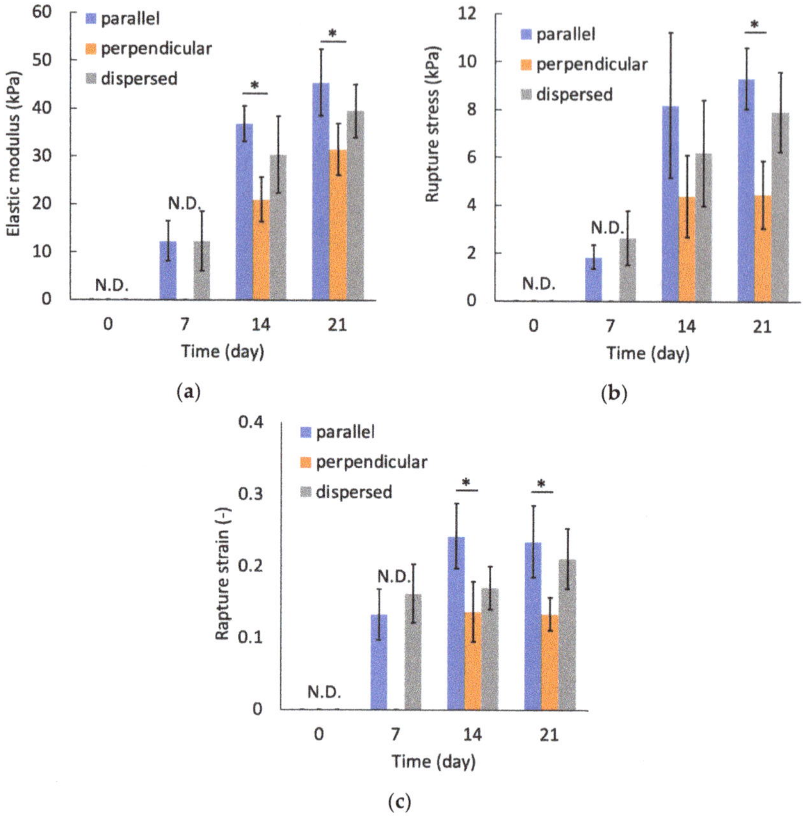

Figure 14. Relationships between culture periods and (**a**) elastic modulus, (**b**) rupture stress, and (**c**) rupture strain. Mean +/− S.D., $n = 4$. * indicates a significant difference between parallel and perpendicular groups at each time point, $p < 0.05$.

4. Discussion

In this study, to regenerate cartilaginous tissue with mechanical anisotropy simulating that of "native" articular cartilage, we established a 3-D culture method to accumulate chondrocytes in a parallel-line pattern inside agarose gels by negative dielectrophoresis. There are many reports on the three-dimensional culture of calf chondrocytes using agarose gels as scaffold material, which state that cartilaginous tissue was reconstructed inside the gel [16–20]. In our study, chondrocytes synthesized an extracellular matrix inside the agarose gel and reconstructed cartilaginous tissue along the parallel-line pattern of accumulated cells. There was no significant difference in the sGAG content among the experimental groups where the chondrocytes were patterned, and the group where the chondrocytes were uniformly distributed inside the gel during the 21-day culturing period. This result indicated that the manipulation of cells by dielectrophoresis and cell accumulation did not affect the cell viability or ability to synthesize extracellular matrix.

The elastic modulus of cultured constructs increased with an increase in the culturing period. This increase in elastic modulus showed a similar trend to that reported in previous studies [17,18]. Furthermore, mechanical anisotropy of the regenerated tissue was observed on days 14 and 21. The elastic modulus of the cultured construct with chondrocytes patterned in the same direction as the tensile direction was significantly higher than that of the sample with perpendicularly patterned chondrocytes. The sGAG content in the cultured construct was not significantly different among the experimental groups. However, it is thought that the extracellular matrix reconstructed around the patterned cell arrays became denser and that the tissue structure became more oriented, resulting in a higher stiffness in the direction of cell patterning. In the cultured specimens from the parallel group, the regenerated tissue was not anisotropic in geometry because the tissue was formed to be connected to adjacent cell arrays, whereas significant mechanical anisotropy was observed on day 21. It was assumed that the dense extracellular matrix with orientation along each cell array, which was formed during the early 14-day culture, was still contained in the 21-day cultured specimens. In this study, the gel sheets containing cell-array were stacked to reconstruct the anisotropic biological tissue. To achieve the higher functionality of regenerated tissue, lamination of the cell sheets in an aligned state [38] will be required. Our dielectrophoresis and stacking method will be applicable to align the cell sheets as a pattern and to stack the sheets.

Inside the structure of native articular cartilage, cells and collagen fiber networks are oriented parallel to the articulating surface in the top tissue layer, randomly in the middle layer, and vertically in the deep layer [1,10]. This complex structure enables articular cartilage tissue to withstand shear, tensile, and compressive deformations caused by motion, such as daily walking. To reconstruct this complex structure of articular cartilage, a three-dimensional structure of electrode patterns regarding dielectrophoresis and a more complex assembly technique regarding the gel sheets are required. It has been shown that the mechanical anisotropy of biological tissues represented by articular cartilage can be reconstructed using anisotropic scaffold materials or mechanical stimulation [11–13,39]. However, it is difficult to replicate the orientation of chondrocytes in vivo using these methods. The dielectrophoretic cell patterning technique inside hydrogels proposed in this study can control the cellular orientation in the scaffold material. Furthermore, this cell accumulation method would enable the reconstruction of tissue structures similar to those of "native" articular cartilage at a cellular organization level.

5. Conclusions

In this study, chondrocytes were accumulated to form parallel-line cell clusters in agarose gels via negative-DEP forces. The tensile strength of the chondrocyte-accumulated agarose gel construct, sGAG content, and cell number count increased with increasing culturing time. The stiffness of the constructs in which chondrocytes were patterned in the same direction as the tensile direction was significantly greater than that of the samples with perpendicularly patterned chondrocytes. Finally, the DEP cell-accumulating methodology inside a hydrogel could become a promising approach for regenerating mechanically anisotropic tissues in cartilaginous tissue reconstruction.

Author Contributions: Conceptualization, S.M. and Y.T.; methodology, Y.T.; validation, S.M. and Y.T.; investigation, Y.T.; resources, S.M.; data curation, Y.T. and S.M.; writing—original draft preparation, Y.T. and S.M.; writing—review and editing, S.M.; visualization, Y.T. and S.M.; supervision, S.M.; project administration, S.M.; funding acquisition, S.M. All authors have read and agreed to the published version of the manuscript.

Funding: This research was partially supported by a research grant (Adaptable and Seamless Technology Transfer Program through Target-Driven R&D) from JST, a grant from the Keio Leading-edge Laboratory of Science and Technology (KLL) specified research projects, and the Keio Gijuku Fukuzawa Memorial Fund for the Advancement of Education and Research.

Institutional Review Board Statement: Not applicable.

Acknowledgments: The authors are grateful to Toru Hiwada of the Fuzzy Logic Systems Institute for fabricating the micro transparent electrodes.

Conflicts of Interest: The authors declare no conflict of interest.

References

1. Camarero-Espinosa, S.; Rothen-Rutishauser, B.; Foster, E.J.; Weder, C. Articular cartilage: From formation to tissue engineering. *Biomater. Sci.* **2016**, *4*, 734–767. [CrossRef]
2. Mow, V.C.; Kuei, S.C.; Lai, W.M.; Armstrong, C.G. Biphasic creep and stress relaxation of articular cartilage in compression: Theory and experiments. *J. Biomech. Eng.* **1980**, *102*, 73–84. [CrossRef] [PubMed]
3. Mow, V.C.; Holmes, M.H.; Michael Lai, W. Fluid transport and mechanical properties of articular cartilage: A review. *J. Biomech.* **1984**, *17*, 377–394. [CrossRef]
4. Hunziker, E.B. Articular cartilage repair: Are the intrinsic biological constraints undermining this process insuperable? *Osteoarthr. Cartil.* **1999**, *7*, 15–28. [CrossRef] [PubMed]
5. Willers, C.; Partsalis, T.; Zheng, M.H. Articular cartilage repair: Procedures versus products. *Expert Rev. Med. Devices* **2007**, *4*, 373–392. [CrossRef] [PubMed]
6. Uchio, Y.; Ochi, M.; Matsusaki, M.; Kurioka, H.; Katsube, K. Human chondrocyte proliferation and matrix synthesis cultured in atelocollagen gel. *J. Biomed. Mater. Res.* **2000**, *50*, 138–143. [CrossRef]
7. Shimizu, R.; Kamei, N.; Adachi, N.; Hamanishi, M.; Kamei, G.; Mahmoud, E.E.; Nakano, T.; Iwata, T.; Yamato, M.; Okano, T.; et al. Repair mechanism of osteochondral defect promoted by bioengineered chondrocyte sheet. *Tissue Eng. Part A* **2015**, *21*, 1131–1141. [CrossRef]
8. Hayashi, S.; Kamei, N.; Ikuta, Y.; Shimizu, R.; Ishikawa, M.; Adachi, N.; Ochi, M. Chondrocyte cell-sheet transplantation for treating monoiodoacetate-induced arthritis in rats. *Tissue Eng. Part C Methods* **2017**, *23*, 346–356. [CrossRef]
9. Yang, J.; Zhang, Y.S.; Yue, K.; Khademhosseini, A. Cell-laden hydrogels for osteochondral and cartilage tissue engineering. *Acta Biomater.* **2017**, *57*, 1–25. [CrossRef]
10. Verteramo, A.; Seedhom, B.B. Zonal and directional variations in tensile properties of bovine articular cartilage with special reference to strain rate variation. *Biorheology* **2004**, *41*, 203–213.
11. Steele, J.A.M.; McCullen, S.D.; Callanan, A.; Autefage, H.; Accardi, M.A.; Dini, D.; Stevens, M.M. Combinatorial scaffold morphologies for zonal articular cartilage engineering. *Acta Biomater.* **2014**, *10*, 2065–2075. [CrossRef]
12. McCullen, S.D.; Autefage, H.; Callanan, A.; Gentleman, E.; Stevens, M.M. Anisotropic fibrous scaffolds for articular cartilage regeneration. *Tissue Eng. Part A* **2012**, *18*, 2073–2083. [CrossRef]
13. Lee, J.K.; Huwe, L.W.; Paschos, N.; Aryaei, A.; Gegg, C.A.; Hu, J.C.; Athanasiou, K.A. Tension stimulation drives tissue formation in scaffold-free systems. *Nat. Mater.* **2017**, *16*, 864–873. [CrossRef] [PubMed]
14. Lee, E.J.; Holmes, J.W.; Costa, K.D. Remodeling of engineered tissue anisotropy in response to altered loading conditions. *Ann. Biomed. Eng.* **2008**, *36*, 1322. [CrossRef] [PubMed]
15. Okuda, Y.; Konishi, R.; Miyata, S. Effect of cyclic compressive stimuli on mechanical anisotropy of chondrocyte-seeded agarose gel culture. *Trans. Jpn. Soc. Mech. Eng. C* **2013**, *79*, 1736–1743. [CrossRef]
16. Benya, P.D.; Shaffer, J.D. Dedifferentiated chondrocytes reexpress the differentiated collagen phenotype when cultured in agarose gels. *Cell* **1982**, *30*, 215–224. [CrossRef]
17. Mauck, R.L.; Soltz, M.A.; Wang, C.C.B.; Wong, D.D.; Chao, P.H.G.; Valhmu, W.B.; Hung, C.T.; Ateshian, G.A. Functional tissue engineering of articular cartilage through dynamic loading of chondrocyte-seeded agarose gels. *J. Biomech. Eng.* **2000**, *122*, 252–260. [CrossRef]
18. Mauck, R.L.; Nicoll, S.B.; Seyhan, S.L.; Ateshian, G.A.; Hung, C.T. Synergistic action of growth factors and dynamic loading for articular cartilage tissue engineering. *Tissue Eng.* **2003**, *9*, 597–611. [CrossRef]
19. Kelly, T.A.N.; Ng, K.W.; Wang, C.C.B.; Ateshian, G.A.; Hung, C.T. Spatial and temporal development of chondrocyte-seeded agarose constructs in free-swelling and dynamically loaded cultures. *J. Biomech.* **2006**, *39*, 1489–1497. [CrossRef]
20. Bougault, C.; Aubert-Foucher, E.; Paumier, A.; Perrier-Groult, E.; Huot, L.; Hot, D.; Duterque-Coquillaud, M.; Mallein-Gerin, F. Dynamic compression of chondrocyte-agarose constructs reveals new candidate mechanosensitive genes. *PLoS ONE* **2012**, *7*, e36964.
21. Jones, T.B. *Electromechanics of Particles*; Cambridge University Press: Cambridge, UK, 2005.
22. Rosenthal, A.; Voldman, J. Dielectrophoretic traps for single-particle patterning. *Biophys. J.* **2005**, *88*, 2193–2205. [CrossRef] [PubMed]
23. Albrecht, D.R.; Tsang, V.L.; Sah, R.L.; Bhatia, S.N. Photo- and electropatterning of hydrogel-encapsulated living cell arrays. *Lab Chip* **2005**, *5*, 111–118. [CrossRef] [PubMed]
24. Albrecht, D.R.; Underhill, G.H.; Wassermann, T.B.; Sah, R.L.; Bhatia, S.N. Probing the role of multicellular organization in three-dimensional microenvironments. *Nat. Methods* **2006**, *35*, 369–375. [CrossRef] [PubMed]
25. Albrecht, D.R.; Underhill, G.H.; Mendelson, A.; Bhatia, S.N. Multiphase electropatterning of cells and biomaterials. *Lab Chip* **2007**, *7*, 702–709. [CrossRef] [PubMed]
26. Lin, Y.-H.; Yang, Y.-W.; Chen, Y.-D.; Wang, S.-S.; Chang, Y.-H.; Wu, M.-H. The application of an optically switched dielectrophoretic (ODEP) force for the manipulation and assembly of cell-encapsulating alginate microbeads in a microfluidic perfusion cell culture system for bottom-up tissue engineering. *Lab Chip* **2012**, *12*, 1164–1173. [CrossRef]
27. Menad, S.; Franqueville, L.; Haddour, N.; Buret, F.; Frenea-Robin, M. nDEP-driven cell patterning and bottom-up construction of cell aggregates using a new bioelectronic chip. *Acta Biomater.* **2015**, *17*, 107–114. [CrossRef]

28. Takahashi, Y.; Miyata, S. Continuous ES/feeder cell-sorting device using dielectrophoresis and controlled fluid flow. *Micromachines* **2020**, *11*, 734. [CrossRef]
29. Sugimoto, Y.; Miyata, S. Multi-layered cell assembling technology using dielectrophoresis and construction of skin tissue microelement. *Trans. Jpn. Soc. Mech. Eng. C* **2017**, *83*, 16-00387.
30. Ojima, Y.; Miyata, S. Discrimination methodology of living-cells and microbeads using dielectrophoresis and fluid-induced shear force. *J. Biorheol.* **2015**, *29*, 42–50. [CrossRef]
31. Takeuchi, Y.; Miyata, S. Micro cell patterning technology by dielectrophoresis and application to regenerated cartilage. *Trans. Jpn. Soc. Mech. Eng. C* **2010**, *76*, 3015–3020. [CrossRef]
32. Miyata, S.; Furukawa, K.S.; Ushida, T.; Nitta, Y.; Tateishi, T. Static and dynamic mechanical properties of extracellular matrix synthesized by cultured chondrocytes. *Mater. Sci. Eng. C* **2004**, *24*, 425–429. [CrossRef]
33. Miyata, S.; Numano, T.; Homma, K.; Tateishi, T.; Ushida, T. Feasibility of noninvasive evaluation of biophysical properties of tissue-engineered cartilage by using quantitative MRI. *J. Biomech.* **2007**, *40*, 2990–2998. [CrossRef]
34. Miyata, S.; Tateishi, T.; Ushida, T. Influence of cartilaginous matrix accumulation on viscoelastic response of chondrocyte/agarose constructs under dynamic compressive and shear loading. *J. Biomech. Eng.* **2008**, *130*, 051016. [CrossRef] [PubMed]
35. Miyata, S.; Homma, K.; Numano, T.; Tateishi, T.; Ushida, T. Evaluation of negative fixed-charge density in tissue-engineered cartilage by quantitative MRI and relationship with biomechanical properties. *J. Biomech. Eng.* **2010**, *132*, 071014. [CrossRef]
36. Farndale, R.W.; Buttle, D.J.; Barrett, A.J. Improved quantitation and discrimination of sulphated glycosaminoglycans by use of dimethylmethylene blue. *BBA Gen. Subj.* **1986**, *883*, 173–177. [CrossRef]
37. Kim, Y.J.; Sah, R.L.Y.; Doong, J.Y.H.; Grodzinsky, A.J. Fluorometric assay of DNA in cartilage explants using Hoechst 33258. *Anal. Biochem.* **1988**, *174*, 168–176. [CrossRef]
38. Tanaka, N.; Ota, H.; Fukumori, K.; Miyake, J.; Yamato, M.; Okano, T. Micro-Patterned Cell-Sheets Fabricated with Stamping-Force-Controlled Micro-Contact Printing. *Biomaterials* **2014**, *35*, 9802–9810. [CrossRef]
39. Meinert, C.; Schrobback, K.; Hutmacher, D.W.; Klein, T.J. A novel bioreactor system for biaxial mechanical loading enhances the properties of tissue-engineered human cartilage. *Sci. Rep.* **2017**, *7*, 1–14. [CrossRef] [PubMed]

micromachines

MDPI

Article

One-Dimensional Flow of Bacteria on an Electrode Rail by Dielectrophoresis: Toward Single-Cell-Based Analysis

Yukihiro Yamaguchi and Takatoki Yamamoto *

Department of Mechanical Engineering, School of Engineering, Tokyo Institute of Technology, 2-12-1 Ookayama, Meguro-ku, Tokyo 152-8550, Japan; yamaguchi.y.ay@m.titech.ac.jp
* Correspondence: yamamoto.t.ba@m.titech.ac.jp; Tel.: +81-3-5734-3182

Abstract: Many applications in biotechnology and medicine, among other disciplines, require the rapid enumeration of bacteria, preferably using miniaturized portable devices. Microfluidic technology is expected to solve this miniaturization issue. In the enumeration of bacteria in microfluidic devices, the technique of aligning bacteria in a single line prior to counting is the key to an accurate count at single-bacterium resolution. Here, we describe the numerical and experimental evaluation of a device utilizing a dielectrophoretic force to array bacteria in a single line, allowing their facile numeration. The device comprises a channel to flow bacteria, two counter electrodes, and a capture electrode several microns or less in width for arranging bacteria in a single line. When the capture electrode is narrower than the diameter of a bacterium, the entrapment efficiency of the one-dimensional array is 80% or more within 2 s. Furthermore, since some cell-sorting applications require bacteria to move against the liquid flow, we demonstrated that bacteria can move in a single line in the off-axial direction tilted 30° from the flow direction. Our findings provide the basis for designing miniature, portable devices for evaluating bacteria with single-cell accuracy.

Keywords: dielectrophoresis; microfluidics; single cell; single bacterium; flow cytometer

Citation: Yamaguchi, Y.; Yamamoto, T. One-Dimensional Flow of Bacteria on an Electrode Rail by Dielectrophoresis: Toward Single-Cell-Based Analysis. *Micromachines* **2021**, *12*, 123. https://doi.org/10.3390/mi12020123

Academic Editor: Antonio Ramos
Received: 23 December 2020
Accepted: 21 January 2021
Published: 24 January 2021

1. Introduction

Technologies for measuring the number of bacteria in a sample are important in biotechnology, medicine, the food industry, and hygiene management. For example, in the field of medical care, measuring the number of bacteria in the mucosa of a patient can be used to detect an abnormality in the body. In the food industry, the number of bacteria is monitored to control fermentation and to prevent food poisoning [1,2]. Such applications require rapid and precise bacterial measurement techniques, preferably using devices that are small and portable and thus suitable for on-site analysis such as environmental monitoring and point-of-care [3,4]. This need has led to the recent development and widespread use of small bacterial measurement devices using microfluidics technology and particularly miniaturized flow cytometers for separating and counting cell types at the single-cell level. Various methods have been developed based on optical detection [5,6], electrical impedance response [7,8], the Coulter counter method (nanopore measurement) [9], Raman spectroscopy [10,11], and a magnetic method involving binding magnetic beads to bacteria [12,13]. All these methods require the separation of bacteria into a single line which is fed to the measurement portion of the flow cytometer where the bacteria are processed individually. This is realized in a simple manner by limiting the channel diameter to that of a bacterium [14], but such channels tend to get clogged with bacteria. Conventionally, the flow diameter is limited to that of a bacterium using a sheath flow method, wrapping the sample flow with another flow stream [15,16]. This approach allows high-throughput counting but is difficult to miniaturize due to the complicated liquid delivery system. Stable sheath flow requires precise control of the flow rate and sample dilution by the sheath flow. Furthermore, problems such as damage to the cells caused by exposure to high pressure and shear forces caused by the sheath flow have been pointed out [17]. Sheathless methods have also been investigated, such as surface acoustic waves [18,19], the use of inertial force [20,21], and the electrostatic force of dielectrophoresis (DEP). However, the equipment for generating surface acoustic waves is expensive and non-disposable, and the equipment for using inertial force must be redesigned for each bacterial type and is not ideal for submicron particles such as bacteria. There are many reports on arranging mammalian cells in a single line using DEP [22–24] but few reports on arranging bacteria. In addition, most previous work used negative DEP [25]. Negative DEP acts toward the weak portion of an electric field, which has the advantage of causing less damage to the cells due to the electric field. However, this approach has the disadvantage that a steric and complex electrode structure is required to stably generate the force [26]. Furthermore, the conductivity of the solvent must be higher than that of the cell, causing wear of the electrode due to the electrode reaction and the high amount of heat generated, making it difficult to generate a strong dielectrophoretic force [27]. In contrast, a positive DEP acts toward the electrode edge where the electric field is strongest. However, the potential drawback of this approach is that it may damage cells due to the strong electric field and because the conductivity of the solvent must be lower than that of the cell [28].

At the same time, the low conductivity of the solution provides the advantage that a strong DEP can be applied because heat generation and the electrode reaction are suppressed more than in a negative DEP. The volume of a bacterial cell is less than 1/1000th that of mammalian cells, and thus, a positive dielectrophoretic force is preferable to effectively utilize the volumetric force of the dielectrophoretic force. However, there have been few reports on bacterial alignment by positive DEP and no reports of a technique in which bacterial cells are continuously arranged in a single line and fed individually downstream, although one example of generating a single stream of a cluster consisting of a small number of cells has been reported [29].

Here, we propose an electrode-rail structure to realize a technique to arrange bacterial cells in a single line using positive DEP. Our approach alleviates damage caused by the electric field by avoiding the electrode edges, where the greatest damage is caused by the electric field. We demonstrate the generation of a single line of bacteria on the electrode-rail by first analyzing the proposed device using the finite element method, then by fabricating the device, and by experimentally evaluating it based on the conditions obtained by the numerical analysis.

2. Materials and Methods

2.1. Design of the Device

A schematic diagram of the device is shown in Figure 1a. This device delivers the bacteria in a single line downstream to a flow cytometer and other bacterial single-cell analysis devices. The device comprises a flow channel to flow bacteria, two counter electrodes, and a capture electrode (electrode rail) several microns or less in width for arranging bacteria in a single row.

The electric field is generated using one capture electrode and two counter electrodes located outside the microchannel. Only the capture electrode is in the microchannel, and thus, the electric field gradient toward the capture electrode is increased in the microchannel. This is achieved by an electric field applied through an insulator via the use of a high-frequency electric field in DEP (Figure 1b). Therefore, bacteria flowing along the flow path are collected by the capture electrode via DEP and flow downstream, resulting in the bacteria flowing in a single line in the vicinity of the downstream outlet. Since the channel is used only for flowing a sample, clogging by bacteria can be prevented by using a channel with a sufficiently large cross-sectional area yet compatible with the range of the dielectrophoretic force.

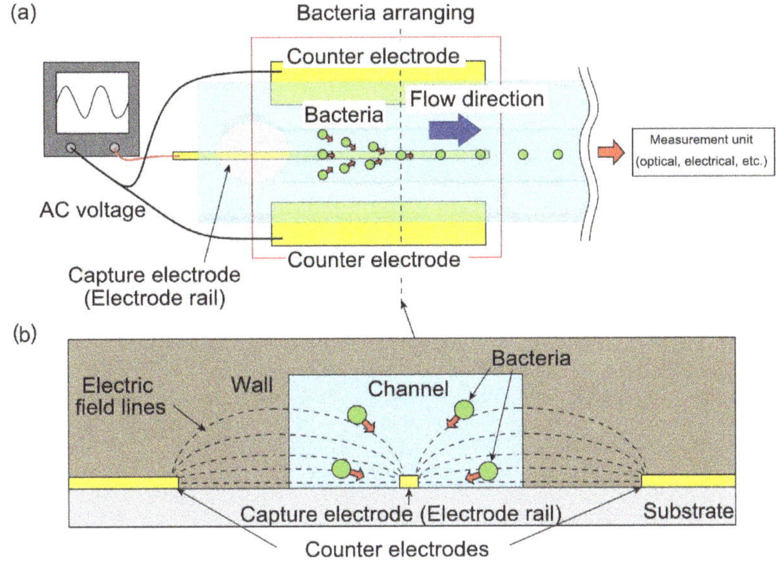

Figure 1. Schematic of the device used to form a line of bacterial cells: (**a**) top view of the device and connection of the AC power source to the electrodes and (**b**) cross-sectional view of the device. The dotted lines indicate the electric field lines which pass through the microchannel wall, creating a large electric field gradient to generate dielectrophoresis (DEP). Bacteria are attracted on the capture electrode (electrode rail).

The dielectrophoretic force is an electrostatic force acting on a particle and arises from the dipole moment induced in the particle by an applied external heterogeneous electric field [30,31].

Equation (1) represents the dielectrophoretic force F_{DEP} acting on particles of radius r_p:

$$F_{DEP} = 2\pi r_p^3 \varepsilon_m Re[K(\omega)] \nabla \Big| E^2 \tag{1}$$

where ε_m is the dielectric constant of the solvent, ∇ is the vector derivative operator, and $Re[K(\omega)]$ is the real part of the Clazius–Mossotti equation represented by Equation (2):

$$K(\omega) = \frac{\varepsilon_p^* - \varepsilon_m^*}{\varepsilon_p^* + 2\varepsilon_m^*} \tag{2}$$

where $\varepsilon_p{}^*$ and $\varepsilon_m{}^*$ are the complex dielectric constants of particles and solutions and are represented by Equation (3):

$$\varepsilon^* = \varepsilon - \frac{2\sigma}{\omega} j = \varepsilon - \frac{\sigma}{\pi f} j \tag{3}$$

where ε is the dielectric constant, σ is the conductivity, ω is the angular frequency, and f is the frequency of the applied alternating current.

From Equations (2) and (3), since $Re\ [K(\omega)]$ depends on frequency, the force exerted on the particle also varies with frequency. If $Re[K(\omega)]$ is positive, a positive dielectrophoretic force is applied and the particles are forced in the direction of increasing electric field strength. In the proposed device, as shown in Figure 1b, the electric field is concentrated on the capture electrode, and the positive dielectrophoretic force is utilized to collect bacteria in the flow path onto the capture electrode on which the electric field is concentrated.

2.2. Numerical Analysis

We designed the shapes and arrangements of the electrodes and microchannel by analyzing the movement of particles by the dielectrophoretic force in the microchannel using the commercially available finite element method analysis software package COMSOL (COMSOL 5.6, COMSOL Inc., Stockholm, Sweden). The applied electric potential is assumed to remain constant along the capture and the counter electrodes. Therefore, the numerical calculation can be reduced to a two-dimensional problem of any cross section. The calculation was conducted to calculate the trapping behavior by considering only the dielectrophoretic force and the Stokes drag force in the absence of fluid flow. A model of the analysis space is shown in Figure 2, and the physical properties are shown in Table 1. The conductivity (σ_m) and dielectric constant (ε_m) values are those of water. The conductivity (σ_p) and dielectric constant (ε_p) values of the particles were based on the physical properties of *Escherichia coli* [32], and the external forces acting on the particles are the dielectrophoretic force given by Equation (1) and the Stokes drag force F_D given by Equation (4).

$$\mathbf{F}_D = \frac{18\mu}{\rho_p d_p^2} m_p(\mathbf{u} - \mathbf{v}) \tag{4}$$

where μ is the kinematic viscosity, m_p is the mass of a bacterium, $\mathbf{v}$ is the velocity of a bacterium, $\mathbf{u}$ is the velocity of the liquid, ρ_p is the density of a bacterium, and d_p is the diameter of a bacterium.

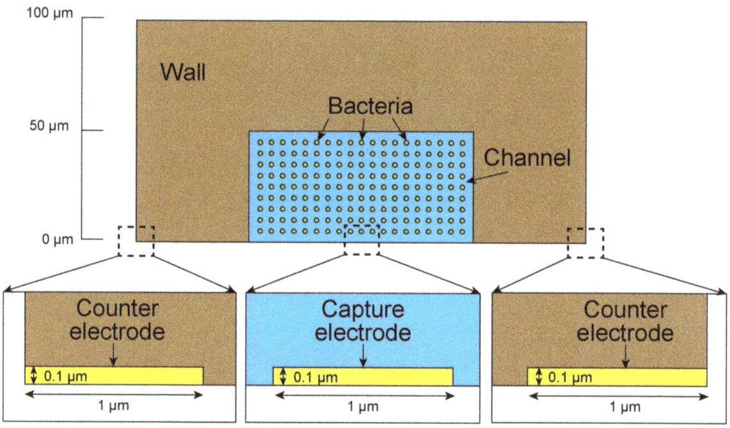

Figure 2. Analysis space in the numerical analysis.

Table 1. Physical properties for numerical analysis.

Parameter	Value
r_p	0.5 (μm)
ε_p	60
ε_m	80
σ_p	0.44 (S/m)
σ_m	5.56 (μS/m)
μ	1 (mPa)
ρ_p	1000 (kg/m^3)
d_p	1 (μm)

The physical properties of the channel structure were based on those of polydimethylsiloxane (PDMS), a silicone rubber [33,34]. The thickness of the electrode was 100 nm, and an AC voltage was applied. As an initial condition, the bacteria were placed in the 9 × 19 lattice pattern at 5 μm intervals in the channel without fluid flow. Under the above boundary, physical property, and initial conditions, we determined the electrode width dependence of the capture state at the capture electrode.

The numerical analysis in COMSOL for the electric field and dielectrophoresis were carried out in electric current interface using Equations (5) to (7).

$$\nabla \cdot \mathbf{J} = Q \tag{5}$$

$$\mathbf{J} = \sigma\mathbf{E} + j\omega\mathbf{D} + \mathbf{J_e} \tag{6}$$

$$\mathbf{E} = -\nabla V \tag{7}$$

where σ, ω, D, V, J, and Q are conductivity, angular frequency, electric displacement, voltage, external current density, and charge density, respectively. Electrical insulating boundary conditions were set at all boundaries, except for the electrodes. All three electrodes are modeled as perfect conductors. The boundary condition of the capture electrodes was set as the voltage terminal with a voltage of ± 100 V_{pp}, while 0 V was applied to the two counter electrodes. Particle tracing interface was coupled with the electric current interface and the laminar flow interface to predict the external forces F_D exerted on the particles. Particle tracing interface calculated the particle momentum and trajectory by solving the motion given by Equation (8).

$$\frac{d(m_p\mathbf{v})}{dt} = \mathbf{F}_t \tag{8}$$

The freezing wall boundary conditions were set at all boundary walls inside the channel.

The model of the analysis space was meshed using an extremely fine free triangular mesh having maximum and minimum element sizes of 2 μm and 0.004 μm, respectively, for the calculation domain, with a total of 32,743 mesh elements. The electric current interface was initially solved in the frequency domain. The applied electric potential V was approximated by a polynomial of low order at each mesh point in the calculation. The values of the parameters describing the dielectrophoretic behavior of the particles, such as the electrical field E and the gradient E^2, can be obtained from the numerical solution. We then calculated the particle tracing interface in time-dependent mode to evaluate the distribution of the electric field to predict the DEP force and Stokes drag force acting on the particles to obtain the trajectory of the particles inside the microchannel.

2.3. Fabrication

The device was fabricated by conventional photolithography and micro-molding.

This device contains a channel chip and an electrode chip bonded together. The fabrication method is conventional and thus described briefly. The channel-chip was fabricated from a PDMS silicone elastomer (KE-106, Shin-Etsu Chemical, Tokyo, Japan). First, a negative resist of SU-8 (SU-8 3050, KAYAKU Advanced Materials, Tokyo, Japan) was spin-coated on a 4-inch silicon wafer at 3000 rpm for 30 s. It was then pre-baked on a hot plate at 95 °C for 15 min and exposed through the photomask of the microchannel using UV light at a wavelength of 365 nm at 240 mJ. The channel was subsequently heated on a hot plate at 95 °C for 5 min and developed (SU-8 developer, KAYAKU Advanced Materials, Tokyo, Japan) to form the negative structure of the microchannel. A release agent (DURASURF831-TH, Daikin, Oosaka, Japan) was coated on the surface of the mold by dipping for several seconds. KE-106 was mixed with the polymerization agent at 10:1 volume ratio, poured onto the mold, degassed in a desiccator, and then heat-cured at 120 °C for 2 h using an oven. The dimensions of the fabricated channel chip were channel width 100 μm, channel height 50 μm, and channel length 20 mm.

The electrode chip was fabricated by patterning Cr (about 100 nm thick) deposited on a glass substrate using conventional photolithography. First, the photoresist AZP-1350 was spin-coated on the vacuum-evaporated Cr surface on a quartz glass slide at 3000 rpm for 30 s. The photomask in the shape of the electrode was transferred to the photoresist by exposing 365 nm UV light, 60 mJ, through the photomask. Next, the photoresist was developed with a developer (NMD-3, TOKYO OHKA KOGYO, CO., LTD., Kanagawa, Japan) and Cr was patterned to the electrode shape by etching with a Cr etchant. A fluorine-based coating agent (SFE-DP02H, AGCs) was spin-coated at 2000 rpm for 30 s on the fabricated electrode to prevent nonspecific adhesion of bacteria. Finally, the fabricated channel chip was aligned and hermetically bonded to the electrode chip and used as the device. An overall photograph of the fabricated device and an enlarged view around the capture electrode are shown in Figure 3b.

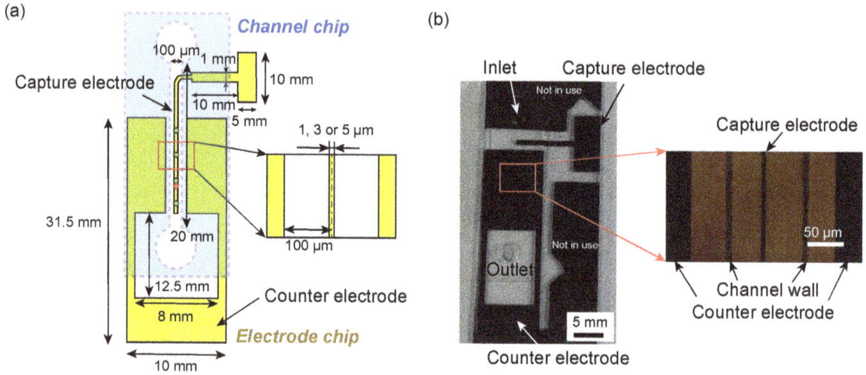

Figure 3. Single-bacteria arraying device: (**a**) an illustration of the device, and (**b**) a photograph of the device and an enlarged view around the capture electrode.

2.4. Materials

K-12 strain of *Escherichia coli* (20135, RIKEN BRC) and the lactic acid bacterium *Pediococcus pentosaceus* (AOK-L4140, Akita Konno Co., Ltd., Akita, Japan) were used as bacterial samples. LB medium (12780-052, Thermo Fisher, MA, USA) was used to cultivate *E. coli*, which was cultured in an incubator at 37 °C for 4 h. MRS medium (288130, Becton Dickinson, NJ, USA) was used for *P. pentosaceus*, which was cultured at 25 °C for 24 h. After incubation, a 1 mL portion of culture was centrifuged at 5000 *g* for 2 min, the pellet was suspended in pure water, and then, the cells were stained with SYBR Gold (S11494, Thermo Fisher) at a final concentration of 0.5%. Finally, the surfactant Triton-X (9002-93-1, Sigma-Aldrich, MI, USA) was added to the sample at a final concentration of 0.05% to prevent bacterial adhesion onto the inner surface of the microchannel.

2.5. Experimental Setup

The flow rate of bacteria in the channel was controlled by applying negative pressure to the outlet using a micropump (RP-HX01S-1A-DC3VS, Aquatech, Oosaka, Japan), and then, the behavior of the bacteria in the microchannel was observed when the dielectrophoretic force was applied. The electrical parameters were set based on preliminary experiments and the simulation results. The voltage used was within the range where neither electrolysis by the electrode occurred nor heat was generated. The frequency chosen was 1 MHz, as this frequency generates a sufficiently strong positive dielectrophoretic force within the bandwidth of the amplifier. The voltage was generated (waveform of 1 MHz, 10 V_{pp}) using a function generator (SG4322, IWATSU ELECTRIC CO., LTD., Tokyo Japan) and amplified to 200 V_{pp} using an amplifier (BA4825, NF CORPORATION, Tokyo Japan). Bacterial movement was observed under a fluorescence microscope (LG-PS2, Olympus Corporation, Tokyo, Japan) to determine how bacteria are transported to the electrode by DEP while flowing in the microchannel. The whole view of the setup is shown in Supplementary Figure S1.

3. Results

3.1. Numerical Analysis of the Motion of Bacteria

The driving voltage and frequency of the DEP force used in the numerical analysis and experiments were obtained based on the preliminary experiments and calculation results. The applied voltage 200 V_{pp} used was the maximum voltage within the range where neither electrolysis by the electrode occurred nor heat was apparently generated in the preliminary experiments. The frequency 1 MHz was determined as high as possible in the range of positive dielectrophoresis to suppress electrolysis of the electrode from the numerical calculation, as shown in Figure 4. We also confirmed that 1 MHz was the best in the preliminary experiment and used it for the numerical analysis and experiment.

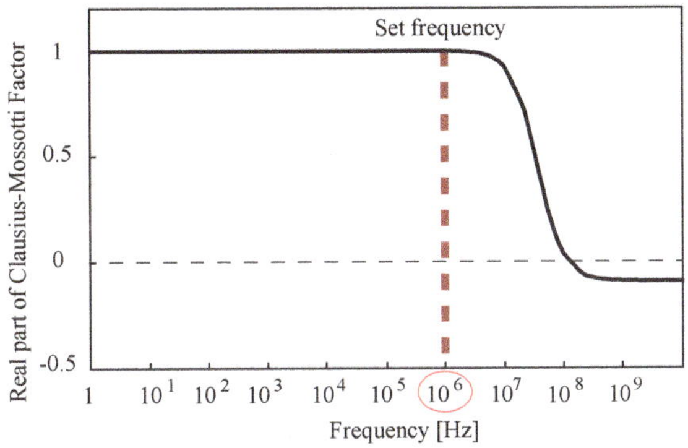

Figure 4. Calculated values of the Clasius-Mossotti equation using the parameters in Table 1: the frequency of 1 MHz was determined as high as possible in the range of positive dielectrophoresis to suppress electrolysis of the electrode from this.

The results of the finite element analysis are shown in Figure 5, which represents the potential energy of the dielectrophoretic force (U_D), represented by Equation (9).

$$U_D = -\int F_{DEP} \cdot dr = -2\pi r_p^3 \varepsilon_m Re[K(\omega)] \left| E^2 \right| \quad (9)$$

Bacteria in the channel are eventually trapped on the electrode because they move to the point of lowest potential. Figure 5b shows the potential energy of the DEP force in the vicinity of the electrode, where the electric field is strongest at the edge of the electrode, as shown in Figure 5c. Therefore, if the capture electrode is wide, bacteria are arranged in two rows on both edges of the electrode, as shown in Figure 5d. On the other hand, if the electrode is narrow, the spacing between the edges decreases and the bacteria are arranged in one row on the electrode. We also calculated the cross-sectional size of the microchannel where DEP acts effectively. As is apparent from Figure 5a, since the local minimum potentials are formed at the center, the left, and the right corners of the channel ceiling, the bacteria in these vicinities cannot be collected by the capture electrode because the bacteria are attracted toward the ceiling. This makes it difficult to collect bacteria in 100% of the channel cross-sectional area: about 80% of the area was available for collecting bacteria experimentally. Figure 6 shows the collection ratio of bacteria collected when initially arranged 171 bacteria in the 9 × 19 lattice pattern at 5 μm intervals in the microchannel for each width of capture electrode. The results indicate that the trend in the number of captured bacteria remains essentially unchanged regardless of the width of the capture electrode, with more than 80% of the bacteria in the channel collected in about 10 s for *P. pentosaceus* and about 16 s for *E. coli* when using a capture electrode with a cross section 100 μm wide and 50 μm high.

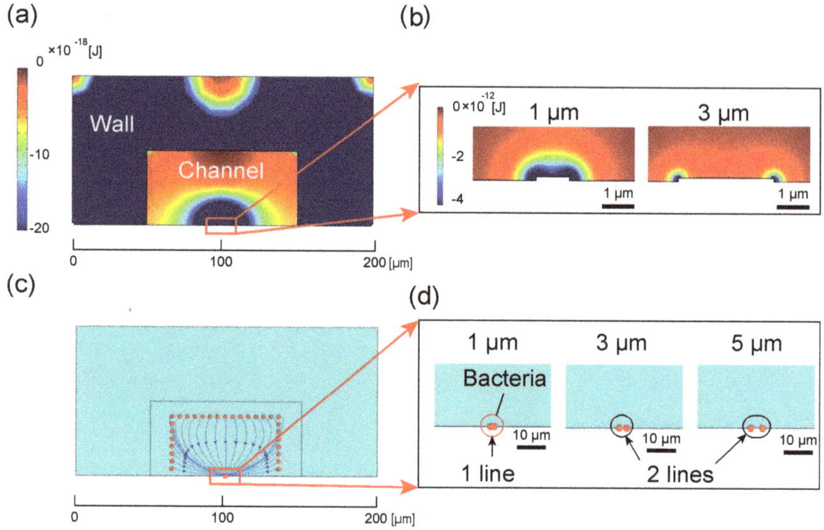

Figure 5. Analysis results: (**a**) the potential energy of the dielectrophoretic force in the cross section of the microchannel, (**b**) an enlarged view of the vicinity of the capture electrode, (**c**) the trajectory of particle tracing, and (**d**) a view of the capture results for each electrode width. Poor contrast between the particles, background, and electrode made it difficult to distinguish these components, and thus, the background was dyed light blue.

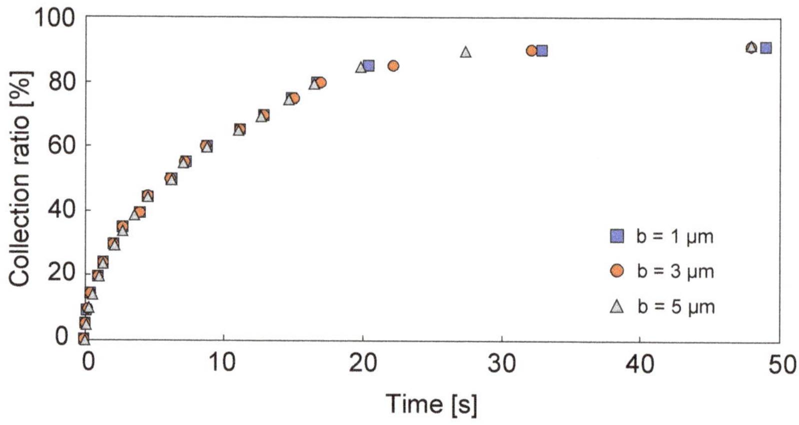

Figure 6. Time evolution of the bacterial capture rate by numerical simulation: the time trend of the capture rate remains essentially unchanged regardless of the electrode width.

3.2. One-Dimensional Arrangement of Bacteria

We measured the time evolution of the number of bacteria trapped on the capture electrode using an external dielectrophoretic force under no-flow conditions to evaluate capture efficiency. The number of bacteria collected over time is shown in Figure 7. Numerical analysis was calculated by the equivalent diameter of lactic acid bacteria (1.3 µm for lactic acid bacteria and 1 µm for *E. coli*) [35]. The times required to capture 80% of the bacterium were about 10 s for *P. pentosaceus* and about 16 s for *E. coli* in the numerical analysis, whereas the required times were about 1 s for *P. pentosaceus* and about 2 s for *E. coli* in the experiments. In comparison to the experimental and calculated values in Figure 7, the trend in time response of the capture rate is similar for both the numerical analysis and the experiments. However, the time required to capture 80% of the bacteria experimentally was more than 8 times faster than the simulation. Despite the differences between the numerical analysis and actual experiments, we nonetheless demonstrated the underlying concept.

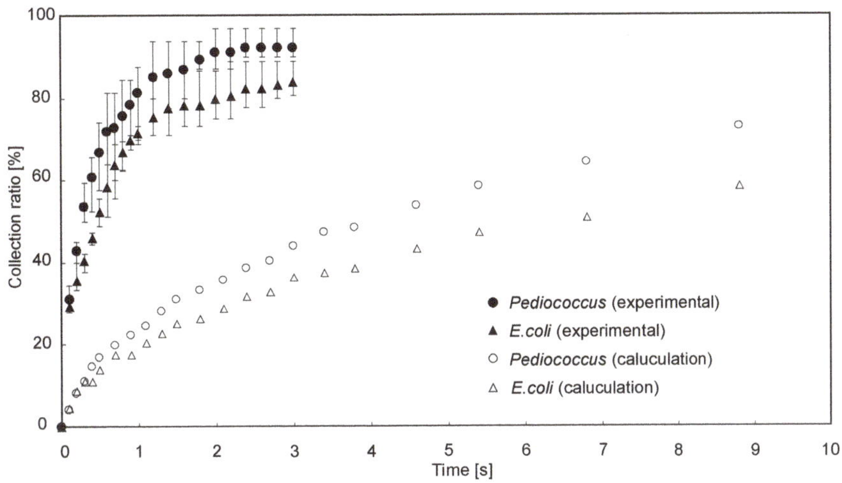

Figure 7. Time variation in the number of captured bacteria: measurements were performed at least three times to confirm reproducibility. The trapping amount saturates at a value of more than 80% in about 2 s.

Figure 8 shows how bacteria collected on the electrode as the width of the capture electrode varied. Figure 8 was obtained at the flow rate of 590 µm/s for *E. coli* and 560 µm/s for *P. pentosaceus*. Both the Gram-negative bacterium *E. coli* and the Gram-positive bacterium *P. pentosaceus* were similarly arranged in two rows at the edges of both side of the electrode when the electrode was thick (3 µm to 5 µm). On the other hand, when the electrode width was 1 µm or less, the bacteria were arranged in one row on the electrode, as shown in the figure. Consequently, bacteria can be ordered in a one-dimensional arrangement if the electrode is approximately 1 µm wide, which is equal to or less than the size of a bacterium.

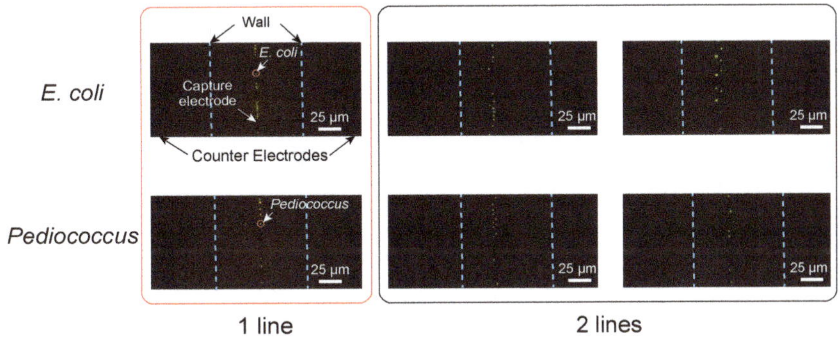

Figure 8. Electrode width dependence of the captured bacteria: Both *E. coli* at a flow rate of 590 µm/s and *P. pentosaceus* at a flow rate of 590 µm/s are similarly arranged in one row when the electrode was about 1 µm or less in width and are arranged in two rows at both edges of the electrode when the electrode was thick (3 µm to 5 µm).

3.3. *Off-Axis Manipulation of Bacteria*

Some applications, such as cell sorting, require bacteria moving against liquid flow for cell fractionation. We thus evaluated the device with bacteria moving in the off-axis direction by flowing bacteria while trapping them on an electrode bent to the right at 30° to the flow direction. The results verified that the bacteria could move in directions other than the flow direction along the edge of the electrode. Figure 9 shows the off-axis manipulation of *E. coli* and *P. pentosaceus*. The bacteria could flow either parallel to the flow or 30° oblique on the electrode. Although the relationship between flow angle, flow velocity, and applied voltage is currently unclear, bacterial movement can clearly be controlled in off-axis directions.

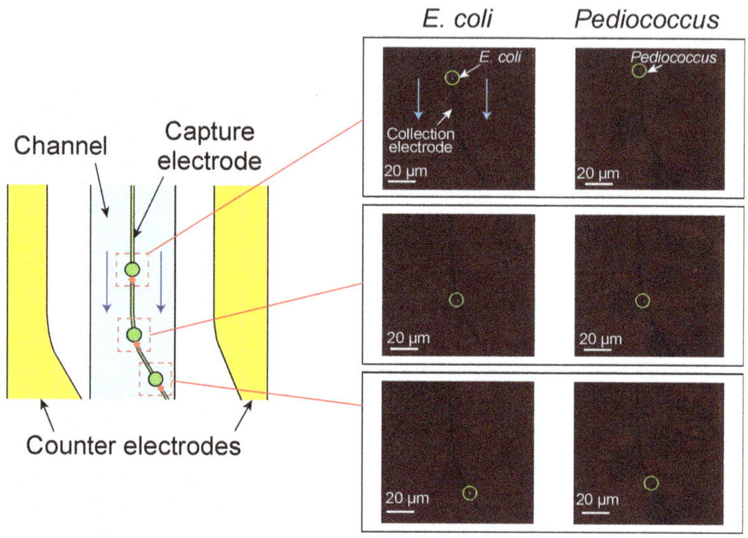

Figure 9. Off-axis manipulation of the bacteria: the bacteria move along the electrode in a direction oblique to the flow.

4. Discussion

Experimentally, we succeeded in collecting bacteria much faster than suggested by numerical analysis. Although it is desirable to collect bacteria faster, it is problematic that the prediction of the collection rate by numerical calculation was unsuccessful. Experimental results may include the effects of electric double layer (EDL) and electrohydrodynamic (EHD) convection [36,37]. Microparticles in fluids, such as cells, bacteria, and charged non-bioparticles, are surrounded by an EDL. Counterion relaxation and conduction in an EDL affect the behavior of particles in a dielectrophoretic force significantly. As the particle size decreases, the EDL thickness becomes larger relative to the particle size, especially when the conductivity of the solvent is low, such as the low concentration of electrolyte used in positive DEP in this work. Numerical analysis of the dielectrophoretic collection did not take into account the effects of EDL and EHD convection, thereby leading to a large difference in the calculated and experimentally determined collection rates. The experiments were carried out at voltages and frequencies where no apparent convection occurred, but there could have been weak EHD convection undetectable by microscopic observation. Moderate EHD convection is known to increase the collection efficiency of submicron particles by dielectrophoretic force [38], further suggesting that EHD convection could have helped increase the collection efficiency. Evaluating the effect of EHD convection is a future challenge since it could not be observed microscopically, even though it apparently improved collection efficiency.

Detailed experimental characterization of the device will be reported in the future. The present results nonetheless demonstrate that bacteria could flow either parallel to the liquid flow or 30° oblique on the electrode, suggesting that the method can also be used for separation procedures, e.g., sorting the outlets to collect them depending on the type of bacteria.

Although dielectrophoretic forces have the disadvantage that the force exerted decreases as the particle size decreases because they are volumetric forces, they are also known to act effectively on nanometer-scale particles such as proteins. Therefore, it may be possible to move viral and exosome bionanoparticles arranged in a single line using a smaller capture electrode. Because such bionanoparticles are difficult to manipulate with existing flow cytometers, this approach would not only miniaturize conventional methodologies but also would find value as a platform technology for bionanoparticle manipulation.

5. Conclusions

Here, we proposed and demonstrated a microfluidic device for achieving the manipulation of one-dimensional arrays of bacteria using the dielectrophoretic force generated on an electrode rail. We first showed the feasibility of our method by numerical analysis using the finite element method. Based on the various parameters obtained from the numerical analysis, we experimentally evaluated the device and found that, when the electrode width was narrower than the diameter of a bacterium, as indicated by the numerical analysis, the solution can be delivered downstream while arranging the bacteria in one row on the capture electrode. The capture efficiency of the one-dimensional array was 80% or more within 2 s. In addition, since some cell-sorting applications require bacteria to move against the liquid flow, such as during fractionation, we demonstrated that bacteria can move in the axial direction tilted 30° from the flow direction. In the future, we will develop a device for evaluating bacteria with single-cell accuracy by incorporating a sensor downstream of this device and by applying it to bacterial sorters and related devices.

Supplementary Materials: The following are available online at https://www.mdpi.com/2072-666X/12/2/123/s1, Figure S1: Experimental setup.

Author Contributions: Y.Y. conducted the experimental work and analyzed the data, and T.Y. designed the experiment and wrote the paper. All authors have read and agreed to the published version of the manuscript.

Funding: This work was supported by a grant-in-aid for scientific research (KAKENHI) (No. 20H02160) from the Japan Society for the Promotion of Science (JSPS).

Acknowledgments: Part of this work was conducted in association with the Four-University Nano-Micro Fabrication Kawasaki Consortium (Fab-4U Kawasaki), supported by the Ministry of Education, Culture, Sports, Science, and Technology (MEXT), Japan. This work was partially supported by the LNest Grant program. *Escherichia coli* was provided by the RIKEN BRC through the National BioResource Project of the MEXT/AMED, Japan.

Conflicts of Interest: The authors declare no conflict of interest.

References

1. Lazcka, O.; Campo, F.J.D.; Munoz, F.X. Pathogen detection: A perspective of traditional methods and biosensors. *Biosens. Bioelectron.* **2007**, *22*, 1205–1217. [CrossRef] [PubMed]
2. Leonard, P.; Hearty, S.; Brennan, J.; Dunne, L.; Quinn, J.; Chakraborty, T.; O'Kennedy, R. Advances in biosensors for detection of pathogens in food and water. *Enzym. Microb. Technol.* **2003**, *32*, 3–13. [CrossRef]
3. Foudeh, A.M.; Fatanat Didar, T.; Veres, T.; Tabrizian, M. Microfluidic designs and techniques using lab-on-a-chip devices for pathogen detection for point-of-care diagnostics. *Lab Chip* **2012**, *12*, 3249–3266. [CrossRef] [PubMed]
4. Law, J.W.-F.; Ab Mutalib, N.-S.; Chan, K.-G.; Lee, L.-H. Rapid methods for the detection of foodborne bacterial pathogens: Principles, applications, advantages and limitations. *Front. Microbiol.* **2015**, *5*, 770. [CrossRef]
5. Piyasena, M.E.; Graves, S.W. The intersection of flow cytometry with microfluidics and microfabrication. *Lab Chip* **2014**, *14*, 1044–1059. [CrossRef]

6. Bene, M.C. Microfluidics in flow cytometry and related techniques. *Int. J. Lab. Hematol.* **2017**, *39*, 93–97. [CrossRef]
7. Bernabini, C.; Holmes, D.; Morgan, H. Micro-impedance cytometry for detection and analysis of micron-sized particles and bacteria. *Lab Chip* **2011**, *11*, 407–412. [CrossRef]
8. Petchakup, C.; Tay, H.M.; Li, K.H.H.; Hou, H.W. Integrated inertial-impedance cytometry for rapid label-free leukocyte isolation and profiling of neutrophil extracellular traps (NETs). *Lab Chip* **2019**, *19*, 1736–1746. [CrossRef]
9. Apetrei, A.; Ciuca, A.; Lee, J.-k.; Seo, C.H.; Park, Y.; Luchian, T. A Protein Nanopore-Based Approach for Bacteria Sensing. *Nanoscale Res. Lett.* **2016**, *11*, 501. [CrossRef]
10. Walter, A.; Marz, A.; Schumacher, W.; Rosch, P.; Popp, J. Towards a fast, high specific and reliable discrimination of bacteria on strain level by means of SERS in a microfluidic device. *Lab Chip* **2011**, *11*, 1013–1021. [CrossRef]
11. Lyu, Y.; Yuan, X.; Glidle, A.; Fu, Y.; Furusho, H.; Yang, T.; Yin, H. Automated Raman based cell sorting with 3D microfluidics. *Lab Chip* **2020**, *20*, 4235–4245. [CrossRef] [PubMed]
12. Lin, G.; Makarov, D.; Schmidt, O.G. Magnetic sensing platform technologies for biomedical applications. *Lab Chip* **2017**, *17*, 1884–1912. [CrossRef]
13. Chicharo, A.; Martins, M.; Barnsley, L.C.; Taouallah, A.; Fernandes, J.; Silva, B.F.B.; Cardoso, S.; Dieguez, L.; Espina, B.; Freitas, P.P. Enhanced magnetic microcytometer with 3D flow focusing for cell enumeration. *Lab Chip* **2018**, *18*, 2593–2603. [CrossRef] [PubMed]
14. Song, Y.; Zhang, H.; Chon, C.H.; Chen, S.; Pan, X.; Li, D. Counting bacteria on a microfluidic chip. *Anal. Chim. Acta* **2010**, *681*, 82–86. [CrossRef] [PubMed]
15. Lopez, P.A.; Hulspas, R. Special Issue on Enhancement of Reproducibility and Rigor. *Cytom. Part A* **2020**, *97*, 105–106. [CrossRef] [PubMed]
16. Kuan, D.-H.; Huang, N.-T. Recent advancements in microfluidics that integrate electrical sensors for whole blood analysis. *Anal. Methods* **2020**, *12*, 3318–3332. [CrossRef]
17. Huang, C.-T.; Weng, C.-H.; Jen, C.-P. Three-dimensional cellular focusing utilizing a combination of insulator-based and metallic dielectrophoresis. *Biomicrofluidics* **2011**, *5*, 044101. [CrossRef]
18. Zhao, S.; Wu, M.; Yang, S.; Wu, Y.; Gu, Y.; Chen, C.; Ye, J.; Xie, Z.; Tian, Z.; Bachman, H.; et al. A disposable acoustofluidic chip for nano/microparticle separation using unidirectional acoustic transducers. *Lab Chip* **2020**, *20*, 1298–1308. [CrossRef]
19. Witte, C.; Reboud, J.; Wilson, R.; Cooper, J.M.; Neale, S.L. Microfluidic resonant cavities enable acoustophoresis on a disposable superstrate. *Lab Chip* **2014**, *14*, 4277–4283. [CrossRef]
20. Zhang, J.; Yan, S.; Yuan, D.; Alici, G.; Nguyen, N.-T.; Ebrahimi Warkiani, M.; Li, W. Fundamentals and applications of inertial microfluidics: A review. *Lab Chip* **2016**, *16*, 10–34. [CrossRef]
21. Zhou, Y.; Ma, Z.; Ai, Y. Dynamically tunable elasto-inertial particle focusing and sorting in microfluidics. *Lab Chip* **2020**, *20*, 568–581. [CrossRef] [PubMed]
22. Fiedler, S.; Shirley, S.G.; Schnelle, T.; Fuhr, G. Dielectrophoretic Sorting of Particles and Cells in a Microsystem. *Anal. Chem.* **1998**, *70*, 1909–1915. [CrossRef] [PubMed]
23. Choongho, Y.; Vykoukal, J.; Vykoukal, D.M.; Schwartz, J.A.; Li, S.; Gascoyne, P.R.C. A three-dimensional dielectrophoretic particle focusing channel for microcytometry applications. *J. Microelectromech. Syst.* **2005**, *14*, 480–487. [CrossRef]
24. Muller, T.; Gradl, G.; Howitz, S.; Shirley, S.; Schnelle, T.; Fuhr, G. A 3-D microelectrode system for handling and caging single cells and particles. *Biosens. Bioelectron.* **1999**, *14*, 247–256. [CrossRef]
25. Xuan, X.; Zhu, J.; Church, C. Particle focusing in microfluidic devices. *Microfluid. Nanofluidics* **2010**, *9*, 1–16. [CrossRef]
26. Chu, H.; Doh, I.; Cho, Y.-H. A three-dimensional (3D) particle focusing channel using the positive dielectrophoresis (pDEP) guided by a dielectric structure between two planar electrodes. *Lab Chip* **2009**, *9*, 686–691. [CrossRef]
27. Cheng, I.-F.; Chang, H.-C.; Hou, D.; Chang, H.-C. An integrated dielectrophoretic chip for continuous bioparticle filtering, focusing, sorting, trapping, and detecting. *Biomicrofluidics* **2007**, *1*, 021503. [CrossRef]
28. Ateya, D.A.; Erickson, J.S.; Howell, P.B.; Hilliard, L.R.; Golden, J.P.; Ligler, F.S. The good, the bad, and the tiny: A review of microflow cytometry. *Anal. Bioanal. Chem.* **2008**, *391*, 1485–1498. [CrossRef]
29. Kim, M.; Jung, T.; Kim, Y.; Lee, C.; Woo, K.; Seol, J.H.; Yang, S. A microfluidic device for label-free detection of Escherichia coli in drinking water using positive dielectrophoretic focusing, capturing, and impedance measurement. *Biosens. Bioelectron.* **2015**, *74*, 1011–1015. [CrossRef]
30. Morgan, H.; Hughes, M.P.; Green, N.G. Separation of Submicron Bioparticles by Dielectrophoresis. *Biophys. J.* **1999**, *77*, 516–525. [CrossRef]
31. Pethig, R. Review Article-Dielectrophoresis: Status of the theory, technology, and applications. *Biomicrofluidics* **2010**, *4*, 022811. [CrossRef] [PubMed]
32. Holzel, R. Non-invasive determination of bacterial single cell properties by electrorotation. *Biochim. Et Biophys. Acta Mol. Cell Res.* **1999**, *1450*, 53–60. [CrossRef]
33. National Astronomical Observatory of Japan. *Chronological Scientific Tables*; Maruzen: Tokyo, Japan, 2020.
34. Molberg, M.; Leterrier, Y.; Plummer, C.J.G.; Walder, C.; Lowe, C.; Opris, D.M.; Nuesch, F.A.; Bauer, S.; Manson, J.-A.E. Frequency dependent dielectric and mechanical behavior of elastomers for actuator applications. *J. Appl. Phys.* **2009**, *106*, 054112. [CrossRef]
35. Jones, R.B. *Electromechanics of Particles*; Cambridge University Press: Cambridge, UK, 2008.

36. Washizu, M.; Suzuki, S.; Osamu, K.; Nishizaka, T.; Shinohara, T. Molecular dielectrophoresis of biopolymers. *IEEE Trans. Ind. Appl.* **1994**, *30*, 835–843. [CrossRef]
37. Muller, T.; Gerardino, A.; Schnelle, T.; Shirley, S.G.; Bordoni, F.; Gasperis, G.D.; Leoni, R.; Fuhr, G. Trapping of micrometre and sub-micrometre particles by high-frequency electric fields and hydrodynamic forces. *J. Phys. D Appl. Phys.* **1996**, *29*, 340–349. [CrossRef]
38. Dash, S.; Mohanty, S. Dielectrophoretic separation of micron and submicron particles: A review. *Electrophoresis* **2014**, *35*, 2656–2672. [CrossRef]

micromachines

MDPI

Article

A Study of Dielectrophoresis-Based Liquid Metal Droplet Control Microfluidic Device

Lu Tian [1,2,†], Zi Ye [2,3,†] and Lin Gui [2,3,*]

1 Beijing Smart-Chip Microelectronics Technology Company, Ltd., Beijing 100192, China; tianlu1@sgitg.sgcc.com.cn
2 Key Laboratory of Cryogenics, Technical Institute of Physics and Chemistry, Chinese Academy of Sciences, 29 Zhongguancun East Road, Haidian District, Beijing 100190, China; yezi_1992@mail.ipc.ac.cn
3 School of Future Technology, University of Chinese Academy of Sciences, Beijing 100039, China
* Correspondence: lingui@mail.ipc.ac.cn; Tel.: +86-010-82543483
† These two authors contribute equally to this work.

Abstract: This study presents a dielectrophoresis-based liquid metal (LM) droplet control microfluidic device. Six square liquid metal electrodes are fabricated beneath an LM droplet manipulation pool. By applying different voltages on the different electrodes, a non-uniform electric field is formed around the LM droplet, and charges are induced on the surface of the droplet accordingly, so that the droplet could be driven inside the electric field. With a voltage of ±1000 V applied on the electrodes, the LM droplets are driven with a velocity of 0.5 mm/s for the 2.0 mm diameter ones and 1.0 mm/s for the 1.0 mm diameter ones. The whole chip is made of PDMS, and microchannels are fabricated by laser ablation. In this device, the electrodes are not in direct contact with the working droplets; a thin PDMS film stays between the electrodes and the driven droplets, preventing Joule heat or bubble formation during the experiments. To enhance the flexibility of the chip design, a gallium-based alloy with melting point of 10.6 °C is used as electrode material in this device. This dielectrophoresis (DEP) device was able to successfully drive liquid metal droplets and is expected to be a flexible approach for liquid metal droplet control.

Keywords: dielectrophoresis; microfluidics; metal droplet driving

Citation: Tian, L.; Ye, Z.; Gui, L. A Study of Dielectrophoresis-Based Liquid Metal Droplet Control Microfluidic Device. *Micromachines* **2021**, *12*, 340. https://doi.org/10.3390/mi12030340

Academic Editors: Rodrigo Martinez-Duarte, Xiangchun Xuan and Nam-Trung Nguyen

Received: 7 February 2021
Accepted: 19 March 2021
Published: 23 March 2021

Publisher's Note: MDPI stays neutral with regard to jurisdictional claims in published maps and institutional affiliations.

1. Introduction

Droplet-based microfluidics, also known as "digital microfluidics", has become a hot spot in the chemical and biological sciences [1,2], benefitting from its advantages, such as no cross-contamination, small sample size, and potential of high throughput analysis [3]. Droplet manipulation is the base of digital microfluidics in which the droplet control is of most importance. Approaches of droplet control can usually be categorized as active and passive methods [4], where passive control is of less interest due to its lack of flexibility. Among all active manipulation methods, electric-based ones are most popular because there are diverse and convenient ways to apply the electrical field inside microchannels. Normal electrical active droplet driving methods include electrostatic-charging actuation [5,6], electrowetting-on-dielectric (EWOD) [7], and dielectrophoresis (DEP) [8,9].

In most droplet-based microfluidics studies, droplets are often made of deionized water or aqueous solution and flow inside the oily or gaseous medium, with cells [10], particles [11], or multiphase droplets [12] wrapped inside. Recently, room temperature liquid metal (LM) has been introduced into droplet microfluidics as well [13]. Some metals, such as mercury, or alloys, most known as gallium-based alloys, have melting points lower than the room temperature and could form droplets at room temperature [14,15]. Among all these metal types, the gallium-based alloy is now of most interest due to its low toxicity, low cytotoxicity, and benign biocompatibility [16,17]. Droplets of this kind of metal have high electrical and thermal conductivities; hence, compared with ordinary aqueous solutions, they could perform different functions and applications. Due to their particular characteristics, the driving approaches are also slightly different. The most common principles of driving an LM droplet are electrowetting-on-dielectric and electrostatic. One of the most popular applications of LM droplet control is to use its high electrical and thermal conductivity in microfluidics as a micromechanical switch inside microchannels for thermal or electrical conduction [18–20]. Basic microfluidics components, such as micropumps [21] and mixers [22], are achieved by electrowetting (EW) controlled liquid metal droplets as well. Furthermore, Wu et al. [23] have recently developed a wheelchair mobile robot with electrical movement control of LM droplets, thereby largely expanding its application scope.

Among all the liquid metal droplet control studies mentioned above, DEP principle-based studies are barely reported. The DEP phenomenon is described as follows. If dielectric particles or droplets are put into non-uniform electric fields, these particles or droplets could move due to a force exerted from an electric field [24]. Instead of driving the droplet, DEP is commonly used in LM droplet generation [25]; however, DEP is also capable of driving LM droplets. In DEP microfluidics devices, the electric field and electrode material or design are of most concern, since the DEP's driving effect largely depends on the electric field around the droplet. Standard electrode types include solid electrodes, deposition or sputtering electrodes, liquid electrodes, and salt bridge electrodes [26]. Room temperature liquid metal, as mentioned above, is a new choice of electrode material in microfluidics [27–29]. Compared with traditional solid electrodes, liquid electrodes have the advantages of easy fabrication, softness; and compared with salt bridge liquid electrodes, liquid metal has much a higher electrical conductivity. Unlike volatile and toxic mercury, gallium-based alloys have been widely used as electrodes in microfluidic devices because of their biocompatibility and safety.

Here, a new DEP-based LM droplet control microfluidic device is presented. An open channel structure is introduced into this device to simplify the manipulating process and waste recycling. A gallium-based alloy at room temperature ($Ga_{66}In_{20.5}Sn_{13.5}$, melting point: 10.6 °C) is used as the electrode material. This device could drive liquid metal droplets moving between electrodes. Instead of the standard soft lithography process, laser ablation is used in this study to achieve simpler, faster, and cheaper microchannel fabrication. The LM droplet driving principle and the result with this DEP device is presented and discussed.

2. Design and Methods

2.1. DEP Chip Design and Fabrication

Figure 1a shows the microchip structure of the DEP droplets control system. There are three layers in the microchip. The upper layer is 5.0 mm thick and has a rectangular through hole as the manipulating pool for the droplets. The middle layer is 600 μm thick and has six square island electrode paddles connected with the electrode microchannels of the lower layer. The lower layer is 5.0 mm thick and has six pairs of electrode microchannels. The upper and lower layers are determined to be 5.0 mm thick because they work as manipulation pool and bottom of the device, respectively. A thick PDMS block is stiff enough to support the whole device. The middle layer is thinner because the upper surface of this layer is in direct contact with the manipulated LM droplet, while voltages are applied on the square island electrode paddles, which are on the bottom side of the middle layer. The PDMS thickness between the electrode and the LM droplet could largely affect the electric field on the LM droplet, and finally determine the driving efficiency of this device. So, a thinner middle layer is chosen. These six pairs of microchannels are patterned and aligned with the six "island" electrode paddles in the middle layer. These six pairs of microchannels are used as the LM injection inlet and outlet of the island electrode paddles of the middle layer. The microstructures of the middle layer and lower layer are bonded face-to-face to make them connected with each other directly. Thus, the island electrode paddles in the middle layer and the paired microchannels in the lower layer are matched and connected to form six complete electrode channels, and work as six independent electrodes. In fabricating this two-layer structure alignment attention should be paid to making sure that each electrode paddle in the middle layer should be connected to the paired microchannel beneath it, and not to the one adjacent to it. As with the upper layer, the rectangular through cavity should cover all six complete electrode channels in the two-layer structure, i.e., both "island" electrode paddles and paired microchannels. Thus, the sidewalls of a rectangular cavity of the upper layer and the upper surface of the middle layer constitute the open area of the DEP device, working as the manipulation pool. When voltages are added to these electrodes, there is an inhomogeneous electric field generated in the manipulation pool, and the droplets there could be driven. For the convenience of the experimental operation, all the inlets and outlets of the microchannels are set and led out from the top, shown as through holes in Figure 1a. In the design, there is no direct contact between the working droplet and the electrodes, avoiding cross-contamination, electrode corrosion, and bubble formation during usage.

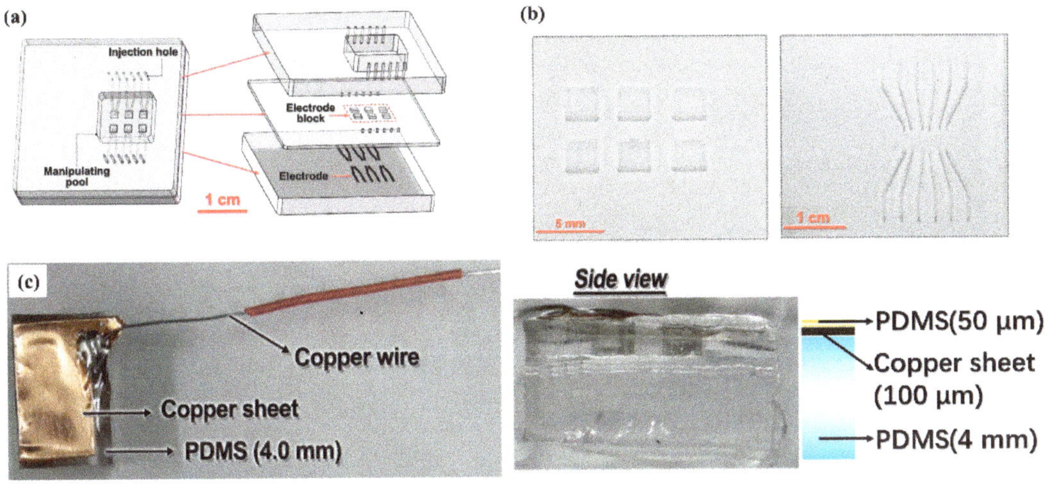

Figure 1. Design schematic and photo of the droplet driving device. (**a**) Three layers of the channel structure; (**b**) photo of the laser ablation channel; (**c**) photo of the copper sheet electrode.

Fabrication of this DEP microchip included microchannel fabrication and the assembling process. Microchannels on both middle and lower layers are made by laser ablation. After drawing the designed channel structure with the L-edit software, the structure was imported to a laser ablation machine (Diatools, Shanghai, China), modeled LE-3, with laser power 1000 mW. An empty PDMS (10:1 mixture ratio by weight, Dow Corning, Midland, MI, USA) block, covered with black oily pigment to enhance the ablating strength, was then put under the laser beam. After calibrating and adjusting the beam position, the ablation began. After the ablation on the PDMS layer finished, residues should be wiped off along the microchannels to finish the microchannel fabrication process. The ablated channels are shown in Figure 1b. The "island" electrode paddles in the middle layer are squares with a side length of 2.0 mm, and the electrode channels in the lower layers are 200 µm in width. The height of all these channels is 80 µm. The upper layer was fabricated with a PDMS slide, where a rectangular area (1.3 cm × 1.5 cm) was fabricated by thrusting four thin copper sticks at its four corners first, and then cutting the sidelines with a knife. Finally, through holes were punched in the upper layer at all inlets and outlets of the electrode channels.

Assembling of this DEP microchip was achieved by the oxygen plasma bonding technology with a plasma cleaner (YZD08-2C, Yanzhao technology, Tangshan, China). The middle layer and lower layer were first aligned and bonded face-to-face as mentioned above, and then the two-layer structure was bonded with upper layer using the same bonding technique. Liquid metal electrodes were then made by directly injecting liquid metal (gallium-based alloy) into the electrode channels using a syringe. A silver-plated copper wire was put into the inlets and outlets of the electrode channels to connect the LM electrodes with the outer voltage supplier. All channel inlets and outlets were finally sealed with silicone rubber.

Because the liquid metal is easy to oxidize in the air, a layer of oxide film with high viscosity is formed on its surface and this film tends to adhere to PDMS. In order to avoid the liquid metal sticking to the bottom PDMS layer, a thin layer of insulating silicone oil was spread at the bottom of the manipulation pool for lubrication and insulation purposes.

2.2. Droplet Driving Experiment

The driving performance of LM droplets was tested with this DEP microchip. During the experiment, LM droplets were directly sucked up and dripped into the manipulation pool. The volume of the droplets used in the experiment was controlled by a pipettor.

The electric field during the experiment was realized by applying D.C. voltages on liquid metal electrodes in the microchip through a high voltage sequencer (HVS448 6000D, LabSmith, Inc., Livermore, CA, USA). Unless specified, the voltage applied during the experiment is ±1000 V and marked as "+" or "−" in Figures 2–4.

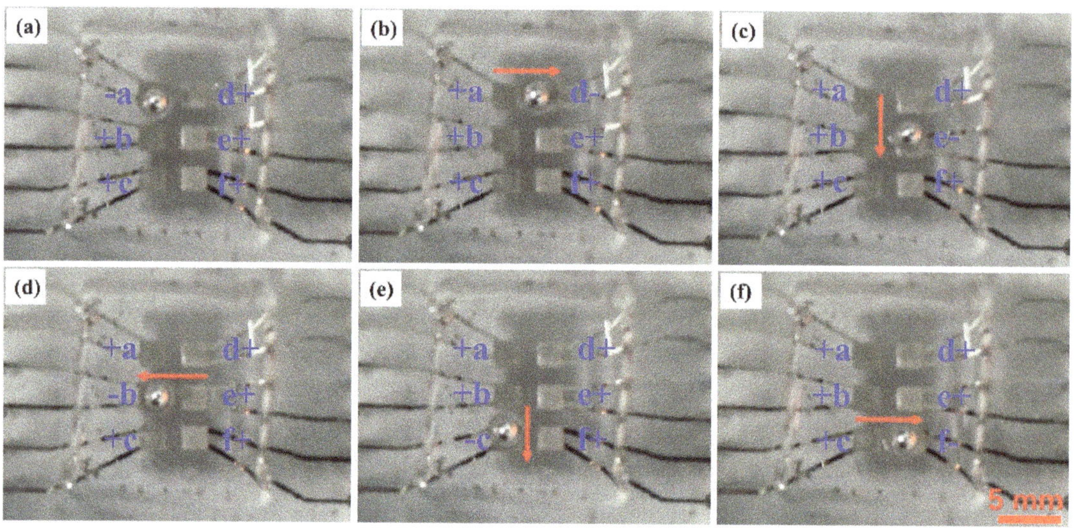

Figure 2. Experimental results for large liquid metal (LM) droplet driving. Six electrodes are numbered as a–d. "+" represents +1000 V, and "−" represents −1000 V. (**a**–**f**) Droplet position and travelling trace at different electrode conditions.

2.3. Plate Electrode Experiment

To explore the mechanism of LM droplet driving, an additional experiment was performed with a large plate electrode. The large plate electrode was made of a thin copper sheet (1.0 cm × 0.5 cm) with a silver-plated copper wire welded to it, as shown in Figure 1c. For the convenience of the stable placement of the plate electrode, the copper sheet was stuck to a 4 mm thick PDMS block. In order to prevent direct contact between the electrode and the liquid metal, the surface of the copper sheet was spin-coated with a 50 micron thick PDMS membrane.

This experiment aims to see how an LM droplet is driven under different electric field strengths. During the experiment, the PDMS-attached plate electrode was put vertically in the manipulation pool, as shown in Figure 5. A +1000 V D.C. voltage was applied on the plate electrode, and the electrodes in the DEP microchip remained unpowered. An LM droplet was put into the manipulation pool and its position was manually controlled with a pair of tweezers. Experimental results will be presented later.

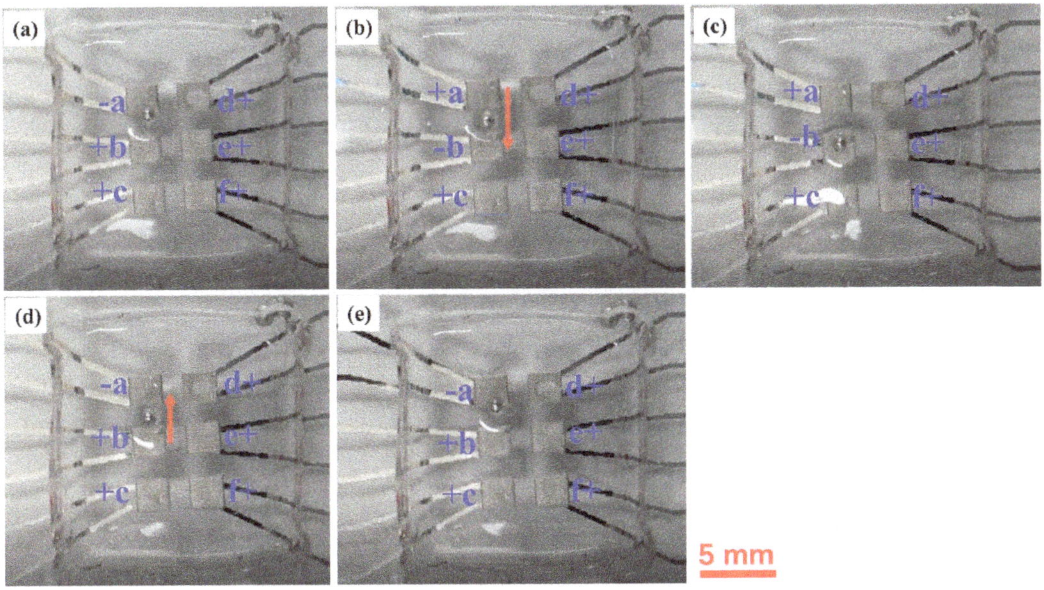

Figure 3. Experimental results for small LM droplet driving. Six electrodes are numbered as a–f. "+" represents +1000 V, and "−" represents −1000 V. The small droplet is controlled to travel between electrodes "a" and "b". (**a**) Droplet initial position at electrode "a"; (**b**) droplet traveling towards electrode "b"; (**c**) stabilized position of droplet at electrode "b"; (**d**) droplet traveling back to electrode "a"; (**e**) stabilized position of droplet at electrode "a".

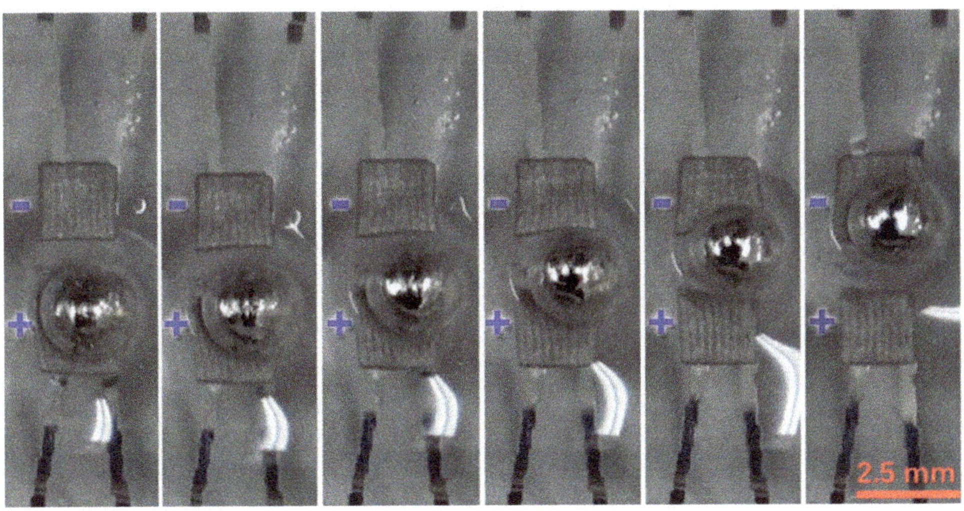

Figure 4. Details of large droplet moving from one electrode to another. Only two electrodes are included. "+" represents +1000 V, and "−" represents −1000 V.

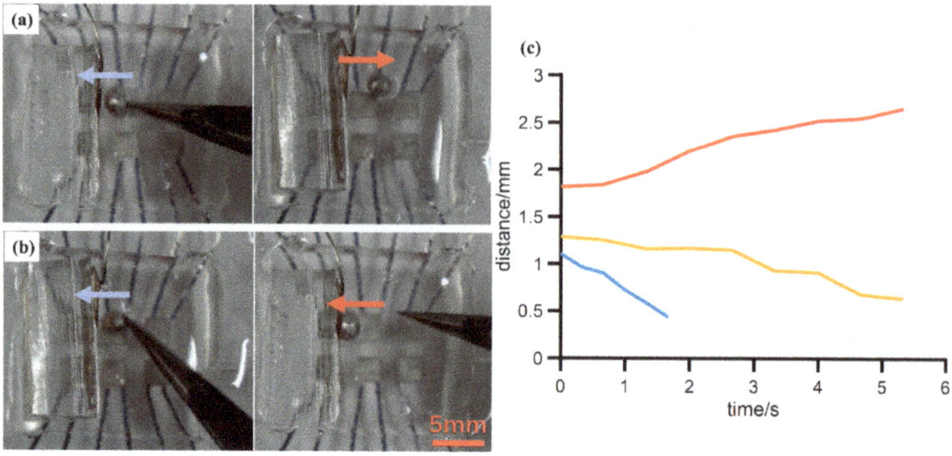

Figure 5. Experimental results of LM droplet moving towards a copper board electrode. Photo of droplet's performance when it is further from (**a**) or nearer (**b**) the copper sheet electrode (+1000 V). (**c**) Test of droplet movement with different initial distance. Different colored lines represent different tests. x-axis: recording time; y-axis: distance between droplet center and board electrode. Zero time point is voltage application time.

3. Results

This section will be divided into two parts. It provides a concise and precise description of the experimental results, their interpretation, as well as the experimental conclusions that can be drawn.

In the first part of the experiment, the performance of this LM droplet driving device was tested. Voltages were applied on six electrodes on the microchip, labeled as "a" to "f", and a droplet with 2.0 mm diameter, which is the same size as the electrode side length, was controlled to move as "S" tracks by a designed voltage sequence, as shown in Figure 2. The "+" and "-" signs in the figure represent +1000 V and −1000 V, respectively. The results show that the LM droplet tends to be attracted to the "−" voltage electrode. In the beginning, the electrode "a" was in "−" state, while other electrodes were in "+" state, and the LM droplet was attracted and trapped upon the electrode "a", as shown in Figure 2a. In the second step, when the electrode "a" turned back to the "+" state, and the electrode "d" to the right of the electrode "a", turned to the "−" state, the droplet moved to the right and was trapped upon the electrode "d", as shown in Figure 2b. By turning the electrodes "e", "b", "c", "f" to the "−" state, in turn, the droplet was controlled to move upon these electrodes by turns, and its track formed an opposite "S". The whole experimental process was recorded with a camera. The LM droplet driving velocity is then calculated by calculating the time period while the droplet traveled between the two electrodes from the Video S1 (in mp4 format, liquid metal droplet manipulation process), and with the known distance, the velocity can be acquired. At the voltage of ±1000 V, the velocity is 0.5 mm/s.

To study how the droplet size affects its movement velocity, a 1.0 mm diameter droplet, which is much smaller than the electrode size, was further tested, and the results are shown in Figure 3. All experimental conditions were the same as mentioned above. The result indicates that the smaller droplet moves faster than larger ones under the same voltage, in this case, the 1.0mm diameter droplet is driven with 1.0 mm/s velocity. However, smaller droplets would only move across the nearest electrode edges. As shown in Figure 3a, where the electrode "a" was "−" at the beginning, the LM droplet was trapped at the edge of the electrode and unable to go to the electrode center like the 2.0 mm diameter droplet. In the second step, when the electrode "a" was turned back to "+" state and "b" to the "−" state, the small droplet was driven to move towards the electrode "b", and finally stopped at the electrode edge, as shown in Figure 3b,c. When the electrodes "a" and "b" turned back to "−" and "+", respectively, again, the small droplet was driven back to the electrode "a", and again, stabled at the electrode edge, as shown in Figure 3d,e. This indicates that the edges of the electrode provide larger driving forces or electric field gradients than the electrode inner area could. In actual application of this droplet driving device, the size of droplet and electrode should be reasonably chosen according to the actual need.

In addition to the driving velocity, the way of the droplet movement is worth studying. During the process of the movement of the LM droplets between the bottom electrodes, LM droplets were found to move in a rolling manner, rather than a translative manner, as shown in Figure 4. This phenomenon indicates that the upper and lower parts of a droplet are subjected to forces in different directions under the electric field. The calculated rotation speeds for the 2 mm and 1 mm diameter droplets are 4.78 rpm and 19.11 rpm, respectively. A detailed analysis will be discussed later.

Except for the 1000 V driving voltage, the performance of this device to other voltages were tested. The lowest voltage can be also seen as an effective voltage in the system and was found to be +1000 V/−500 V (and reversed). When the voltages went up to +2500 V/−2500 V, because the interval between neighboring copper wires (those used to connect the electrode on the device to the outer high voltage source) is about 2 mm at the outlets of the electrodes, the air between these two wires was broken down and a short circuit happened, and the whole device was invalidated. So, the range of voltages that can be used for this experiment is +1000 V/−500 V (and reversed) to ±2500 V.

To further study the driving mechanism of the LM droplet driving, a driving experiment based on the copper sheet plate electrode was performed. During the experiment, a 2.0 mm diameter LM droplet was put into the manipulation pool, and the PDMS-attached plate electrode was put vertically in the manipulating pool, as shown in Figure 5. The six "island" electrodes beneath the manipulation pool were not connected to any power supply and remained at a floating voltage. A +1000 V voltage was applied to the copper sheet electrode. As shown in Figure 5a, when the LM droplet stayed relatively far from the copper sheet and was slowly pushed against the copper sheet manually by a pair of tweezers, the droplet would spontaneously move away from the copper electrode. However, when the droplet was pushed harder and got closer to the electrode, the droplet would suddenly be attracted by the electrode and voluntarily moved towards and finally "stuck" to the copper electrode, as shown in Figure 5b. We quantified this process by repeating the experiment with different initial distances between the electrode and droplet, and the results are shown in Figure 5c, where the y-axis indicates the distance between electrode interface and the center of the droplet, and the x-axis indicates the recording time. The zero-time point is the time when voltage is applied on an electrode, and the distance at this point represents the initial position of the droplet. In the three repeats shown here, when the droplet was initially set nearer to the electrode, as shown in yellow and blue lines, it would be attracted and finally stuck to the electrode. The shorter the initial distance was, the faster the droplet moved. If the initial distance was larger, like the one shown by the red line, the droplet was pushed away from the electrode. After sticking to the electrode, the shape of the LM droplet was changed from a ball to an ellipsoid, as shown in Figure 5b. So, the final distance between the droplet center and the electrode could be smaller than the droplet radius. This phenomenon might be explained by the fact that the droplet induces different charge distributions when it is at different distances from the electrode. The details will be provided in Section 4.

4. Discussion

In this study, a DEP device was developed and tested for driving liquid metal (LM) droplets. Directional movements of a LM droplet in this work were achieved by applying DEP forces. By controlling the electrode field direction around those droplets, the LM droplets were driven with a velocity of 0.5 mm/s for 2.0 mm-diameter droplets, and 1.0 mm/s for 1.0 mm-diameter droplets at 1000 V. Theoretical analysis of the driving phenomena will be discussed below. This LM droplet driving device is made of PDMS, where laser ablation was used in its microchannel fabrication process. Room temperature liquid metal, gallium-based alloy ($Ga_{66}In_{20.5}Sn_{13.5}$) is used as electrode material in this device. In all, this device drives LM droplets by DEP forces, giving an alternative in liquid metal droplet manipulation, and has potential in microfluidics devices like pumps, valves, and soft robots.

First, to explore the theoretical basis of this LM droplet driving device, a simplified numeral model is established. According to the force analysis of the charge in the electrostatic field, the potential value of any point P in space should be found when there exists an ungrounded metal ball with radius R_0 and charge q and a point charge Q, as shown in Figure 6a. The ball is not grounded, so that charges are induced by the point charge Q. According to the mirror image method, the induced charge on the ball can be equivalently expressed as an image charge Q'. As shown in Figure 6a, according to reference [30], the electric potential φ_0 at P can be expressed as:

$$\varphi_0 = \frac{1}{4\pi\varepsilon_0}\left\{\frac{Q}{(R^2+a^2-2Racos\theta)^2} - \frac{\frac{R_0}{a}Q}{\left(R^2+\frac{R_0^4}{a^2}-2R\frac{R_0^2}{a}cos\theta\right)^{\frac{1}{2}}} + \frac{q+\frac{R_0}{a}Q}{R}\right\} \tag{1}$$

where a is distance between the image charge and the point charge Q, R is distance between the point P and the center of the ball. The image charge Q' is expressed as:

$$Q' = -\frac{R_0 Q}{a} \tag{2}$$

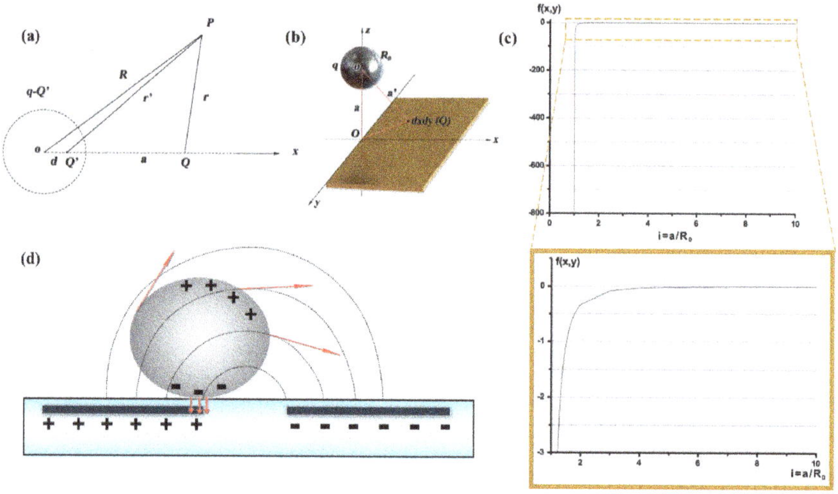

Figure 6. (**a**) Geometrical model of theoretical analysis; (**b**) geometric model of copper board electrode and LM droplet; (**c**) calculation result of forces on LM droplet from charged board with different droplet-board distance; (**d**) induced charge in a LM droplet inside electric field and schematic of droplet rolling principle.

The force between the metal ball and the charge Q is calculated by:

$$\vec{F} = Q\vec{E} \tag{3}$$

where

$$\vec{E} = -\frac{\partial\varphi_0}{\partial n}\hat{n}|_{R=a} = -\frac{\partial\varphi_0}{\partial R}\vec{e}_x|_{R=a} \tag{4}$$

Combining Equations (1)–(4), $\vec{F}$ can be expressed as:

$$\vec{F} = \frac{1}{4\pi\varepsilon_0}\left[\frac{Q(q+\frac{R_0}{a}Q)}{a^2} - \frac{Q^2\frac{R_0}{a}}{\left(a^2-\frac{R_0^2}{a}\right)^2}\right]\vec{e}_x = \frac{1}{4\pi\varepsilon_0}\frac{Q}{a^2}\left[q - \frac{QR_0^3(2a^2-R_0^2)}{a(a^2-R_0^2)^2}\right]\vec{e}_x \quad (5)$$

Then, Equation (5) can be used to derive the electric force exerted on the metal ball by a plate electrode. The point charge Q can be regarded as the charge of a small element $dxdy$ on the plate electrode, as shown in Figure 6b. According to Newton's third law, the metal ball is also subjected to an opposing force $-\vec{F}$ from Q. Thus, to calculate the total force acting on the ball for a given area of the plate electrode, $\vec{F}$ can be integrated as follows:

$$\vec{F}_{all} = \iint \frac{1}{4\pi\varepsilon_0}\frac{Q}{\left(\sqrt{x^2+y^2+a^2}\right)^2}\left\{q - \frac{QR_0^3\left[2\left(\sqrt{x^2+y^2+a^2}\right)^2 - R_0^2\right]}{\sqrt{x^2+y^2+a^2}\left[\left(\sqrt{x^2+y^2+a^2}\right)^2 - R_0^2\right]}\right\}dxdy \quad (6)$$

Equation (6) can be used to calculate the force applied on the LM droplet from a charged plate electrode. It is assumed that it is a ball-shaped droplet with radius 1.0 mm. The size of the charged plate is similar to the copper sheet used in the experiment (0.5 cm × 1.0 cm). Assuming $b = q/Q$, $i = a/R_0$, Equation (6) can be written as:

$$\vec{F}_{all} = \frac{Q^2}{4\pi\varepsilon_0}\int_{y_1}^{y_2}\int_{x_1}^{x_2} f(x,y)dxdy \quad (7)$$

where

$$f(x,y) = \frac{1}{\left(\sqrt{x^2+y^2+a^2}\right)^2}\left\{b - \frac{R_0^3\left[2\left(\sqrt{x^2+y^2+a^2}\right)^2 - R_0^2\right]}{\sqrt{x^2+y^2+a^2}\left[\left(\sqrt{x^2+y^2+a^2}\right)^2 - R_0^2\right]}\right\} \quad (8)$$

The result of $\vec{F}_{all}$ is plotted in Figure 6c. In the experiment, the LM droplets were uncharged, so $b = 0$. When $i \leq 2$, $\vec{F}_{all}$ is negative, meaning that the force on the droplet is attractive when the droplet is close to the plate electrode. As i becomes smaller, $\vec{F}_{all}$ decreases sharply, indicating the attractive force on the droplet becomes larger as well. As i becomes larger, $\vec{F}_{all}$ slowly increases and finally becomes positive, and forces on the droplet become repulsive accordingly.

During the experiment, when the LM droplet was pushed to the charged copper sheet, repulsive forces were first observed when the distance between the droplet and the copper sheet was large. This is in good agreement with the calculation results with larger i. When the droplet was pushed closer, the repulsive force turned into an attractive force, and the droplet spontaneously moved towards the charged electrode, until they stuck together. This is similar to the calculation results with smaller i. So, the theoretical analysis and the experimental results are consistent.

As described in the former part, when moving between the square electrode paddles, the droplet moved in a rolling manner, instead of a translative manner. This could be explained by the fact that the droplet is driven by induced charges in the electric field from the electrodes below. As shown in Figure 6d, when the LM droplet was put into an electric field, charges inside the droplet would move according to the electric field. Take the electric field in Figure 6d as an example, the negative charges inside the droplet would be attracted to the bottom by the positive electrode beneath it; positive charges inside the droplet will then be repelled to the upper part of the droplet. As shown in Figure 6d, when a LM droplet was placed between two opposite electrodes, the upper part of the droplet was attracted to the electrode ahead and started to move forward, while the lower part, with negative charge induced, was attracted to the electrode beneath it, and tended to stay still. Thus, the droplet would roll forward instead of sliding forward.

In most reported LM droplet manipulation studies, using electricity to control is the most convenient method. In these devices, LM droplets, not only gallium-based alloys, but mercury as well, are driven by a pair of electrodes to move or transform to perform as switches, pumps, or mixers [19,20]. In most cases, LM droplets have to be wholly wrapped with electrolyte solution, such as NaOH solution, so that the droplet could be easily electrically driven [29,30]. In the DEP device proposed in this work, on the other hand, no electrolyte is necessary. A thin layer of silicone oil, as mentioned above, is applied in the manipulation pool, simply to prevent an oxide film on gallium-based alloy droplets from sticking to PDMS. Due to the high viscosity (350 ± 10 mPa·s at 25 °C) and wettability of silicone oil to PDMS, there would be a very thin trace of oil left when it flows through the PDMS surface of the manipulation pool. This layer is thick enough to prevent LM droplets from sticking to the PDMS surface, but thin enough to maintain low flow resistance during LM droplet movement. Besides, in this DEP device, there is no need to surround the droplet with fluid, like silicone oil or electrolyte solution; LM droplets could be driven in air, and only a thin protection layer on the PDMS surface is needed. The device proposed here gets rid of this limitation and shows strong potential in wider applications of LM droplet movement.

As for metal droplet driving studies based on the electrostatic principle, contact electrodes are used in most cases, i.e., electrodes generating an electric field are in direct contact with the LM droplet itself, or with electrolyte around the droplet. Except for sample contamination from contact electrodes, bubble generation and heating problems during use are of more concern. The electrode materials involved in this kind of device are mainly inert metals, like gold or platinum, to prevent their reaction with electrolyte and electrode corrosions. However, high voltage and current applied on electrodes while working, would cause hydrolysis, and bubbles performed from hydrolysis might block microchannels and invalidate the whole device; the occurrence of squeezing bubbles needs additional channel design or surface modification of the microchannel walls. To avoid bubble formation, voltages used in electrostatic droplet manipulation are limited, so that the driving efficiency is restricted. In the device mentioned in this work, there is a thin PDMS membrane between the droplet and the driving electrode, thereby avoiding any contact of the electrodes with the working droplet. Thus, all problems that might happen in contact-electrode devices are perfectly prevented.

Except for removing the restriction of electrolyte solution during manipulation and application of non-contact electrodes, this device is advantageous over currently existing LM droplet manipulation platforms by its flexible electrode design. In most reported platforms for LM droplet control, only one pair of electrodes is used, so that the moving route of an LM droplet is restricted, mostly controlled by the shape of microchannels; usually one structure could provide only one motion path for droplets. In the device reported here, on the other hand, a multi-electrode structure is used. The position and moving routes of droplets are controlled by the electric field provided from electrodes, so that this device could realize a series of moving paths for droplets within one design. Electrodes are set beneath the manipulation pool, making it possible to bring more complicated electrode patterns to this device, and to realize more complicated and flexible movement patterns of LM droplets than most of the reported works.

The simplest and most direct application of this device is a surface microswitch for electrical, thermal, or even optical control. With proper electrode design and applied voltage sequence, LM droplets could be controlled to move to the needed position to "open" or "close" a circuit or heater. With more design, this device could also be used to build valves, pumps, or even soft robots. However, due to the low electrical conductivity of PDMS, the driving voltage of this device is much higher than previously reported electrical devices. Reducing the thickness of the PDMS membrane between electrodes and droplets could reduce the voltage needed in droplet driving. Except for the driving voltage, more attempts are expected with this design and driving principle in the future.

5. Conclusions

Here, a microfluidic device for liquid metal (LM) droplet manipulation is presented, tested, analyzed, and discussed. By applying different voltages on six square electrodes in this device, a non-uniform electric field is formed inside the manipulation pool; LM droplets inside the pool are then induced and partially charged, and driven according to the electric field. With ±1000 V voltages applied on electrodes, LM droplets are driven 0.5 mm/s for 2.0 mm-diameter ones, and 1.0 mm/s for 1.0 mm-diameter ones. The whole chip is formed by PDMS, and microchannel structures are fabricated by laser ablation, a simpler and alternative approach compared to standard soft lithography. In this device, there is a thin PDMS film between the electrodes and the manipulation pool, so that the electrodes are not in direct contact with the working droplets, preventing Joule heat or bubble formation during working. Gallium-based alloy with melting point 10.6 °C is used as electrode material in this device, to simplify the fabrication process. This DEP metal droplet driving device is expected to be a flexible approach for metal droplet driving.

Supplementary Materials: The following are available online at https://www.mdpi.com/2072-666X/12/3/340/s1, Video S1: liquid metal droplet manipulation process.

Author Contributions: Conceptualization, L.T., Z.Y. and L.G.; methodology, L.T. and Z.Y.; software, L.T.; validation, L.T. and Z.Y.; formal analysis, L.T.; investigation, L.T. and Z.Y.; resources, L.T.; data curation, L.T. and Z.Y.; writing—original draft preparation, Z.Y. and L.T.; writing—review and editing, Z.Y. and L.G.; visualization, L.T.; supervision, L.G.; project administration, L.G.; funding acquisition, L.G. All authors have read and agreed to the published version of the manuscript.

Funding: This work is supported by the National Key Research and Development Program of China (2019YFB2204903): "The Study of medical pressure sensor, micropump and chip-level heterogeneous integration" and the Science and Technology Program from State Grid Corporation of China (No. 5700-201955318A-0-0-00): "Principle, Structural Design and Functional Verification of Multi-parameter Parallel Sensing with Liquid Metals".

Conflicts of Interest: The authors declare no conflict of interest.

References

1. Basova, E.Y.; Foret, F. Droplet microfluidics in (bio)chemical analysis. *Analyst* **2015**, *140*, 22–38. [CrossRef]
2. Mashaghi, S.; Abbaspourrad, A.; Weitz, D.A.; van Oijen, A.M. Droplet microfluidics: A tool for biology, chemistry and nanotechnology. *Trends Anal. Chem.* **2016**, *82*, 118–125. [CrossRef]
3. Teh, S.Y.; Lin, R.; Hung, L.H.; Lee, A.P. Droplet microfluidics. *Lab Chip* **2008**, *8*, 198–220. [CrossRef]
4. Shang, L.; Cheng, Y.; Zhao, Y. Emerging droplet microfluidics. *Chem. Rev.* **2017**, *117*, 7964–8040. [CrossRef]
5. Roux, J.; Fouillet, Y.; Achard, J. 3D droplet displacement in microfluidic systems by electrostatic actuation. *Sens. Actuators A Phys.* **2007**, *134*, 486–493. [CrossRef]
6. Ahn, B.; Lee, K.; Louge, R.; Oh, K.W. Concurrent droplet charging and sorting by electrostatic actuation. *Biomicrofluidics* **2009**, *3*, 1–8. [CrossRef] [PubMed]
7. Nelson, W.C.; Kim, C.J.C. Droplet actuation by electrowetting-on-dielectric (EWOD): A review. *J. Adhes. Sci. Technol.* **2012**, *26*, 1747–1771. [CrossRef]
8. Wang, K.L.; Jones, T.B.; Raisanen, A. Dynamic control of DEP actuation and droplet dispensing. *J. Micromech. Microeng.* **2007**, *17*, 76–80. [CrossRef]
9. Ahmed, R.; Jones, T.B. Optimized liquid DEP droplet dispensing. *J. Micromech. Microeng.* **2007**, *17*, 1052–1058. [CrossRef]
10. Joensson, H.N.; Svahn, A.H. Droplet microfluidics—A tool for single-cell analysis. *Angew. Chem. Int. Ed.* **2012**, *51*, 12176–12192. [CrossRef] [PubMed]
11. Tenje, M.; Fornell, A.; Ohlin, M.; Nilsson, J. Particle manipulation methods in droplet microfluidics. *Anal. Chem.* **2018**, *90*, 1434–1443. [CrossRef]
12. Shui, L.; Eijkel, J.C.T.; van den Berg, A. Multiphase flow in microfluidic systems—Control and applications of droplets and interfaces. *Adv. Colloid Interface Sci.* **2007**, *133*, 35–49. [CrossRef]
13. Zhu, L.; Wang, B.; Handschuh-Wang, S.; Zhou, X. Liquid metal-based soft microfluidics. *Small* **2020**, *16*, 1–32. [CrossRef]
14. Liu, R.; Yang, X.; Jin, C.; Fu, J.; Chen, W.; Liu, J. Development of three-dimension microelectrode array for bioelectric measurement using the liquidmetal-micromolding technique. *Appl. Phys. Lett.* **2013**, *103*, 1–4. [CrossRef]

15. Wang, Q.; Yu, Y.; Yang, J.; Liu, J. Fast Fabrication of flexible functional circuits based on liquid metal dual-trans printing. *Adv. Mater.* **2015**, *27*, 7109–7116. [CrossRef]
16. Sun, X.; Yuan, B.; Rao, W.; Liu, J. Amorphous liquid metal electrodes enabled conformable electrochemical therapy of tumors. *Biomaterials* **2017**, *146*, 156–167. [CrossRef]
17. Yi, L.; Jin, C.; Wang, L.; Liu, J. Liquid-solid phase transition alloy as reversible and rapid molding bone cement. *Biomaterials* **2014**, *35*, 9789–9801. [CrossRef] [PubMed]
18. Sen, P.; Kim, C.J. A fast liquid-metal droplet microswitch using EWOD-driven contact-line sliding. *J. Microelectromech. Syst.* **2009**, *18*, 174–185. [CrossRef]
19. Yang, T.; Kwon, B.; Weisensee, P.B.; Kang, J.G.; Li, X.; Braun, P.; Miljkovic, N.; King, W.P. Millimeter-scale liquid metal droplet thermal switch. *Appl. Phys. Lett.* **2018**, *112*, 063505. [CrossRef]
20. Kim, J.; Shen, W.; Latorre, L.; Kim, C. A micromechanical switch with electrostatically driven liquid-metal droplet. *Sens. Actuators A Phys.* **2002**, *98*, 672–679. [CrossRef]
21. Tang, S.; Khoshmanesh, K.; Sivan, V.; Petersen, P.; Mullane, A.P.O.; Abbott, D. Liquid metal enabled pump. *Proc. Natl. Acad. Sci. USA* **2014**, *111*, 3304–3309. [CrossRef] [PubMed]
22. Hu, Q.; Ren, Y.; Liu, W.; Chen, X.; Tao, Y.; Jiang, H. Fluid flow and mixing induced by AC continuous electrowetting of liquid metal droplet. *Micromachines* **2017**, *8*, 119. [CrossRef]
23. Wu, J.; Tang, S.; Fang, T.; Li, W.; Li, X.; Zhang, S. A wheeled robot driven by a liquid-metal droplet. *Adv. Mater.* **2018**, *30*, 1805039. [CrossRef]
24. Pethig, R. Dielectrophoresis: Status of the theory, technology, and applications. *Biomicrofluidics* **2010**, *4*, 022811. [CrossRef] [PubMed]
25. Tian, L.; Gao, M.; Gui, L. A microfluidic chip for liquid metal droplet generation and sorting. *Micromachines* **2017**, *8*, 39. [CrossRef]
26. Çetin, B.; Li, D. Dielectrophoresis in microfluidics technology. *Electrophoresis* **2011**, *32*, 2410–2427. [CrossRef]
27. Wang, R.; Zhang, L.; Gao, M.; Wang, Q.; Deng, Z.; Gui, L. A liquid-metal-based dielectrophoretic microdroplet generator. *Micromachines* **2019**, *10*, 769. [CrossRef]
28. Zhang, L.; Zhang, P.; Wang, R.; Zhang, R.; Li, Z.; Liu, W.; Wang, Q.; Gao, M.; Gui, L. A performance-enhanced liquid metal-based microheater with parallel ventilating side-channels. *Micromachines* **2020**, *11*, 133. [CrossRef]
29. Ye, Z.; Zhang, R.; Gao, M.; Deng, Z.; Gui, L. Development of a high flow rate 3-D electroosmotic flow pump. *Micromachines* **2019**, *10*, 112. [CrossRef]
30. Guo, S. *Electrodynamics*; Higher Education Press: Beijing, China, 1980.

micromachines

MDPI

Article

Characterization and Separation of Live and Dead Yeast Cells Using CMOS-Based DEP Microfluidics

Honeyeh Matbaechi Ettehad [1,*] and Christian Wenger [1,2]

[1] IHP—Leibniz-Institut für Innovative Mikroelektronik, Im Technologiepark 25, 15236 Frankfurt (Oder), Germany; wenger@ihp-microelectronics.com
[2] BTU Cottbus-Senftenberg, 03046 Cottbus, Germany
* Correspondence: matbaechi@ihp-microelectronics.com; Tel.: +49-335-5625 (ext. 663)

Abstract: This study aims at developing a miniaturized CMOS integrated silicon-based microfluidic system, compatible with a standard CMOS process, to enable the characterization, and separation of live and dead yeast cells (as model bio-particle organisms) in a cell mixture using the DEP technique. DEP offers excellent benefits in terms of cost, operational power, and especially easy electrode integration with the CMOS architecture, and requiring label-free sample preparation. This can increase the likeliness of using DEP in practical settings. In this work the DEP force was generated using an interdigitated electrode arrays (IDEs) placed on the bottom of a CMOS-based silicon microfluidic channel. This system was primarily used for the immobilization of yeast cells using DEP. This study validated the system for cell separation applications based on the distinct responses of live and dead cells and their surrounding media. The findings confirmed the device's capability for efficient, rapid and selective cell separation. The viability of this CMOS embedded microfluidic for dielectrophoretic cell manipulation applications and compatibility of the dielectrophoretic structure with CMOS production line and electronics, enabling its future commercially mass production.

Keywords: cell characterization; cell separation; dielectrophoresis (DEP); interdigitated electrodes (IDEs); microfluidics; CMOS-based lab-on-a-chip

Citation: Matbaechi Ettehad, H.; Wenger, C. Characterization and Separation of Live and Dead Yeast Cells Using CMOS-Based DEP Microfluidics. *Micromachines* **2021**, *12*, 270. https://doi.org/10.3390/mi12030270

Academic Editor: Rodrigo Martinez-Duarte

Received: 4 February 2021
Accepted: 5 March 2021
Published: 6 March 2021

1. Introduction

Characterization and manipulation of biological cells are critical in biomedical and environmental applications. Cells contain crucial information about biological processes and environmental situations [1]. Cell separation, a subcategory of cell manipulation, is vital in clinical applications. Therefore, the secondary aim of a full sample-to-result lab-on-a-chip (LoC) relies on its capability to separate or isolate a cell kind from a cell mixture. For example, separating and sorting live and dead cells is crucial for early stage disease diagnosis [2,3]. Microfluidic LoC devices offer significant benefits in terms of efficiency, accuracy, and cost compared to other macroscopic counterparts [4–6]. There are many ways to characterize and separate cells using a microfluidic device, such as filtration, centrifugation, optical and magnetic tweezers, chromatography, dielectrophoresis (DEP), etc., [7–18]. However, among these techniques, DEP has been widely utilized in microfluidics for such biomedical applications [12,13,19–28]. This electrode-based AC electrokinetic technique requires label-free sample preparation, making the sample processing very simple. It also proposes excellent advantages regarding cost, operational power, efficiency, speed, sensitivity, selectivity, and ease of electrode integration with the device architecture [29]. The combination of microfluidic and electrokinetic actuation platforms leads to a promising direction towards complex sample handling procedures.

The advent of complementary metal-oxide-semiconductor (CMOS) technology and its integration with the LoC system enabled the fully functional sample-to-result LoC devices, which aids in device portability even out of the laboratory or hospitals. CMOS integrated LoC device can manage the data from microfluidics, sensors, and actuators. The integration of microfluidic channel with CMOS can scale down multiple-stage laboratory procedures in a single chip and process micro and nano-liters samples within a fully isolated manner. Implementing CMOS logic into practice is simple and consumes little to no current in an idle state [30]. Based on the More Than Moore approach, the CMOS devices' size is getting smaller and smaller. This approach, on the one hand, allows more functionality to fit in a smaller area. On the other hand, it can provide the right platform for miniaturized hand-held and versatile microfluidic devices [30]. Thus far, a lot of setups have been introduced for such applications. These setups are often bulky and desktop-sized that limits the functionality of the microfluidics to only laboratory-based applications. However, the lack of a promising miniaturized system remains a challenge, and realizing a fully functional device is under research.

This work aims toward the realization of the completely CMOS miniaturized portable and versatile biomedical devices capable of performing various processing laboratory steps (cell preparation, characterization, detection, and separation) using a single device. Monolithic integration of this device and its potentiality for both sample preparation and analyzation on a single platform, as well as operational simplicity and employing a single set of electrodes, offers excellent benefits from a cost and commercialization perspective. The device also provides significant potentialities in terms of sensing, read-out the trapped cells, and performing both sample preparation and analysis (manual or automated using machine learning) on a single device. The simplicity of the device design and its extended functionality for separation as well as detection applications, combined with its flexibility to use different cell types, can place this LoC device among the prominent systems developed to date for commercialization. There are exciting possibilities for the future development of this LoC platform, e.g., the development of multi-functional LoC to be used for clinical diagnostics or as part of the point of care (PoC) or near-patient portable testing systems outside hospitals. The state-of-the-art of this work relies on employing the simplest, most convenient, reliable, and accurate methods and geometries reported so far in one platform.

The combination of CMOS-based microfluidics with dielectrophoresis (DEP) has been demonstrated as one of the most outstanding devices in diagnostics applications for screening, drug delivery, and disease identification. DEP generates a non-uniform electric field (EF) that can be applied to any biological [31] and non-biological [32], charged, or neutral particles. Particle polarization occurs as a result of charge (negative and positive) migration towards the opposite directions. The electric dipole of the particle is defined by the separation of the opposing charges [33]. This technique was first introduced by Pohl in 1951 [34]. AC field using inhomogeneous electromagnetic fields can create a trap of cells. DEP-based microfluidic has been explored to address various needs in biological, medical, and chemical areas. This technique is currently limited to the 1–10 nm accuracy [35], which is sufficient to manipulate and detect micron and nanometer-sized cells.

Cells have complex structures and consist of a cell wall, cell membrane, and nucleus with various proteins, lipid molecules, etc., making their dielectric characteristics and surface charges (i.e., electrical phenotype) of one kind unique from other types. Using the cells' distinct differences and employing DEP electrokinetic technique, cells can be characterized, manipulated, and separated. Based on the Clausius-Mossotti (CM) function differing, the desired cells move to high EF intensity regions using pDEP and undesired cells to areas of low-intensity EFs by nDEP. This method exploits the intrinsic dielectric properties (relative permittivity and electrical conductivity) of the cell and its surrounding medium [31–33]. Therefore, it can be selectively used for the characterization and separation of cells. The first use of DEP for the separation of live and dead yeast cells with the aim of microbe determination was reported in 1966 [36]. In this method, they employed a rudimentary system, including a chamber with two electrodes. Later on, with microelectronics development, DEP's application was advanced to nanometric particles such as viruses.

Interdigitated electrodes (IDEs) were the initial development in DEP microelectrodes' miniaturization process after developing the DEP concept with wire electrodes [37]. IDE-based microfluidic devices are the most convenient DEP devices and usually consist of a microfluidic chamber or channel and IDEs patterned on the bottom surface of the channel [38]. The design of this microelectrode type is based on the two parallel adjacent bar microelectrodes with two poles. These electrodes might have equal or unequal finger width and spacing [38]. The finger width and spacing size can be in the nano to the micro-meter range [39]. The IDE-based DEP devices can be an open chamber or flow-through for cell suspension introduction or injection [40,41]. Due to the fabrication ease of the IDEs with many existing references, they are widely used in conjugation with DEP investigations. IDEs have been used to capture bacteria [42], blood erythrocytes [28], DNA [43], polystyrene beads [28,44] and capture and release ribosomal RNA (rRNA) [45]. Moreover, it was utilized to separate live and dead yeast cells [46] and Listeria innocua cells [47] and micro/nano-particles [43,48], human colon cancer cells from RBCs [49], characterize and separate C. muris, G. lambia, and C. parvum [50], and pattern colloid particles [51]. With technology development, IDEs have also been integrated with other on-chip components for different biological applications over the years. Suehiro et al. [52] developed a DEP impedance measurement (DEPIM) technique for detecting biological cells and bacteria, utilizing positive DEP force to immobilized biological particles in suspension onto an IDE array. Furthermore, IDEs are also integrated with other on-chip components for various applications. Gadish et al. [44] combined IDEs with a chaotic mixer to develop a micro-concentrator to measure the concentration of beads, B. subtilis, and spores. In other work, Vykoukal et al. [53] introduced a DEP field-flow fractionation (DEP-FFF) micro-separator with flex IDEs to enrich stem cells from enzyme-digested adipose tissue. Shim et al. [54,55] used the same DEP-FFF method for the isolation of circulating tumor cells (CTCs) from the blood.

Over recent decades, many studies have been conducted on the characterization and manipulation of the biological species on a monolithic device. Performing many functions on a single chip with a minimum amount of reagents and high efficiency increases the throughput, improves sensitivity, accuracy, and decreases operational complexity and cost. Moreover, using a single microfluidic chip allows high operation frequencies, increasing sensitivity and enabling easy mass production. Thus far, various chip materials were used for the fabrication of such a DEP microfluidic chip. Table 1 describes different DEP-on-a-chip approaches that have been tried for the fabrication of microfluidic-based LoC to enable label-free, fast, high-precision manipulation techniques and setups. Most of these systems have been integrated on various rigid and robust platforms such as silicon, glass, polymer (e.g., elastomer and PDMS), and PCB board [56–59]. For example, PDMS is one of the most common materials used for chip and microfluidic channel fabrication because it is transparent and enables optical observations. However, cost, scalability, and even ecological conditions have changed the manufacturing technologies towards disposable devices. With this background, CMOS-based LoC has been repeatedly reported as one of the most reliable platforms, capable of merging microfluidic with microelectronics, sensors, actuators, and filters as well as micro-electrochemical and microstructure systems into a monolithic device [60,61]. Compared to other DEP-on-a-chip devices, the combination of CMOS technology with DEP has been demonstrated as one of the most prominent devices in diagnostics applications for screening, drug delivery, and disease identification due to their low power consumption, scalability to larger systems like PoC, and high noise immunity [30].

Previously within our group, Guha et al. [62–65] has extensively worked on developing single CMOS LoC platforms for biological cell sensing and detection using different techniques. In all of these works, this sensor's sensing principle is based on the relative permittivity change of the material-under-test (MUT). Variation in the biological cell resulted in the sensor's fringing field capacity change, which caused capacitance variations to be detected by an associated silicon high-frequency read-out circuit. A non-uniform electric field between the adjacent fingers is generated by applying an electric potential to the electrodes. This method is used to detect the dielectric permittivity of the MUT. In other work [63], he proposed integrating the silicon microfluidic channel with a CMOS sensor circuit for cytometric applications to detect the cell concentrations using dielectric spectroscopy. In this method, the IDE arrays were placed on top of the microfluidic channel. As a result of IDE excitation, the fringing electric fields penetrate the fluid flowing through the channel. Other work [62] presented a CMOS-based high-frequency sensor with the capability of distinguishing the blood sample with fat and calcium from the normal blood sample. The sensor is placed at the top and inside the catheter wall and exposed to the blood.

Furthermore, he developed a self-calibrating highly sensitive dynamic IDE sensor in a BiCMOS-based PDMS microfluidic platform that can be applied for particle counting and single-particle sensing in a fluidic system [65]. The steady flow of the particles suspended in the fluid results in capacitive pulses from the sensor embedded in the oscillator. Eventually, these pulses translate to frequency modulation using an integrated phase-locked loop demodulator. They also investigated the relative viscosity variation in an aqueous solution [64] using a radio frequency (RF) CMOS chemo-bio sensor. The fringing field between the adjacent fingers is utilized to detect the dielectric permittivity of the MUT.

Table 1. Various DEP-on-a-chip platform examples for cell analysis applications.

Ref.	Purpose	Analyte	Microfluidic Material	DEP Microelectrode	DEP Parameters	Substrate
[56]	characterization, identification	Stem cells and their differentiated progeny	PDMS	IDE [1]	25 kHz—10 MHz at 8 V	silicon wafers or glass slides
[57]	detection	Live Jurkat's cytoplasm	PDMS	coplanar waveguide (CPW)	10 MHz at 3 V	quartz
[58]	trapping, rotating, detecting	Hela cells and polystyrene particles	PDMS	IDE [1]	1 MHz at 76/80 Vpp	PCB
[59]	detection, separation	Micro-nano particles 500 nm–10 μm	Glass	L-shaped electrode	0–1 MHz at >−1.3 V and <1.4	glass on PCB
[60]	trapping, manipulation	yeast cells	PDMS	3D octa-pole	100 Hz—5 MHz at 5 Vpp	CMOS
[61]	characterization, discrimination	cancer stem cells	PDMS	quadrupole electrode	50–500 MHz at 2–4 Vpp	CMOS

[1] Interdigitated electrode.

Several DEP integrated CMOS platforms have been developed for bio-particle characterization and manipulation (cell trapping, sorting, separation, differentiation, purification, etc.) [60,61,66–69]. However, they use relatively large-scale systems with polymeric microfluidic channels that are not compatible with the standard CMOS process flows. Silicon is a reliable replacement for polymer microfluidics in terms of high integration robustness with CMOS electronics and high precision channel alignments [70]. The bulky polymer-based LoC setups limit the device performance by introducing parasites to the system. Moreover, PDMS microfluidics is more convenient for laboratory applications than for industrial ones [71].

Here we described a CMOS-based microfluidic device that employs DEP via IDEs embedded in a silicon microfluidic channel for characterization and separation of live and dead yeast cells. This IDE platform can be integrated with circuits and microfluidic channel by the CMOS process line of IHP, for simultaneous immobilization, sensing, and detection of biological and non-biological particles. This device's ability to be integrated with circuitry and performing many processes using a single platform for sample preparation, characterization, detection, separation, and analysis, as well as portability, makes it distinct from other setups. This study's primary focus is to create a platform for the characterization and manipulation of cells as the preliminary step for detection and separation applications. Sensing circuitry and read-out of the cells are in the scope of this paper.

This device was successfully used to immobilize yeast cells as a means for detection application [72]. In this study, we demonstrated the applicability of the same device for frequency-dependent DEP characterization of live and dead yeast cells and for cell separation from a cell mixture. DEP responses of live and dead yeast cells were first characterized with respect to the AC field frequency and medium conductivity effects. DEP's different polarities (pDEP and nDEP) were used to selectively separate live or dead yeast cells from a mixture of live and dead cells. Current results showed that the cell separation action was achievable using the same IDEs used for cell immobilization without any additional or specialized electrode structure. This achievement ensures maintaining the simplicity of the active devices and, as a result, the simplicity of the microfluidic chip structure. Moreover, the opportunity to employ various laboratory steps using a single device like characterization, detection, and separation makes the device analysis very prominent.

2. Materials and Methods

2.1. Microfluidic Device Design and DEP Operation

The DEP microfluidic device used in this work was introduced in our recent article [72] for cell immobilization. To analyze biological suspensions, a CMOS device was combined with a microfluidic channel on a single chip. This device is composed of a CMOS integrated microfluidic channel with embedded Al-based IDEs at the bottom of the silicon channel. For simultaneous electrical measurement and optical observations, the microfluidic channel was closed with a glass layer. The fabrication was performed based on the standard 250 nm high-performance SiGe BiCMOS technology of IHP. Figure 1a shows the microfluidic device, which consists of one channel with 2.75 mm length, 0.33 mm width, and 0.075 mm depth with two ports for the inlet and outlet. There are six multi-fingered IDEs with various geometrical dimensions, spaced 250 µm apart, fabricated on the topmost metal level of the back-end-of-line stack of the CMOS process. Figure 1b presents the schematic cross-sectional view of the microfluidic device. The fabrication details and procedures have been previously explained in [72,73]. The microfluidic channel delivers the liquid to the electrodes, and the IDEs model the DEP structures. In this work, for separation experiments, we focused on the results achieved from one of the best operational IDEs [72,74]. The finger width and gap of this IDE are 45 µm and 5 µm, respectively.

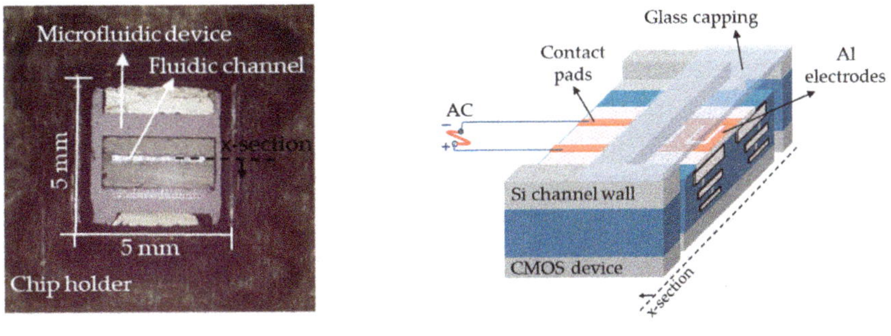

(a) (b)

Figure 1. (**a**) Microfluidic device with isolated fluidic and electrical interfaces; (**b**) Schematic cross-sectional view of the device with one IDE structure.

For cell characterization, the microfluidic channel was first filled with either live or dead cell suspensions, and then for the separation experiments media was replaced by live and dead cell mixture suspension. The liquid suspension was generated in the microfluidic channel through the inlet with 1 µm s^{-1} flow rate, where the IDEs are present to induce the DEP effect by imposing the AC signal (Figure 2).

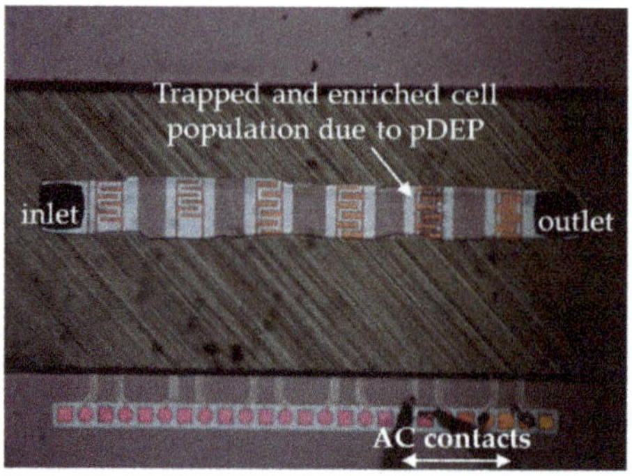

Figure 2. Microfluidic device in operation: cell entrapment at the IDEs after imposing AC and generating DEP.

Cells experienced positive DEP (pDEP) attracted toward the IDEs and trapped there. Meanwhile, in separation operations, cells that undergo negative DEP (nDEP) did not overcome the hydrodynamic drag force to be immobilized at the IDEs and washed away from the channel in the stream of fluid flow. This, resulted in the enrichment of one cell kind from the other one.

2.2. Sample Preparation

Cells of Saccharomyces cerevisiae RXII, Commercial fresh baker's yeast, were used in this study. The samples were suspended in four various medium solutions with different conductivities for DEP experiments. These medium were such as DI-water (DIW), Tap water, Potassium chloride (KCL) solution (Supelco, Merck KGaA, Darmstadt, Germany), Phosphate-buffered saline (PBS) solution (Sigma-Aldrich, Merck KGaA, Darmstadt, Germany), and diluted PBS (D-PBS) with DIW in a volume ratio of 1:25. These suspending mediums were used to compare the DEP behavior of the cells. The choice of media was made so that conductivities of different orders of magnitude ranging from 10^{-4} to 1 were investigated. Once solutions were prepared with desired dilutions, their conductivities were measured at room temperature using a conductivity meter (Accumet AET30 Conductivity Tester, Thermo Fisher Scientific, Inc., Waltham, MA, USA). The yeast cells were considered to be robust enough to survive any osmotic stress that can be caused by the non-physiological media used. To prepare stock yeast suspensions, 30 mg of baker's yeast (fresh form) was diluted in 20 mL of medium solutions. The samples were equally transferred into 2mL micro centrifuge tubes. To achieve dead cells, a yeast sample tube was heated at 100 °C for 30 min in a dry bath block (AccuBlock Mini-Compact Dry Bath, Labnet International, Inc., Edison, NJ, USA).

For characterization, ten suspensions were prepared and applied separately. To this end, live and dead cells were individually suspended in 0.0002 S/m DIW, 0.08 S/m Tap water, 0.2 S/m KCL 20 mM, 1.39 S/m PBS 0.1 M and 0.1566 S/m D-PBS 0.1 M solutions. The PH values of the DI-water, tap-water, KCL, PBS, and D-PBS were measured 7, 7.04, 7.05, 7.8, and 7.1, respectively, using PH meter (Checker Plus, Hanna Instruments Deutschland GmbH, Germany). For separation, live and dead cell suspensions were mixed in a 1:1 ratio and stained with methylene blue (MB) dye (Sigma-Aldrich, Merck KGaA, Darmstadt, Germany). Due to the high cell concentration of cell mixture suspension, the prepared stock solutions were further diluted with their respective medium solutions in the volume ratio of 1:10 for separation experiments, cell density, and the viability test measurements. The 4 mL of 1:10 cell mixtures were stained with 1 mL MB for 20 min.

The induced DEP force by the IDE electrodes was generated by AC voltage (V) supplied by a signal generator (Agilent-33220A, Agilent Technologies/Keysight Technologies, Santa Clara, CA, USA). Cell tracing and motions were monitored and recorded by a CCD camera (Nikon-DS-Fi2, Nikon GmbH, Tokyo, Japan) mounted on an optical microscope (Nikon Eclipse-LV100ND, Nikon GmbH, Tokyo, Japan). Cell suspension solutions were pumped through the channel using a syringe pump (NEMESYS, CETONI GmbH, Korbußen, Germany). A hemocytometer (BLAUBRAND® Neubauer improved, BRAND GMBH + CO KG, Germany) was used to estimate the cell viability and density by counting the cells per volume concentration unit (cells/mL). The viability of the cell mixture was determined using a counting chamber (hemocytometer) through optical microscope observation. The heat-killed cells were stained and showed a blue-color cytoplasm, and, conversely, live cells were not stained. Therefore, the blue spheres represent dead cells, while the white ones represent live cells. Counting the ratio of the viable cells to the total number of the cells gives an estimation about the viability of the sample before and after the measurement.

The cellular density of the stock solutions were estimated to be in the average range of 1.45×10^7 cells mL^{-1}. The cellular density of the mixtures was estimated to be in the average range of 1.3×10^5 cells mL^{-1}. The conductivity value of the live and dead yeast suspension samples is shown in Table 2.

Table 2. Conductivity value of cell suspension samples.

Sample	DI-Water	Tap Water	KCL	PBS	D-PBS
Live yeast cells	1.18×10^{-3} S/m	7.82×10^{-2} S/m	0.262 S/m	1.430 S/m	0.201 S/m
Dead yeast cells	3.87×10^{-3} S/m	7.85×10^{-2} S/m	0.266 S/m S/m	1.435 S/m	0.206 S/m

3. Theoretical Background

3.1. Dielectrophoresis Theory

Dielectrophoresis (DEP) is the motion of a polarizable cell in a dielectric medium subjected to a non-uniform AC electric field [1]. The magnitude of the induced dipole depends on the polarizability of the cell and the surrounding media. The DEP force on a spherical cell of radius r is defined as:

$$F_{DEP}(t) = 2\pi\varepsilon_m r^3 \, Re[f_{CM}] \, \nabla \left| E_{rms} \right|_2, \tag{1}$$

here ε_m represents the relative permittivity of the suspending media, $Re[f_{CM}]$ is the real part of the Clausius-Mossotti factor (f_{CM}), and $\nabla|E_{rms}|^2$ is the root-mean-square of electric field strength and is related to the voltage V. The f_{CM} is a complex number:

$$f_{CM} = \frac{\varepsilon_c^* - \varepsilon_m^*}{\varepsilon_c^* + 2\varepsilon_m^*}, \tag{2}$$

$$\varepsilon^* = \varepsilon - i\frac{\sigma}{\omega}, \tag{3}$$

where ε_m represents the relative permittivity of the suspending media, ε_m^* is the complex permittivity of the fluid and, ε_c^* is the complex permittivity of the cell. Complex permittivity is the function of the conductivity (σ) and angular frequency (ω) of the electric field. The electric field relationship illustrates the local electric field E, which is linked to the potential field V:

$$E = -\nabla V \tag{4}$$

The intensity of E at a point is the gradient of potential V at that point after sign change, in the x-direction.

As can be seen from (1), DEP force is strongly dependent on the cells' size, electrical and dielectric properties, their surrounding media (f_{CM}), voltage, frequency, and electric field vectors. When $\sigma_p < \sigma_m$ and $\varepsilon_p > \varepsilon_m$, f_{CM} is negative at lower frequencies and positive at higher frequencies, and when $\sigma_p > \sigma_m$ and $\varepsilon_p < \varepsilon_m$, f_{CM} becomes positive at lower frequencies and negative at higher frequencies. Therefore, the positive and negative values of f_{CM} results in either positive DEP (pDEP) or negative DEP (nDEP), respectively. When $Re[f_{CM}] > 0$, cells attract toward the high electric field intensity regions. In contrast, when $Re[f_{CM}] < 0$, cells repel from these regions. The transition point from positive to negative or vice versa is known as crossover frequency (fc) or zero force-frequency [75]. This is a specific point at which the real part of the cell's effective polarizabilities and its suspending media are equal (i.e., $Re[f_{CM}] = 0$), thus making the DEP force zero. Creation of the electric field gradient in each of the induvial cells as a result of their electric dipole can cause the attraction of the cells to their neighboring cells. Therefore, the polarized cells creates pearl-chains that are formed along the electric field lines [76]. Based on the selective DEP forces, desired cells can be trapped and isolated from a cell mixture for purifying processes [25]. Furthermore, dead cells can be removed from live cells [23].

Applying AC to these electrodes, a non-uniform EF is created. This EF is the strongest at the finger edges and the weakest in the adjacent finger gaps (i.e., electrode intervals) and the electrode finger's center. The field's magnitude is decreased with the height from the electrode surface, and thus DEP force is significantly reduced in a vertical direction. Therefore, the DEP force is inversely proportional to the distance from the IDEs generating EF [29]. Depending on the relative permittivity of the cells and medium flown over IDEs as well as amplitude and angular frequency of the AC, cells can experience a translational DEP force in two opposite directions. Positive DEP force moves the cells toward the strongest EF locations (i.e., finger edges), and as a result, cells can be trapped there. Negative DEP repels the cells from the IDEs and moves them towards the lowest EF regions.

3.2. *Numerical Determinations of Clausius-Mossotti Factor*

Figure 3 illustrates the CM factor's ovulation with the frequency of the applied electric field for live and dead yeasts suspended in DIW (0.0002 S/m), Tap-water (0.08 S/m), KCL solution (0.2 S/m), PBS (1.39 S/m), and D-PBS (0.1566 S/m). The numerical determinations were calculated using myDEP software [77], based on the two-shell model [78], where cells are assumed to possess two concentric layers of various electric and dielectric properties.

Based on these calculations, as shown in Figure 3a, the CM factor's real part of live cells suspended in DIW was bounded between 0.9 and −0.13. For dead cells, this value was specified between 0.82 and −0.13. For cell suspension in DIW fc was predicted at around 65.4 MHz and 2.5 MHz for live and dead yeast, respectively. At frequencies lower than these crossover frequencies, the DEP force was positive on both live and dead cells. For live-cell suspension in tap-water (Figure 3b) CM factor was bounded between 0.28 and −0.47. In contrast, for dead-cell suspension f_{CM} was bounded entirely in the negative region. For the KCL suspending media (Figure 3c), with a dilution of 20 mM, the same trend as tap-water can be seen for the live and dead cells. For both live and dead cells suspended in tap-water and KCL, the maximum value of the CM factor for nDEP was around −0.49, whereas that for pDEP was around 0.28 and 0.03, respectively. This indicates that forces generated by pDEP are weaker than nDEP forces. For KCL compared with the tap-water, not only the maximum value of CM factor was less, but also the pDEP spectrum was limited to smaller frequency ranges. Figure 3d, shows the CM factor for the highest conductance suspending media (PBS, 0.1M), where the real part for both live and dead cells was under nDEP for all frequencies. However, for diluted PBS live-cell suspension, a pDEP spectrum is expected over a more comprehensive frequency range than KCL (Figure 3e). For this cell suspension the real part of f_{CM} was between 0.1 and 0.48.

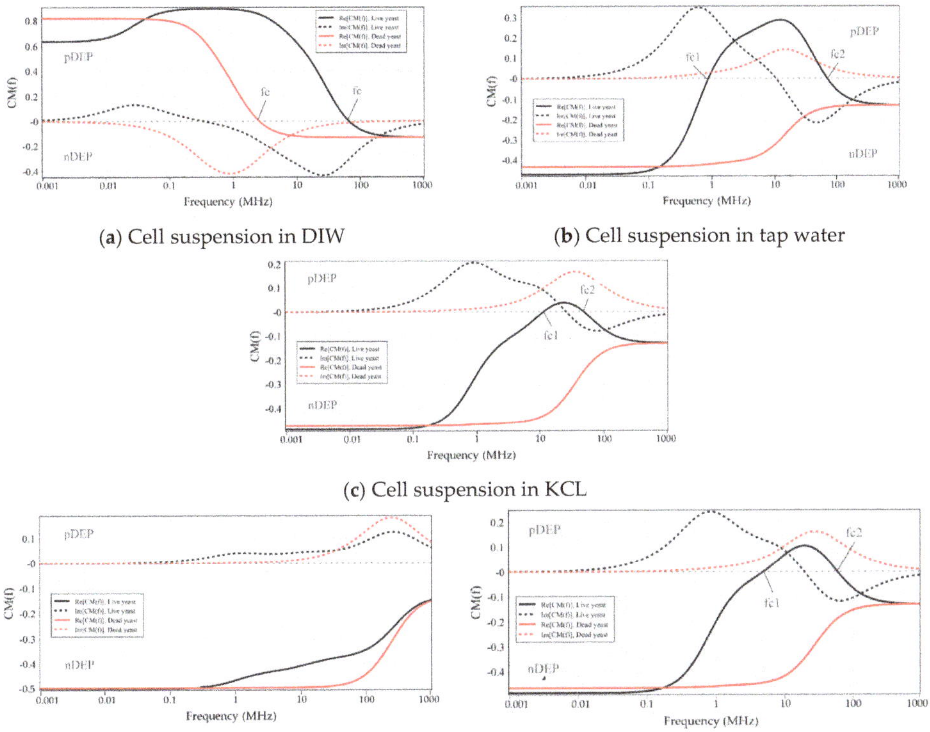

(**a**) Cell suspension in DIW
(**b**) Cell suspension in tap water
(**c**) Cell suspension in KCL
(**d**) Cell suspension in PBS
(**e**) Cell suspension in D-PBS

Figure 3. The Clausius-Mossotti factor as a function of frequency for live and dead yeast cells suspended in: (**a**) DIW; (**b**) Tap-water; (**c**) KCl; (**d**) PBS; (**e**) D-PBS. Electrical and geometrical properties of the live and dead yeasts were taken from [79]. Solid lines and dotted lines represent the real and imaginary parts of the Clausius-Mossotti factor, respectively.

4. Results and Discussion

The DEP separation of cell mixtures using the same operating conditions and experimental configurations was simulated using COMSOL Multiphysics®. For these simulations, we used the same model described in our previous publication [73]. The related parameters and boundary conditions are explained in detail in [72,73]. Figure 4 illustrates the DEP isolation of live cells from dead cells suspended in KCL at 20Vpp, 6 MHz, and 1 $\mu m\ s^{-1}$.

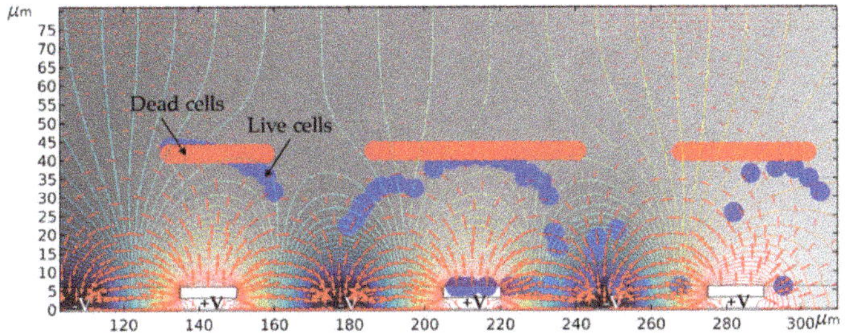

Figure 4. FEM simulation results for cell separation. DEP generated by IDEs, shown in black and white segments (marked by −V and +V, respectively). The line contour illustrates the electric potential applied to the IDEs, and red arrows represent the electric field distribution.

As it was expected from the calculations, live cells experienced pDEP and attracted by the higher electric field intensity regions, and finally got trapped at the IDEs. Meanwhile, dead cells were not influenced by pDEP and moved towards the lower electric field intensity regions, which eventually led to their elution from the channel and separation from the live cells.

4.1. *Characterization of Live and Dead Yeast Cells*

The frequency-dependent DEP behavior of the live and dead cells was first determined empirically by observing cells' behaviors when the given frequency was altered to see whether cells move towards the IDEs or away from them. To this end, 20 Vpp was applied with ranging frequencies between 10 kHz to 20 MHz (maximum frequency allowed by the signed generator). As the numerical calculations predicted, both live and dead yeasts suspended in DIW exhibited pDEP and nDEP behaviors. Figure 5 compares the numerical predictions and the empirically proven pDEp regime. Results showed that, contrary to CM factor calculations, the pDEP spectrum for live yeasts in DIW was restricted between 100 kHz–12 MHz. The frequency range at which cells experienced enough DEP force to get trapped was between 300 kHz–10 MHz for live cells, and the experimental fc point was different from the numerical predictions.

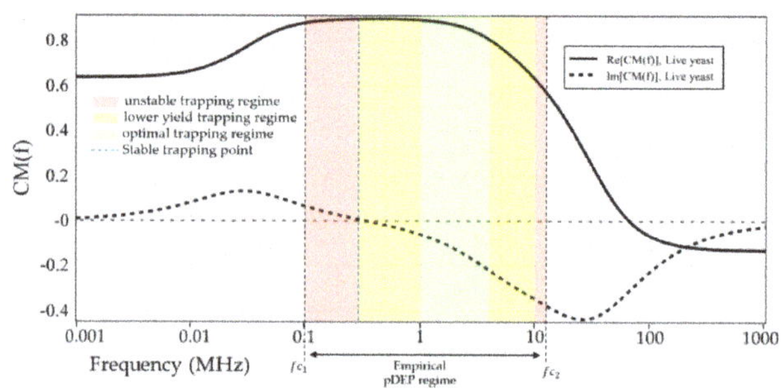

Figure 5. Empirically proven pDEP regime and crossover frequencies (*fc*) vs. numerical predictions for live yeasts suspended in DIW. Colored frequency bands represent trapping behavior based on microscopical observations.

Moreover, within a specific frequency range (green region), the trapping yield was optimum. The trapping yield was lower in yellow regimes than in the green areas. In the pink regions, unstable trappings were observed. Upon applying these ranges, entrapped cells tended to desorb from the electrodes after few seconds. The blue dashed line represents the point at which trapping was getting stable, and desorption of entrapped cells reached the minimum. These variations were observed by the microscopical magnification. Several images were taken at various steps to analyze and estimate the trapping yield for given frequencies. The images were then analyzed using ImageJ software. Each picture had a specific cell coverage on the entire IDE area. The trapping percentage was estimated by calculating the area covered by cells to the total area, as demonstrated in Figure 6a. These measurements were done at the frequency range of 10 kHz–20 MHz. For these live cells, the optimum frequency range was in the field of 1–3 MHz. Figure 6b–g describes the trapping behavior upon changing frequency. For frequencies below 300 kHz and beyond 10 MHz (within the mentioned frequency band), the DEP force could only impact the cell trajectory and not trapping.

It was observed that the velocity reduction of cells reaching the IDEs due to partial absorption toward the IDE region was followed by spontaneous desorption. This could be related to an insufficient amount of pDEP force to trap the cells stably at the IDEs. Upon changing frequency and applying frequencies between 300 kHz–10 MHz, cell trapping got stable, and optimal trapping was achieved between 1 to 3 MHz.

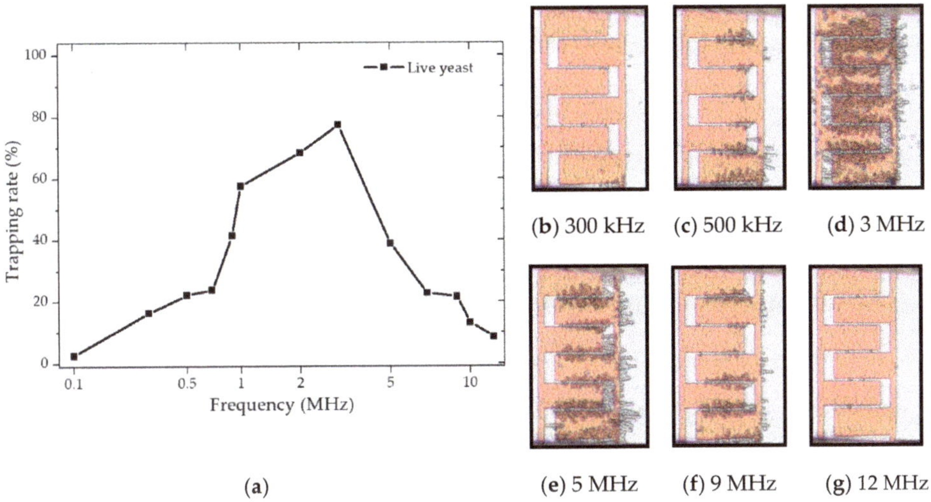

Figure 6. DEP characterization of live yeasts in DIW at 1 μm s^{-1} flow rate: (**a**) Trapping rate (yield) approximation by changing frequency based on covered cell area at the IDEs; (**b**–**g**) the images show the trapping at various frequencies after 1 min.

DEP characterization measurements on dead yeast suspensions showed that the positive scope for dead yeasts in DIW media was between 20 kHz–2.3 MHz. The frequency at which the fc occurred (2.3 MHz) was in good agreement with the numerical simulation results. However, the frequency range at which cells experienced enough DEP force to get trapped was between 70 kHz–2.3 MHz. As shown in Figure 7a, the empirically proven pDEP regime of dead cells occurred at lower frequencies and over a narrower frequency band than live cells. Further observations indicated that the live cells suspended in tap-water and KCL experienced both negative and positive DEP.

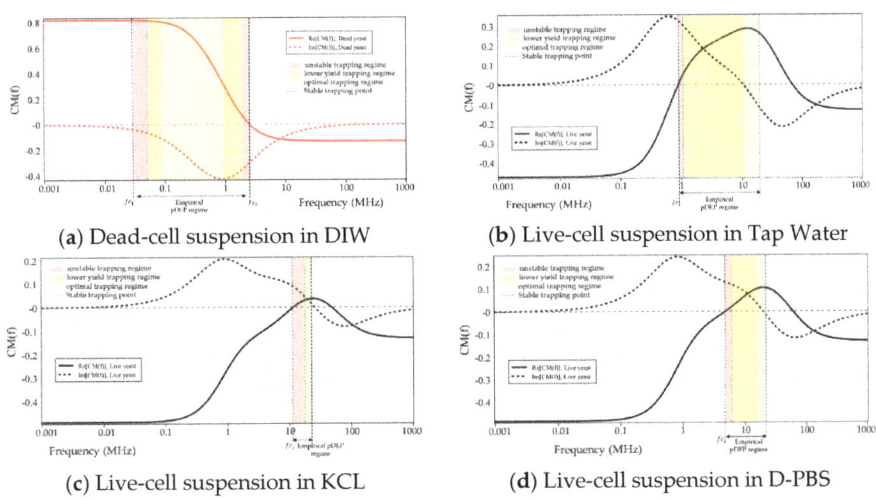

Figure 7. Empirically proven pDEP regime and crossover frequencies (*fc*) vs. numerical predictions for: (**a**) dead yeast suspension in DIW and live cell suspensions in; (**b**) Tap water; (**c**) KCL; (**d**) D-PBS. Colored frequency bands represent empirically proven trapping behavior based on optical observations.

As expected from simulations, live yeast cells in PBS solution experienced nDEP, whereas, in diluted PBS, both negative and positive forces acted on the live cells. Dead cell dilutions in tap-water, KCL, PBS, and D-PBS solutions were solely under nDEP over the entire frequency range. Due to our signal generator's limitation, which was restricted to 20 MHz, investigation of the complete range of pDEP was not possible for all of the live cell suspensions. However, for all the cases, fc1 matched closely with the numerical analysis. DEP force-frequency profile indicated that, keeping all the other parameters constant, changing frequency influences the cell trajectory and gives rise to DEP force in two directions (positive and negative) and impacts the trapping rate (yield) at the IDEs. The trapping trend for all the cases was almost similar. They exhibited a gradual increase in trapping rate by the transition of the real part from negative to positive, and the trapping rate gained peak by increasing frequency. It was seen that the smaller the cell size is, the lower frequency ranges are required for trapping. At lower frequencies, larger electric field gradients were necessary for effective DEP trapping and manipulation. Thus, trapping efficiency can be tuned with increasing electric potential.

Figure 8a–c demonstrates the trajectory of live cells suspended in various solutions (tap water, KCL, PBS, and D-PBS), leading to cell entrapment at the IDEs except for cell suspension in PBS (Figure 8d) that no trapping occurred due to nDEP and cell continued moving in the stream of fluid flow.

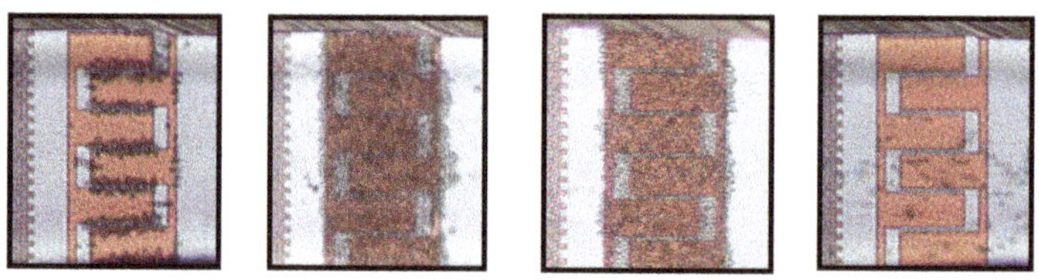

16 MHz (**a**) Tap water; 19 MHz (**b**) KCL; 20 MHz (**c**) D-PBS; 20 MHz (**d**) PBS

Figure 8. DEP behavior of live yeasts in different medium solutions at 20 Vpp and 1 $\mu m\ s^{-1}$ flow rate: (**a–c**) pDEP, which leads to cell trapping and; (**d**) nDEP and no cell trapping.

Comparing all the results, it can be concluded that as the medium conductivity increases, the pDEP spectrum gets smaller, and trapping occurs at higher frequencies. The trapping trend for all the cases was almost similar. The trapping yield gradually increased by the transition of the real part from the negative (dominant conductivity region) to positive (where permittivity starts to dominate) and gained peak with increasing frequency. Moreover, the trapping yield is maximized when $Re[f_{CM}]$ achieves its peak values. Table 3 summarizes the DEP characterization results of viable and dead cells diluted in various medium solutions.

Table 3. DEP characterization results for yeast cell suspensions.

MUT [1]		pDEP Range (Numerical Prediction)		pDEP Range (Empirical)		
Cell	**Media**	**fc1** [2]	**fc2**	**fc1**	**fc2**	f_{opt} [3]
Live Yeast cells	DIW	-	65.4 MHz	100 kHz	12 MHz	1–3 MHz
	Tap water	887 kHz	68.2 MHz	1 MHz	40 MHz [4]	10–14 MHz
	KCL (20 mM)	11.6 MHz	48 MHz	11 MHz	not measured	18–20 MHz
	PBS (0.1 M)	nDEP	nDEP	nDEP	nDEP	-
	D-PBS	4.86 MHz	59.98 MHz	4 MHz	not measured	13–20 MHz
Dead Yeast cells	DIW	-	2.5 MHz	20 kHz	2.3 MHz	90–900 kHz
	Tap water	nDEP	nDEP	nDEP	nDEP	-
	KCL (20 mM)	nDEP	nDEP	nDEP	nDEP	-
	PBS (0.1 M)	nDEP	nDEP	nDEP	nDEP	-
	D-PBS	nDEP	nDEP	nDEP	nDEP	-

[1] Material under test, [2] Low-and high crossover frequency points, [3] Optimum trapping frequency range, [4] Measured through separate experiments not mentioned in the text.

4.2. Cell Separation from a Cell Mixture

Based on numerical predictions and characterization measurements, it was evaluated that each of the live and dead cells has distinct DEP responses with respect to their surrounding media. Considering these particular behaviors, we used the same microfluidic device to separate and isolate a cell type from a cell mixture. Figure 9 combines the experimentally measured effective pDEP ranges for both live (highlighted gray zone) and dead (highlighted pink zone) cells in DIW with the numerical predictions into one plot. Such a graph can be used as a phase diagram for separation regimes of the cell mixture at the microfluidic channel. The labeled 1–3 regions specify the effective frequency ranges for continuous separation of either live or dead cell types.

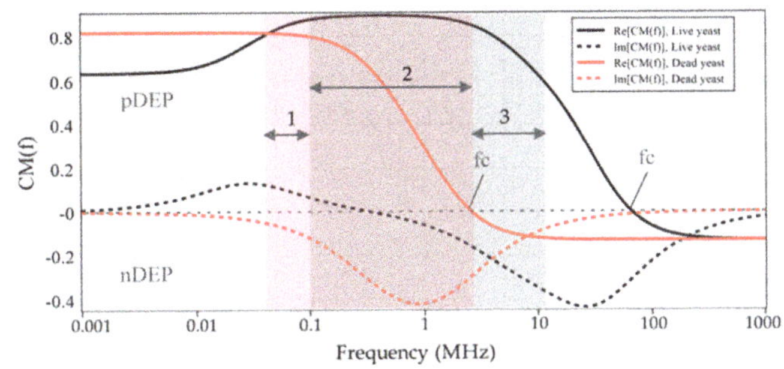

Figure 9. Experimentally measured (pink and gray zones) and numerically predicted (lines) affective pDEP for live and dead cells in 0.0002 S/m DIW at different AC field frequencies at the microfluidic channel using pDEP.

Upon applying any frequencies from zone 1, only the dead cell can get trapped and isolated from the live cells. In contrast, zone 3 was the region where only live cells can be trapped and isolated from the dead cells, while dead yeasts can flow through the channel and be separated from live ones. Upon switching to any frequencies from zone 2, i.e., the highlighted overlapped region, both cell types can get trapped. Therefore, to selectively trap live or dead cells from the mixture, frequencies less than 100 kHz (in zone 1) and more than 2.3 MHz (in zone 3), respectively, were the most effective conditions for differential separation. The cells leaving the microfluidic channel can be collected at the outlet and used for further investigations and analysis, such as viability test.

The isolation efficacy of live cells from a cell mixture was determined during the DEP process for frequencies within zone 1–3, as represented in Figure 10a. This was performed by instantly calculating the viability percentage of entrapped and isolated cells at the IDEs via live and dead cell counting. Viability is the ratio of the number of live (white) cells to the total number of entrapped live and dead (blue-dyed) cells. By switching frequency, the targeted cells were isolated at the IDEs from the cell mixture introduced into the microfluidic channel using the 1 $\mu m.s^{-1}$ flow rate at 20 Vpp. Upon applying frequencies from zone 1, only dead cells could be isolated (Figure 10b). By increasing to frequencies from zone 2, both live and dead cells were trapping, Figure 10c–f. However, the yield of the entrapped live cell population was increasing steadily compared to dead cells, and the isolation efficiency of the live cell population reached 100% at around 5 MHz (Figure 10g). The shift of the higher crossover frequency of dead cells (from 2.3 MHz to about 4.5 MHz) can be attributed to the medium solution's conductivity elevation (from 2.23×10^{-3} S/m to 0.128 S/m) due to staining. Applied frequencies below 100 kHz (within zone 1) diminished the isolation efficiency of live cells drastically to 0%, where the isolation efficiency of dead cells reached 100%.

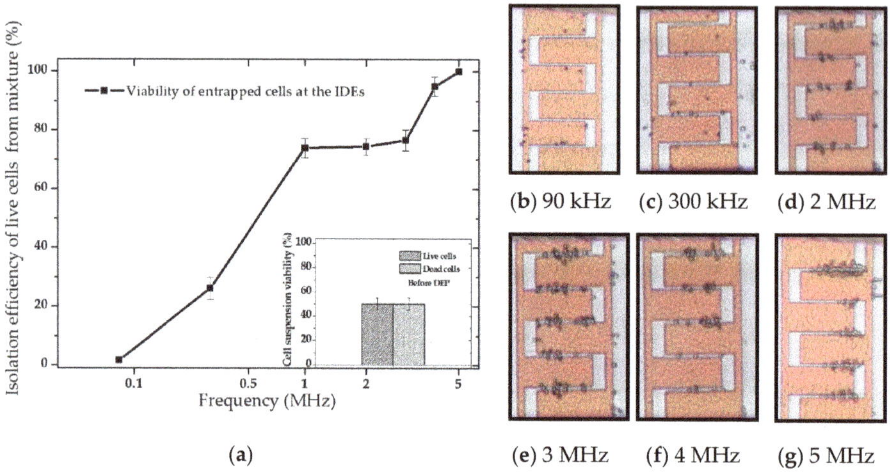

Figure 10. Experimentally recorded results on selectively trapping and separation of live cells from a live and dead yeast mixture (0.1282 S/m) suspended in DIW at a constant voltage of 20 Vpp: (**a**) Isolation efficiency of live cells from a mixture using IDEs; (**b**–**e**) Micrograph examples where the percentage of isolated live cells increase by increasing frequency (**b**) 0%; (**c**) 22.22%; (**d**) 77.35%; (**e**) 80.19%; (**f**) 91.66%; and (**g**) 100%. White cells are Live, and blue-dyed cells are dead cells. The live and dead cells mixture was 1:1 in the suspension prior to separation.

Figure 11 demonstrates the differential separation of live cells from live and dead mixtures in KCL and in D-PBS with almost 100% isolation efficiency at 20 MHz.

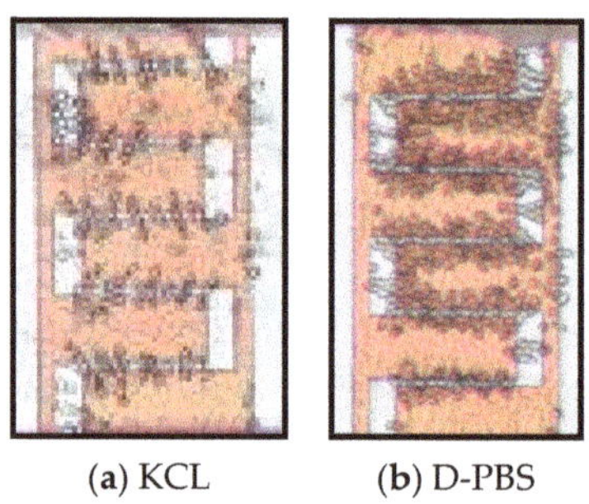

Figure 11. Differential separation of the live yeasts from live and dead (1:1) cell mixtures by DEP at 20 V_{pp}, 20 MHz while dead cells were removed by nDEP: (**a**) cell mixture in KCL; (**b**) cell mixture in D-PBS.

As demonstrated in Figure 12a, analysis before and after DEP showed that the cell solution's viability dropped by 14% after DEP. This viability test approximation showed that at 1 MHz (which was in the overlapped region—zone 2), live yeasts experienced the pDEP force more than the dead yeasts. As a result, more live cells were trapped and isolated from the mixture (see Figure 12b).

It was estimated that upon separation, more than 59% of the live cells and 37% of the dead cells remained in the pDEP at the IDEs when almost 63% of the dead yeasts and less than 41% of live ones were removed from the channel. This is because dead cell response to pDEP is weaker due to the lack of permittivity and conductivity difference with its surrounding buffer. Furthermore, at 1 MHz, the DEP force is closer to fc and weaker than lower frequencies in its effective pDEP region. As a result, the release of dead cells from the channel is 20% more than live ones.

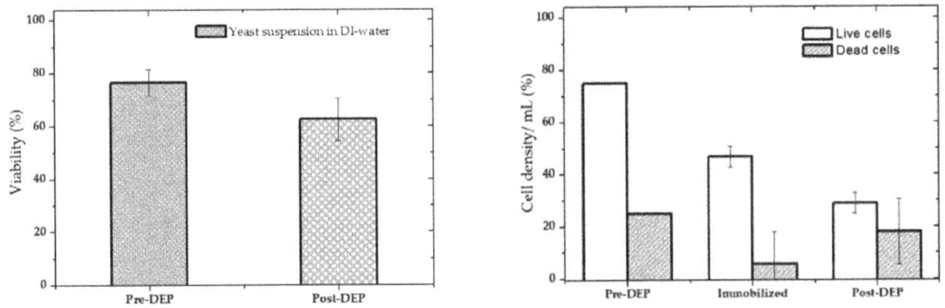

(**a**) (**b**)

Figure 12. Separation estimation of live/dead yeasts in a mixture. The live and dead cells mixture was 3:4 in the suspension before separation: (**a**) Viability estimation before (Pre-DEP) and after measurement (Post-DEP); (**b**) Percentage of trapped and released live and dead cell densities after DEP. The separation condition was: 20 Vpp, 1MHz, and 1 μm·s^{-1} flow rate.

4.3. Isolation Efficiency Assay

For yeast suspension with high cell density in DIW, the isolated cells' efficacy at the IDEs was approximately estimated at 20 Vpp, 1 MHz, via counting of the cells before DEP and after DEP when the AC was still on. 1 mL cell suspension was introduced in the microfluidic channel using a flow rate of 1 μm·s^{-1}. The remaining cells that were not trapped at the IDEs and released from the channel were collected in the tubes at the outlet for further analysis (counting and viability test). Figure 13a shows the cell density assay before and after DEP. Using these approximations, the percentage of the entrapped cells was calculated by deducting the number of the collected cells at the outlets (Post-DEP) from the total number of the cells (Pre-DEP), Figure 13b.

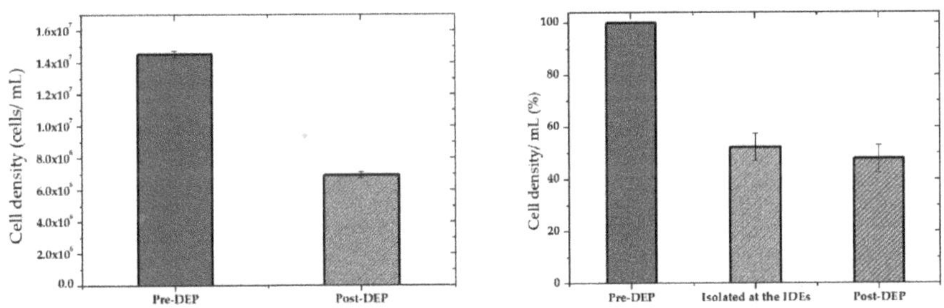

(**a**) (**b**)

Figure 13. Cell density assay before and after DEP. The cells were immobilized using pDEP at 1 MHz, 20 V_{PP}: (**a**) Approximate cell density before (Pre-DEP) and after measurement (Post-DEP); (**b**) Percentage of trapped and released cell density after DEP.

Analyses estimations, for high cell density sample, showed that upon DEP, almost 53% of the cells were trapped at the electrodes. In contrast, around 47% did not overcome drag force and were washed and released from the channel since they were either affected very slightly by pDEP force or did not experience at all. This can be related to the height of the microfluidic channel compared to the cell size and the effective distance of the electric field gradient over the electrodes or the orientation of IDEs in the microfluidic channel. The electric field's magnitude over the IDEs decays with the distance over the IDEs towards the top of the microfluidic channel. Therefore, cells flowing from significantly above the effective electric field region cannot get affected by the electric field gradient because they do not encounter this field and are deflected by any of the DEP forces and only kept flowing in the solution stream eluted from the microfluidic channel. The DEP force decreases quickly as the distance from the planar electrodes increases. Thus, this can likely be improved upon reducing the channel height (75 μm) by at least half of its current size. Moreover, as cell loading continued during the DEP process, the IDEs started to saturate with entrapped cells, and pearl chains structure were formed. This causes a decrease in the trapping yield because the pDEP areas became unavailable to new cells. On the other hand, oversaturation occurred, which led to a significant number of cells desorption with the time passage. However, letting the cells be settled for a more extended period when they reached the channel or adjusting the sample's cell density to lower orders of magnitude ($<10^6$) improved the trapping efficiency (>99%).

Experimental results proved that our microfluidic device has a very fast response, and the trapping starts in a short time after applying AC potential. At optimum AC ranges, trapping occurs spontaneously, in less than 1 s, as soon as the AC was used. Depending on other applied AC ranges, this time increased to 5 s after running the fluid through the microfluidic channel. Cutting the AC, desorption arose very fast.

5. Conclusions

A CMOS integrated silicon-based microfluidic device with embedded arrays of IDE was presented for characterization and separation of live and dead cells and selective separation of one type from the other type in a mixture. This microfluidic device has been used to characterize the DEP responses of these cells in terms of positive and negative DEP and crossover frequency. Various cell suspensions were used during the measurements, and the results were compared with the numerical simulations. Moreover, the selective separation of live yeast cells from dead yeast cells was demonstrated using this device. The differential separation was found to work sufficiently well to be suited for implementing the final design. This device is suitable for the analysis of cells suspended in media with conductivity not more than 10^{-1} S/m order of magnitude. Results showed that as the medium conductivity increases, the pDEP spectrum gets smaller, and trapping occurs at higher frequencies.

Moreover, it was observed that by increasing conductivity, the crossover frequency (transition point of real part from negative to positive) was increased. For highly conductive media (such as PBS), no pDEP response was observed over the cell trajectories, and as a result, no cell entrapment was achieved. However, pDEP trapping and cell separation were tunable for such a buffer solution by changing the medium's electrical conductivity.

Author Contributions: Methodology, H.M.E.; Validation, H.M.E.; Software, H.M.E.; Formal analysis, H.M.E. and C.W.; Data curation, H.M.E.; Writing—Original draft preparation, H.M.E.; Writing—Review and editing, H.M.E. and C.W.; Supervision, C.W.; Funding acquisition, C.W. All authors have read and agreed to the published version of the manuscript.

Funding: This study was supported by the Brandenburg Ministry of Science, Research, and Cultural Affairs for the project within the StaF program in Germany. The publication of this article was funded by the Open Access Fund of the Leibniz Association.

Acknowledgments: The authors would like to thank the technology department of IHP for the fabrication of the sensor chip.

Conflicts of Interest: The authors declare no conflict of interest.

References

1. Demircan, Y.; Yilmaz, G.; Külah, H.; Demirci, U.; Khademhosseini, A.; Langer, R.; Blander, J. Electrophoresis and Dielectrophoresis for Lab-on-a-Chip (LOC) Analyses. In *Microfluidic Technologies for Human Health*; World Scientific: Singapore, 2013; pp. 341–375.
2. Adekanmbi, E.O.; Srivastava, S.K. Dielectrophoretic applications for disease diagnostics using lab-on-a-chip platforms. *Lab. Chip* **2016**, *16*, 2148–2167. [CrossRef] [PubMed]
3. Li, H.; Zheng, Y.; Akin, D.; Bashir, R. Characterization and modeling of a microfluidic dielectrophoresis filter for biological species. *J. Microelectromechanical Syst.* **2005**, *14*, 103–112. [CrossRef]
4. Kersaudy-Kerhoas, M.; Dhariwal, R.; Desmulliez, M.P.Y. Recent advances in microparticle continuous separation. *IET Nanobiotechnology* **2008**, *2*, 1–13. [CrossRef]

5. Tsutsui, H.; Ho, C.-M. Cell separation by non-inertial force fields in microfluidic systems. *Mech. Res. Commun.* **2009**, *36*, 92–103. [CrossRef]
6. Bhagat, A.A.S.; Bow, H.; Hou, H.W.; Tan, S.J.; Han, J.; Lim, C.T. Microfluidics for cell separation. *Med. Biol. Eng. Comput.* **2010**, *48*, 999–1014. [CrossRef] [PubMed]
7. Lee, K.H.; Lee, K.S.; Jung, J.H.; Chang, C.B.; Sung, H.J. Optical mobility of blood cells for label-free cell separation applications. *Appl. Phys. Lett.* **2013**, *102*, 141911. [CrossRef]
8. Zhang, Q.; Yin, T.; Xu, R.; Gao, W.; Zhao, H.; Shapter, J.G.; Wang, K.; Shen, Y.; Huang, P.; Gao, G.; et al. Large-scale immunomagnetic cell sorting of T cells based on a self-designed high-throughput system for potential clinical application. *Nanoscale* **2017**, *9*, 13592–13599. [CrossRef]
9. Quinn, M.M.; Jalalian, L.; Ribeiro, S.; Ona, K.; Demirci, U.; Cedars, M.I.; Rosen, M.P. Microfluidic sorting selects sperm for clinical use with reduced DNA damage compared to density gradient centrifugation with swim-up in split semen samples. *Hum. Reprod.* **2018**, *33*, 1388–1393. [CrossRef]
10. Bishop, D.P.; Blanes, L.; Wilson, A.B.; Wilbanks, T.; Killeen, K.; Grimm, R.; Wenzel, R.; Major, D.; Macka, M.; Clarke, D.; et al. Microfluidic high performance liquid chromatography-chip hyphenation to inductively coupled plasma–mass spectrometry. *J. Chromatogr. A* **2017**, *1497*, 64–69. [CrossRef] [PubMed]
11. Xiang, N.; Shi, Z.; Tang, W.; Huang, D.; Zhang, X.; Ni, Z. Improved understanding of particle migration modes in spiral inertial microfluidic devices. *RSC Adv.* **2015**, *5*, 77264–77273. [CrossRef]
12. Gascoyne, P.R.C.; Shim, S. Isolation of Circulating Tumor Cells by Dielectrophoresis. *Cancers* **2014**, *6*, 545–579. [CrossRef]
13. Gagnon, Z.R. Cellular dielectrophoresis: Applications to the characterization, manipulation, separation and patterning of cells. *Electrophoresis* **2011**, *32*, 2466–2487. [CrossRef]
14. David, R.; Groebner, M.; Franz, W.-M. Magnetic Cell Sorting Purification of Differentiated Embryonic Stem Cells Stably Expressing Truncated Human CD4 as Surface Marker. *Stem Cells* **2005**, *23*, 477–482. [CrossRef] [PubMed]
15. Said, T.M.; Grunewald, S.; Paasch, U.; Glander, H.-J.; Baumann, T.; Kriegel, C.; Li, L.; Agarwal, A. Advantage of combining magnetic cell separation with sperm preparation techniques. *Reprod. Biomed. Online* **2005**, *10*, 740–746. [CrossRef]
16. Werner, M.; Merenda, F.; Piguet, J.; Salathé, R.-P.; Vogel, H. Microfluidic array cytometer based on refractive optical tweezers for parallel trapping, imaging and sorting of individual cells. *Lab. Chip* **2011**, *11*, 2432–2439. [CrossRef] [PubMed]
17. Schmitz, B.; Radbruch, A.; Kümmel, T.; Wickenhauser, C.; Korb, H.; Hansmann, M.; Thiele, J.; Fischer, R. Magnetic activated cell sorting (MACS)-a new immunomagnetic method for megakaryocytic cell isolation: Comparison of different separation techniques. *Eur. J. Haematol.* **1994**, *52*, 267–275. [CrossRef] [PubMed]
18. Mach, A.J.; Di Carlo, D. Continuous scalable blood filtration device using inertial microfluidics. *Biotechnol. Bioeng.* **2010**, *107*, 302–311. [CrossRef]
19. Khoshmanesh, K.; Zhang, C.; Tovar-Lopez, F.J.; Nahavandi, S.; Baratchi, S.; Kalantar-Zadeh, K.; Mitchell, A. Dielectrophoretic manipulation and separation of microparticles using curved microelectrodes. *Electrophoresis* **2009**, *30*, 3707–3717. [CrossRef]
20. Pommer, M.S.; Zhang, Y.; Keerthi, N.; Chen, D.; Thomson, J.A.; Meinhart, C.D.; Soh, H.T. Dielectrophoretic separation of platelets from diluted whole blood in microfluidic channels. *Electrophoresis* **2008**, *29*, 1213–1218. [CrossRef]
21. Pesch, G.R.; Du, F. A review of dielectrophoretic separation and classification of non-biological particles. *Electrophoresis* **2021**, *42*, 134–152. [CrossRef]
22. Martinez-Duarte, R. Microfabrication technologies in dielectrophoresis applications-A review. *Electrophoresis* **2012**, *33*, 3110–3132. [CrossRef] [PubMed]
23. Cheng, I.-F.; Chang, H.-C.; Hou, D.; Chang, H.-C. An integrated dielectrophoretic chip for continuous bioparticle filtering, focusing, sorting, trapping, and detecting. *Biomicrofluidics* **2007**, *1*, 021503. [CrossRef]
24. Patel, S.; Showers, D.; Vedantam, P.; Tzeng, T.-R.; Qian, S.; Xuan, X. Microfluidic separation of live and dead yeast cells using reservoir-based dielectrophoresis. *Biomicrofluidics* **2012**, *6*, 034102. [CrossRef] [PubMed]
25. Piacentini, N.; Mernier, G.; Tornay, R.; Renaud, P. Separation of platelets from other blood cells in continuous-flow by dielectrophoresis field-flow-fractionation. *Biomicrofluidics* **2011**, *5*, 034122–341228. [CrossRef]
26. Yildizhan, Y.; Erdem, N.; Islam, M.; Martinez-Duarte, R.; Elitas, M. Dielectrophoretic Separation of Live and Dead Monocytes Using 3D Carbon-Electrodes. *Sensors* **2017**, *17*, 2691. [CrossRef]
27. Yafouz, B.; Kadri, N.A.; Ibrahim, F. Dielectrophoretic Manipulation and Separation of Microparticles Using Microarray Dot Electrodes. *Sensors* **2014**, *14*, 6356–6369. [CrossRef]
28. Becker, F.F.; Wang, X.B.; Huang, Y.; Pethig, R.; Vykoukal, J.; Gascoyne, P.R. Separation of human breast cancer cells from blood by differential dielectric affinity. *Proc. Natl. Acad. Sci. USA* **1995**, *92*, 860–864. [CrossRef]
29. Zhang, H.; Chang, H.; Neuzil, P. DEP-on-a-Chip: Dielectrophoresis Applied to Microfluidic Platforms. *Micromachines* **2019**, *10*, 423. [CrossRef] [PubMed]
30. Khan, S.M.; Gümüş, A.; Nassar, J.M.; Hussain, M.M. Personalized Healthcare: CMOS Enabled Microfluidic Systems for Healthcare Based Applications (Adv. Mater. 16/2018). *Adv. Mater.* **2018**, *30*, 1–26. [CrossRef]
31. Nerguizian, V.; Stiharu, I.; Al-Azzam, N.; Diab, B.Y.; Alazzam, A. The effect of dielectrophoresis on living cells: Crossover frequencies and deregulation in gene expression. *Analyst* **2019**, *144*, 3853–3860. [CrossRef]

32. Martínez-López, J.I.; Moncada-Hernández, H.; Baylon-Cardiel, J.L.; Martínez-Chapa, S.O.; Rito-Palomares, M.; Lapizco-Encinas, B.H. Characterization of electrokinetic mobility of microparticles in order to improve dielectrophoretic concentration. *Anal. Bioanal. Chem.* **2009**, *394*, 293–302. [CrossRef]
33. Keeble, L.; Moser, N.; Rodriguez-Manzano, J.; Georgiou, P. ISFET-Based Sensing and Electric Field Actuation of DNA for On-Chip Detection: A Review. *IEEE Sensors J.* **2020**, *20*, 11044–11065. [CrossRef]
34. Pohl, H.A. The Motion and Precipitation of Suspensoids in Divergent Electric Fields. *J. Appl. Phys.* **1951**, *22*, 869–871. [CrossRef]
35. Tuukkanen, S.; Toppari, J.J.; Kuzyk, A.; Hirviniemi, L.; Hytönen, V.P.; Ihalainen, T.; Törmä, P. Carbon Nanotubes as Electrodes for Dielectrophoresis of DNA. *Nano Lett.* **2006**, *6*, 1339–1343. [CrossRef]
36. Polevaya, Y.; Ermolina, I.; Schlesinger, M.; Ginzburg, B.-Z.; Feldman, Y. Time domain dielectric spectroscopy study of human cells. *Biochim. et Biophys. Acta (BBA) Biomembr.* **1999**, *1419*, 257–271. [CrossRef]
37. Buyong, M.R.; Kayani, A.A.; Hamzah, A.A.; Majlis, B.Y. Dielectrophoresis Manipulation: Versatile Lateral and Vertical Mechanisms. *Biosens.* **2019**, *9*, 30. [CrossRef] [PubMed]
38. Yang, L. A Review of Multifunctions of Dielectrophoresis in Biosensors and Biochips for Bacteria Detection. *Anal. Lett.* **2012**, *45*, 187–201. [CrossRef]
39. Yang, L.; Guiseppi-Elie, A. Impedimetric Biosensors for Nano- and Microfluidics. In *Encyclopedia of Microfluidics and Nanofluidics*; Springer International Publishing: Berlin/Heidelberg, Germany, 2015; pp. 1364–1380.
40. Yang, L.; Banada, P.P.; Chatni, M.R.; Lim, K.S.; Bhunia, A.K.; Ladisch, M.; Bashir, R. A multifunctional micro-fluidic system for dielectrophoretic concentration coupled with immuno-capture of low numbers of Listeria monocytogenes. *Lab. Chip* **2006**, *6*, 896–905. [CrossRef]
41. Yang, L. Dielectrophoresis assisted immuno-capture and detection of foodborne pathogenic bacteria in biochips. *Talanta* **2009**, *80*, 551–558. [CrossRef]
42. Miltenyi, S.; Müller, W.; Weichel, W.; Radbruch, A. High gradient magnetic cell separation with MACS. *Cytometry* **1990**, *11*, 231–238. [CrossRef]
43. Gascoyne, P.R.C.; Wang, X.-B.; Huang, Y.; Becker, F.F. Dielectrophoretic separation of cancer cells from blood. *IEEE Trans. Ind. Appl.* **1997**, *33*, 670–678. [CrossRef]
44. Zhu, K.; Kaprelyants, A.S.; Salina, E.G.; Markx, G.H. Separation by dielectrophoresis of dormant and nondormant bacterial cells ofMycobacterium smegmatis. *Biomicrofluidics* **2010**, *4*, 022809. [CrossRef]
45. Choi, W.; Kim, J.-S.; Lee, H.; Lee, K.-K.; Koo, D.-B.; Park, J.-K. Dielectrophoretic oocyte selection chip for in vitro fertilization. *Biomed. Microdevices* **2007**, *10*, 337–345. [CrossRef] [PubMed]
46. Sabuncu, A.C.; Liu, J.A.; Beebe, S.J.; Beskok, A. Dielectrophoretic separation of mouse melanoma clones. *Biomicrofluidics* **2010**, *4*, 021101. [CrossRef] [PubMed]
47. Pethig, R.; Huang, Y.; Wang, X.-B.; Burt, J.P.H. Positive and negative dielectrophoretic collection of colloidal particles using interdigitated castellated microelectrodes. *J. Phys. D Appl. Phys.* **1992**, *25*, 881–888. [CrossRef]
48. Demierre, N.; Braschler, T.; Linderholm, P.; Seger, U.; Van Lintel, H.; Renaud, P. Characterization and optimization of liquid electrodes for lateral dielectrophoresis. *Lab. Chip* **2007**, *7*, 355–365. [CrossRef] [PubMed]
49. Tai, C.-H.; Hsiung, S.-K.; Chen, C.-Y.; Tsai, M.-L.; Lee, G.-B. Automatic microfluidic platform for cell separation and nucleus collection. *Biomed. Microdevices* **2007**, *9*, 533–543. [CrossRef] [PubMed]
50. Rajaraman, S.; Noh, H.-S.; Hesketh, P.J.; Gottfried, D.S. Rapid, low cost microfabrication technologies toward realization of devices for dielectrophoretic manipulation of particles and nanowires. *Sens. Actuators B Chem.* **2006**, *114*, 392–401. [CrossRef]
51. Zhang, C.; Khoshmanesh, K.; Tovar-Lopez, F.J.; Mitchell, A.; Wlodarski, W.; Klantar-Zadeh, K. Dielectrophoretic separation of carbon nanotubes and polystyrene microparticles. *Microfluid. Nanofluidics* **2009**, *7*, 633–645. [CrossRef]
52. Suehiro, J. Fabrication and characterization of nanomaterial-based sensors using dielectrophoresis. *Biomicrofluidics* **2010**, *4*, 022804. [CrossRef]
53. Li, Y.; Dalton, C.; Crabtree, H.J.; Nilsson, G.; Kaler, K.V.I.S. Continuous dielectrophoretic cell separation microfluidic device. *Lab. Chip* **2007**, *7*, 239–248. [CrossRef]
54. Asbury, C.L.; Diercks, A.H.; van den Engh, G. Trapping of DNA by dielectrophoresis. *Electrophoresis* **2002**, *23*, 2658–2666. [CrossRef]
55. Gadish, N.; Voldman, J. High-Throughput Positive-Dielectrophoretic Bioparticle Microconcentrator. *Anal. Chem.* **2006**, *78*, 7870–7876. [CrossRef]
56. Flanagan, L.A.; Lu, J.; Wang, L.; Marchenko, S.A.; Jeon, N.L.; Lee, A.P.; Monuki, E.S. Unique Dielectric Properties Distinguish Stem Cells and Their Differentiated Progeny. *Stem Cells* **2008**, *26*, 656–665. [CrossRef]
57. Ning, Y.; Ma, X.; Multari, C.R.; Luo, X.; Gholizadeh, V.; Palego, C.; Cheng, X.; Hwang, J.C.M. Improved broadband electrical detection of individual biological cells. In Proceedings of the 2015 IEEE MTT-S International Microwave Symposium, Phoenix, AZ, USA, 17–22 May 2015; pp. 1–3.
58. Thomas, R.S.; Morgan, H.; Green, N.G. Negative DEP traps for single cell immobilisation. *Lab. Chip* **2009**, *9*, 1534–1540. [CrossRef] [PubMed]
59. Miled, M.A.; Massicotte, G.; Sawan, M. Dielectrophoresis-Based Integrated Lab-on-Chip for Nano and Micro-Particles Manipulation and Capacitive Detection. *IEEE Trans. Biomed. Circuits Syst.* **2012**, *6*, 120–132. [CrossRef] [PubMed]

60. Park, K.; Kabiri, S.; Sonkusale, S. Dielectrophoretic lab-on-CMOS platform for trapping and manipulation of cells. *Biomed. Microdevices* **2016**, *18*, 1–11. [CrossRef]
61. Manczak, R.; Kaynak, C.B.; Kaynak, M.; Palego, C.; Lalloue, F.; Pothier, A.; Saada, S.; Provent, T.; Dalmay, C.; Bessette, B.; et al. UHF-Dielectrophoresis Crossover Frequency as a New Marker for Discrimination of Glioblastoma Undifferentiated Cells. *IEEE J. Electromagn. RF Microw. Med. Biol.* **2019**, *3*, 191–198. [CrossRef]
62. Guha, S.; Schumann, U.; Jamal, F.I.; Wagner, D.; Meliani, C.; Schmidt, B.; Wenger, C.; Wessel, J.; Detert, M. Integrated high-frequency sensors in catheters for minimally invasive plaque characterization. In Proceedings of the 20th European Microelectronics and Packaging Conference and Exhibition: Enabling Technologies for a Better Life and Future, EMPC, Friedrichshafen, Germany, 14–16 September 2015; pp. 1–6.
63. Guha, S.; Schmalz, K.; Meliani Ch Wenger, C. KW CMOS MEMS based Microfluidic System for Cytometry at 5GHz. In Proceedings of the MFHS, Microfluidic handling System, Entschede, The Netherlands, 10–12 October 2012.
64. Guha, S.; Wenger, C. Radio Frequency CMOS Chem-bio Viscosity Sensors based on Dielectric Spectroscopy. In Proceedings of the 10th International Joint Conference on Biomedical Engineering Systems and Technologies; SCITEPRESS Science and Technology Publications, Porto, Portugal, 21 February 2017; pp. 142–148.
65. Guha, S.; Schmalz, K.; Wenger, C.; Herzel, F. Self-calibrating highly sensitive dynamic capacitance sensor: Towards rapid sensing and counting of particles in laminar flow systems. *Analyst* **2015**, *140*, 3262–3272. [CrossRef] [PubMed]
66. Park, K.; Kabiri, S.; Sonkusale, S. CMOS dielectrophoretic Lab-on-Chip platform for manipulation and monitoring of cells. In Proceedings of the 2015 37th Annual International Conference of the IEEE Engineering in Medicine and Biology Society, EMBCMilan, Italy, 25–29 August 2015; Volume 2015, pp. 7530–7533.
67. Otto, S.; Kaletta, U.; Bier, F.F.; Wenger, C.; Hölzel, R. Dielectrophoretic immobilisation of antibodies on microelectrode arrays. *Lab. Chip* **2014**, *14*, 998. [CrossRef] [PubMed]
68. Manaresi, N.; Romani, A.; Medoro, G.; Altomare, L.; Leonardi, A.; Tartagni, M.; Guerrieri, R. A CMOS Chip for Individual Cell Manipulation and Detection. *IEEE J. Solid-State Circuits* **2003**, *38*, 2297–2305. [CrossRef]
69. Ameri, A.; Zhang, L.; Gharia, A.; Anwar, M.; Niknejad, A.M. Dielectrophoretic-Assisted Biosensor for Single-Cell Characterization at Mmwave Frequencies in CMOS 28nm Technology. In Proceedings of the 2019 20th International Conference on Solid-State Sensors, Actuators and Microsystems & Eurosensors XXXIII (TRANSDUCERS & EUROSENSORS XXXIII, Berlin, Germany, 23–27 June 2019; pp. 174–177.
70. Iliescu, C.; Taylor, H.; Avram, M.; Miao, J.; Franssila, S. A practical guide for the fabrication of microfluidic devices using glass and silicon. *Biomicrofluidics* **2012**, *6*, 016505–1650516. [CrossRef]
71. Becker, H. Mind the gap! *Lab. Chip* **2010**, *10*, 271–273. [CrossRef]
72. Matbaechi Ettehad, H.; Soltani Zarrin, P.; Hölzel, R.; Wenger, C. Dielectrophoretic Immobilization of Yeast Cells Using CMOS Integrated Microfluidics. *Micromachines* **2020**, *11*, 501. [CrossRef] [PubMed]
73. Ettehad, H.M.; Yadav, R.K.; Guha, S.; Wenger, C. Towards CMOS Integrated Microfluidics Using Dielectrophoretic Immobilization. *Biosensors* **2019**, *9*, 77. [CrossRef]
74. Ettehad, H.M.; Guha, S.; Wenger, C. Simulation of CMOS Compatible Sensor Structures for Dielectrophoretic Biomolecule Immobilization. In Proceedings of the COMSOL-Bioscience and Bioengineering, Rotterdam, The Netherlands, 19 October 2017; p. 6.
75. Lei, J.; Wan, J.T.K.; Yu, K.W.; Sun, H. First-principle approach to dielectric behavior of nonspherical cell suspensions. *Phys. Rev. E* **2001**, *64*, 012903. [CrossRef]
76. Thomas, B. *Jones Electromechanics of Particles;* Cambridge University Press in Cambridge: New York, NY, USA, 1995; ISBN 0521431964.
77. Cottet, J.; Fabregue, O.; Berger, C.; Buret, F.; Renaud, P.; Frénéa-Robin, M. MyDEP: A New Computational Tool for Dielectric Modeling of Particles and Cells. *Biophys. J.* **2019**, *116*, 12–18. [CrossRef]
78. Huang, Y.; Holzel, R.; Pethig, R.; Wang, X.-B. Differences in the AC electrodynamics of viable and non-viable yeast cells determined through combined dielectrophoresis and electrorotation studies. *Phys. Med. Biol.* **1992**, *37*, 1499–1517. [CrossRef]
79. Talary, M.S.; Burt, J.P.H.; Tame, J.A.; Pethig, R. Electromanipulation and separation of cells using travelling electric fields. *J. Phys. D Appl. Phys.* **1996**, *29*, 2198–2203. [CrossRef]

micromachines

Article

Sub–100 nm Nanoparticle Upconcentration in Flow by Dielectrophoretic Forces

Maria Dimaki [1,*], Mark Holm Olsen [2], Noemi Rozlosnik [3] and Winnie E. Svendsen [1]

1 DTU Bioengineering, Technical University of Denmark, Søltofts Plads, Bldg 221, 2800 Kongens Lyngby, Denmark; wisv@dtu.dk
2 Center for Plastics and Packaging Technology, Teknologisk Institut, Gregersensvej 1, 2630 Taastrup, Denmark; mols@teknologisk.dk
3 Eir Diagnostics ApS, Muslingevej 36B, 8250 Egå, Denmark; nr@eir-diagnostics.com
* Correspondence: madi@dtu.dk

Abstract: This paper presents a novel microfluidic chip for upconcentration of sub–100 nm nanoparticles in a flow using electrical forces generated by a DC or AC field. Two electrode designs were optimized using COMSOL Multiphysics and tested using particles with sizes as low as 47 nm. We show how inclined electrodes with a zig-zag three-tooth configuration in a channel of 20 μm width are the ones generating the highest gradient and therefore the largest force. The design, based on AC dielectrophoresis, was shown to upconcentrate sub–100 nm particles by a factor of 11 using a flow rate of 2–25 μL/h. We present theoretical and experimental results and discuss how the chip design can easily be massively parallelized in order to increase throughput by a factor of at least 1250.

Keywords: dielectrophoresis; upconcentration; microfluidics; nanoparticles

Citation: Dimaki, M.; Olsen, M.H.; Rozlosnik, N.; Svendsen, W.E. Sub–100 nm Nanoparticle Upconcentration in Flow by Dielectrophoretic Forces. *Micromachines* **2022**, *13*, 866. https://doi.org/10.3390/mi13060866

Academic Editor: Aiqun Liu

Received: 9 May 2022
Accepted: 27 May 2022
Published: 30 May 2022

1. Introduction

The separation of subpopulations of particles is important for many biological, diagnostic, or environmental applications. Such particles can be found in a variety of different fluids, including but not limited to drinking water, surface water, blood, serum, urine, and milk, and in sizes ranging from a few tens of nanometers (e.g., viruses, exosomes, liposomes) up to a few tens of micrometers (e.g., blood cells, bacteria). Some particles are available in large quantities in a sample and can therefore be readily identified and processed, while others are very rare, requiring one or several upconcentration steps for their isolation and identification. While exosomes for example are found in concentrations of about 1×10^8 exosomes/mL in blood [1], circulating tumor cells (CTCs) are only present in concentrations of 1–100 cells/mL [2].

Manipulation of micro- and nanoparticles by passive or active techniques has been the subject of research for at least two decades. Microfluidic technology has been the driving force, particularly due to low sample volume requirements, which are very attractive for many diagnostic applications. Properties such as precise control of flow and the relative ease by which externally applied forces can be integrated into these systems make microfluidic technology ideal for most such applications.

Passive techniques utilize the properties of the flow itself along with geometrical modifications to manipulate particles. Several examples of passive systems can be found in the literature. For example, exosomes (size of 30–200 nm) were sorted with a purity of 92% and a recovery of 81% using a combination of inertial microfluidics and viscoelastic fluids [3]. Inertial microfluidics by itself is a method that has been used for the size sorting of particles in relatively large microfluidic channels and at relatively high Reynolds numbers [4]. Although the method is powerful for the size separation of particles, it operates mostly on particles in the micrometer range. The use of deterministic lateral displacement arrays (DLDs) is another popular passive method for particle separation that can be applied to particles from 20 nm to several tens of micrometers. However, for nanoparticle sorting, it is also necessary to use pillars in the nanometer range, which complicates fabrication and use of the system, and the method is also limited by throughput [5].

For continuous nanoparticle separation, active techniques are the preferred method. Here microfluidics is combined with some sort of external force that acts on the particle and displaces it away from its initial position. External forces are mainly electrical, magnetic, acoustic, or optical. Most of these forces scale with the particle size, which means that they are generally very small for nanoparticles, and have limited throughput [6]. Magnetic forces require the particle to be sorted to either have magnetic properties or be labeled with magnetic particles.

One of the most popular active manipulation methods is the use of electrical forces generated by electrodes immersed in a liquid. Electrical forces acting on uncharged but polarizable particles give rise to a so-called dielectrophoretic (DEP) force. This force depends on particle volume and on the electrical properties of the particles and the medium they are dispersed in (see Section 2.1 for further details). The method requires rather high voltages, particularly in the case where the so-called insulator DEP is used [7]; the high voltages lead to Joule heating and electrothermal flow that can disturb the separation [8].

The sorting of different biological particles, particularly cells, in a continuous flow by DEP is very well studied. One of the main obstacles is the similarity between human cells; they have similar densities and similar turnover frequencies [9]. Therefore, developed systems utilize long channels, multifrequency DEP, and pretreatment protocols to achieve separation. The method is quite efficient for this type of particles because they are rather large; DEP forces scale with the particle radius, and cells are of the order of 5–30 μm.

The DEP force is also highly localized, as it is proportional to the electric field gradient, which declines rapidly the further away from the electrodes the particle is. Therefore, much effort has been put into designing electrodes that can generate significant gradients in as much of the volume of a microfluidic channel as possible, where the DEP force should act. These geometries include top and bottom electrodes; sidewall electrodes, also called 3D electrodes [10]; liquid electrodes, where the channel geometry results in virtual sidewall electrodes, even though the real electrodes are planar [11]; and other techniques, such as nanosized orifices [12], laser-induced DEP [13], and nanopipette-based DEP [14]. Characteristic of all the above is a need for a specialized fabrication process or additional equipment to achieve the large gradients necessary to affect the movement of nanoparticles.

In this paper, we present a simple and scalable microfluidic chip with microsized electrodes generating the electrical field for the continuous sorting of sub–100 nm particles, with the aim of upconcentrating these for further applications. Apart from an AC voltage generator at a single frequency, no other equipment is required for its function, and the fabrication process is simple and modifiable for large-scale production using screen printing and injection molding. We used numerical simulations to find an optimal electrode design and tested the chip using particles with sizes as low as 47 nm, corresponding to the size of small viruses. We further show how this simple design is easily scalable for increasing throughput.

2. Materials and Methods

2.1. General Requirements

The main function of the chip is to concentrate sub-nanometer particles at the center of the channel using dielectrophoretic forces generated by a set of electrodes powered by an AC voltage. The time-averaged dielectrophoretic force is given by Equation (1) for a spherical particle of radius α [15].

$$F_{\mathbf{DEP}} = \pi \cdot \alpha^3 \cdot \varepsilon_m \cdot \varepsilon_0 \cdot Re\left(\frac{\varepsilon_p^* - \varepsilon_m^*}{\varepsilon_p^* + 2\varepsilon_m^*}\right) \cdot \nabla \left| \boldsymbol{E}^2 \right| \quad (1)$$

where ε_m is the relative permittivity of the suspending medium, ε_0 is the permittivity of free space, and $\boldsymbol{E}$ is the applied electric field amplitude. The quantities ε_i^*, where the subscript i can be either p or m, refer to the complex permittivities of the particle (p) and the medium (m). These are given by Equation (2):

$$\varepsilon_i^* = \varepsilon_0\varepsilon - j\frac{\sigma}{\omega} \tag{2}$$

where σ is the conductivity, ε is the relative permittivity, and ω is the angular frequency of the applied electric field.

The term containing the complex permittivities is known as the Clausius–Mossotti factor and determines whether the particles are attracted towards the electrodes (known as positive DEP) or move away from them (negative DEP). Due to surface conductance effects, polystyrene beads have been shown to exhibit positive DEP at frequencies below 1–10 MHz [16].

The term $\nabla|\boldsymbol{E}^2|$ in Equation (1) is dependent on the applied voltage (V) and the distance between the electrodes (d). The quantity can be approximated as V^2/d^3, which means that the force is essentially inversely proportional to the distance between the electrodes to the power of 3. A small electrode distance will generate a localized large field at the tip of the electrodes, but nowhere else in the channel, unless the electrodes are designed in such a way so that they maintain a small spacing at a larger proportion of the channel width, without compromising the inhomogeneity of the electric field. Moreover, to achieve a large field across the channel height, we would ideally need electrodes on both the channel top and bottom. This, however, significantly complicates the fabrication process. Therefore, electrodes were placed only at the bottom of the channel, and the channel height was restricted to 25 µm. The channel width was restricted to 40 µm, for the same reasons. The general chip design is shown in Figure 1a. Each chip contains three individually addressed electrode arrays inside a microfluidic channel. There is one inlet and three outlets for each channel, and the nanoparticles are collected through the central outlet after they have been focused by the DEP forces. A close-up of the outlet region is shown in Figure 1b. Electrodes were placed either directly opposite each other, or in a zig-zag pattern. The angles of the main electrodes were at 60 degrees relative to the channel middle, and the protrusions were spaced 5 µm (face-to-face) and 6 µm (zig-zag) apart, with opposite electrodes having a minimum distance of 2.5 µm (face-to-face) and 2 µm (zig-zag).

Further, the option of using a DC voltage for the same purpose was investigated. In this case, a voltage was applied on a thin electrode running along the entire length of the channel exactly in the middle while several parallel ones provided the grounding voltage. The channel was 100 µm wide, the electrodes were 3 µm wide, and the distance between them was 6 µm. The particles were attracted towards the middle electrode due to electrophoresis, as they are charged. The general design for the DC chip is shown in Figure 1c. The dimensions of the two chips (AC and DC) were the result of a numerical optimization of the upconcentration performance under AC or DC conditions.

Each chip contained 3 cm long channels with electrodes on the bottom. The main channel splits into the three outlets, with the middle outlet designed to carry about 10% of the flow by making it longer than the outer outlets. All three outlets of the DC chips were 50 µm in width, while all three outlets of the AC chips were 20 µm in width. As all the particles coming from the inlet are concentrated in the middle of the channel, the middle outlet will collect all the particles but only in a 10th of the original fluid volume, thus resulting in the upconcentration of the sub-nanometer particles. Chips with more than one row of electrodes were also fabricated in order to evaluate the ability of the process to create identical electrodes along a large wafer surface.

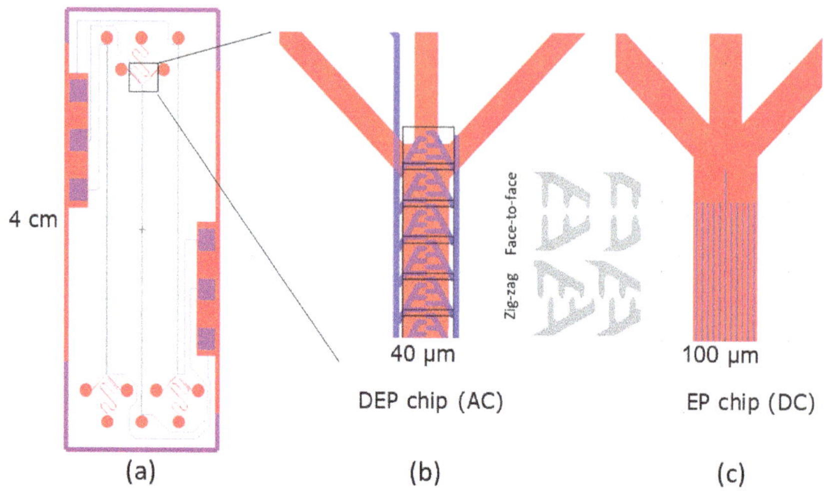

Figure 1. (**a**) The general chip design. (**b**) The channel contains a single channel with electrodes at the bottom. In the case of the AC chip, the electrodes are either in a zig-zag or face-to-face configuration (inset). (**c**) In the case of the DC chip, the electrodes are simple lines running parallel to the channel length.

2.2. Numerical Simulations

COMSOL Multiphysics Version 5.4 was used to calculate the particle trajectories due to the flow and the dielectrophoretic force. To minimize the computation time, only a 500 µm long channel was simulated. The electric field gradient and the dielectrophoretic force were calculated using the AC/DC module, while the flow profile was calculated using the laminar flow module. A voltage of 20 V peak-to-peak was applied to the electrodes in order to calculate the electric field. All other boundaries were set to the default boundary condition of "Electric Insulation" ($n \cdot J = 0$). For the flow simulations, all boundaries apart from the inlet and outlet were set with a no-slip boundary condition. A flow rate of 1 µL/h was applied to the inlet, and a pressure of 0 was used as a boundary condition at the outlet. Once the flow profile and the electric field were calculated, the particle tracing module was used to calculate the trajectories of particles of varying diameters (in the range of 30–100 nm) and a density of 1200 kg/m^3 starting at the channel inlet on a line spanning the channel width at half the channel height. The particles were affected by the flow, gravity (which was, however, negligible for nanoparticles), and dielectrophoresis. The general size of the mesh was set to "coarse", calibrated for fluid dynamics; however, all boundaries were set to a "normal" size, and a mesh control face was added in the middle of the channel (xz plane running along the channel length) with a maximum element size of 2 µm.

The particles starting and ending positions on the y and z axes (Figure 2) of a 500 µm long channel were noted, and the displacement was used to evaluate the performance of each design. Due to computational constraints, Brownian motion was not included in the COMSOL simulations; however, the values of the velocity and electric field in a unit cell of the geometry were exported in matrix form to Matlab (Version 2017b), and an algorithm was developed to calculate the particle trajectories with the thermal force included [17]. The algorithm stops when the particle touches one of the boundaries (sidewalls of the channel or bottom). Due to the complexity of the simulation, bouncing on the boundaries has not been included. The Matlab algorithm can be performed for any flow velocity (by correcting the COMSOL calculated velocity with the correct multiplication factor, MF_vel = (wanted flow rate)/(COMSOL used flow rate). Similarly, the dielectrophoretic force can also be adjusted in the Matlab program by a multiplication factor MFforce = (wanted voltage)^2/(COMSOL used voltage)^2.

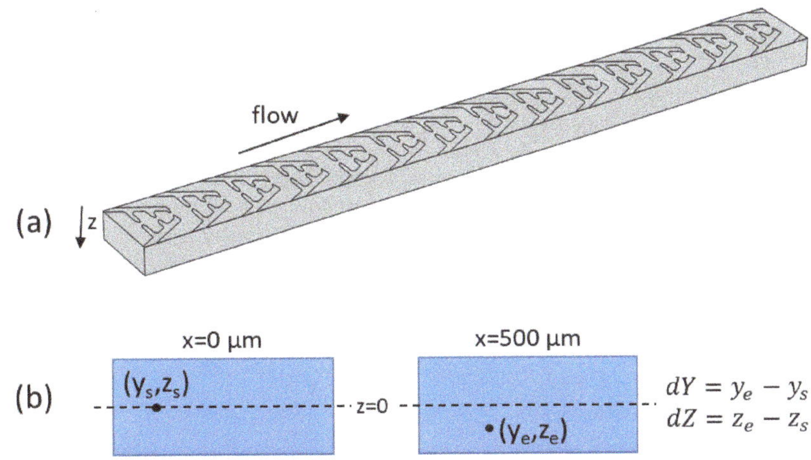

Figure 2. (**a**) The 500 μm long channel with electrodes on the bottom side used for the simulations. (**b**) The starting and ending position of each particle starting at a certain (y,z) point at x = 0 μm is noted after the particle has moved in the channel to x = 500 μm. The displacement along the height and width is used to evaluate how well each electrode design is performing in terms of particle focusing.

The simulations using Matlab were performed with an applied AC voltage of 20 V peak-to-peak and for particles 100 nm in diameter. The particle movement was simulated for flow velocities of 1 μL/h, 10 μL/h, and 25 μL/h. To gain statistical information, 10 to 200 particles were simulated from each position. For the end position graphs, only those particles that had ended their movement on the bottom boundary were included.

We also performed similar simulations for the DC chip (data not shown), which resulted in the chosen chip design and experimental conditions.

2.3. Chip Fabrication Process

The two most promising designs were fabricated using standard photolithography processes on silicon wafers. The fabrication was completed at the National Center for Nanofabrication and Characterization and the Technical University of Denmark. Briefly, 200 nm of silicon oxide was thermally grown on a standard 4-inch silicon wafer with a thickness of 525 μm. Then, 1.5 μm of AZ5214E resist was spun on the wafer and exposed to UV light (365 nm) for 7 s (KS Aligner) with a power of 8 mW/cm^2 through a mask containing the electrodes in hard contact mode. The resist was developed using a standard AZ resist developer (AZ351B). Then, 10 nm of Ti and 100 nm of gold were evaporated on the wafer by e-beam evaporation using a Wordentec QCL 800. This was followed by a lift-off process in acetone for 20 min. SU8-2075 was spun on the wafer and exposed through a second mask containing the microfluidic channels using the KS Aligner. The target height for the SU8 resist was 20 μm. After development in propylene glycol methyl ether acetate (PGMEA) for 10 min, the chips were diced using a diamond saw.

2.4. Measurement Setup

A holder was fabricated using micromilling. The holder contained the silicon chip, a PDMS gasket to close the microfluidic channels, and holes for the inlet and outlet tubing and the electrical connections using spring pins.

An AC signal generator (TG2000, Aim and Thurlby Thandar Instruments, Cambridgeshire, UK) was used to apply an AC voltage of varying amplitude and frequency to the electrodes. Voltages from 1 V to 20 V peak-to-peak were used at a frequency of 200 kHz. This frequency was selected after evaluating the performance of the particle focusing at frequencies ranging from 10 kH to 1 MHz. The flow rate was 2 μL/h. To avoid the particles becoming stuck to the middle electrode, an on–off frequency of 3 Hz was used, although frequencies between 1 Hz and 5 Hz were also investigated.

For the DC case, a DC voltage of 2 V was used. The flow rate in this case was varied between 10 and 50 μL/h. A standard syringe pump (Chemyx Fusion 200) was used for controlling the flow rate.

Polystyrene beads of sizes ranging between 47 nm and 1 μm were used. The beads were fluorescently labeled (Fluoresbrite YG Microspheres, from Polysciences, Inc., Warrington, PA, USA). The bead movement was observed using a fluorescent microscope (Nikon TE2000U). The beads were suspended in tap water with an addition of 0.1% Tween (Sigma Aldrich, St. Louis, MO, USA). The pH of the solution was 7 and the conductivity was 50 mS/m.

2.5. Result Evaluation

ImageJ was used to measure the minimum distance between electrodes for the AC chip, as well as the dimensions of the electrodes in the DC chip. Image J was also used in order to evaluate whether the fluidic channels were designed properly so that only 10% of the flow was collected in the central channel. In this case, recordings of the particle movements were taken, enhanced, and analyzed by manually tracking the position of up to 20 particles per outlet channel. The average velocity of each particle was calculated, and from these results, an average velocity for each outlet was found. The concentration ability of the fluidic system was then calculated as follows:

$$\text{Ratio} = \frac{\text{Middle channel velocity}}{\text{Upper} + \text{Middle} + \text{Lower channel velocity}} \quad (3)$$

Figure 3 outlines the general algorithm of the velocity calculation for each particle.

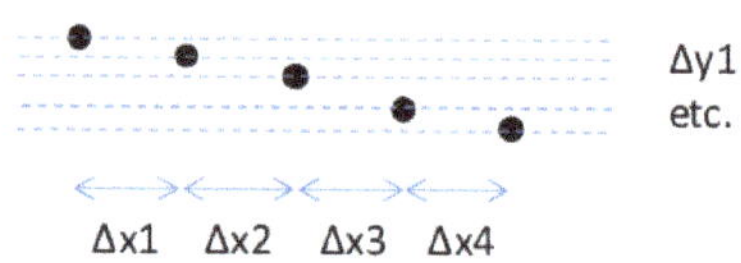

$$Velocity(i) = \frac{\sqrt{\Delta x(i)^2 + \Delta y(i)^2}}{\Delta t} \Longrightarrow [Velocity] = \frac{1}{N}\sum_{i=1}^{N} Velocity(i)$$

Figure 3. The general algorithm for the calculation of the particle velocity by using recordings of particle movement in the channel. To calculate the average velocity of each particle, 5 to 30 positions were tracked per particle.

3. Results

The results of the chip fabrication, the simulations, and the experimental particle upconcentration are presented in this section.

3.1. Chip Fabrication

A close-up of the electrodes and SU8 channels can be seen in Figure 4.

The lift-off process results in well-defined structures. ImageJ was used to measure the distances between the electrode tips in both designs. For the face-to-face design, the mean distance between the electrode tips is 2.53 ± 0.10 μm, while for the zig-zag structure, the minimum distance between the electrode tips is 1.79 ± 0.18 μm. The DC electrodes are 3.24 ± 0.14 μm. All results are based on 10 different measurements.

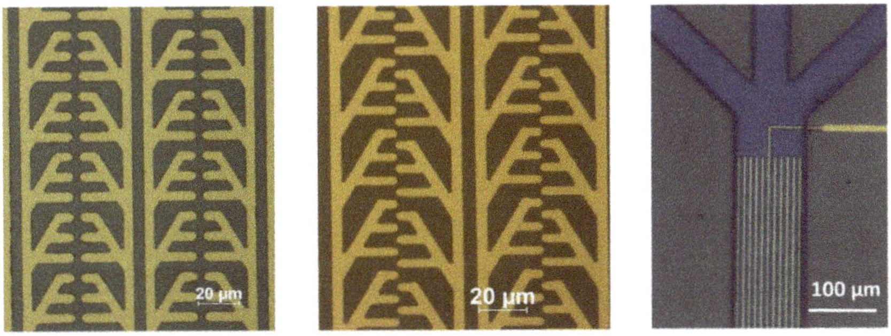

(a) (b) (c)

Figure 4. Close up of the chip. (**a**) Face-to-face electrode design. (**b**) Zig-zag electrode design. (**c**) DC chip design.

3.2. Numerical Simulations

The chip's geometrical requirements presented in Section 2.1 were based on numerical simulations performed on a much wider channel (150 µm) and with electrodes at angles of 30, 45, and 60 degrees relative to the direction of the channel width, located on the bottom and top of the channel. The results of these simulations, performed for particles of 30 nm in diameter, in a channel with a flow of 1 µL/h, and with single, double, or triple face-to-face electrodes per electrode arm, plotted as described in Section 2.2 and Figure 2 (for the case of a triple electrode), are shown in Figure 5. The graphs show that only the particles situated in the central 40–60 µm of the channel width are influenced by the electric field and that electrodes at a 60-degree angle result in a larger displacement than those at 30 or 45 degrees. We can also see that as the channel height becomes smaller (plotted here for h = 5 µm and h = 2 µm), the displacement becomes bigger. However, channel heights below 10 µm are difficult to fabricate and operate without clogging issues, so this solution was not considered as optimal or realistic.

Therefore further simulations were conducted, where the channel width was reduced to 40 µm and only a 60-degree angle was used. Furthermore, the particle diameter was increased to 40 nm. The results are plotted in Figure 6, where not only the displacement along the channel width, but also the displacement along the channel height is shown, for both the face-to-face and the zig-zag designs. The zig-zag design is used to illustrate the placement of the electrodes, for the purpose of guiding the eye. The simulations shown were performed for both types of electrodes.

The results show that the zig-zag pattern is more effective than the face-to-face pattern and that there is a displacement of particles regardless of their position along the channel width. The maximum displacement per 500 µm channel length is 1 µm towards the channel center and about 3 µm towards the channel bottom for the zig-zag pattern.

Figure 7 shows the required length to catch a particle (Figure 7a) as well as the position across the channel width where the particle was caught (Figure 7b). It can be seen that particles starting closest to the electrodes (in height) are reaching the channel bottom furthest away from the center of the channel (dotted line at y = 75 µm). This simulation was performed for three different starting positions across the width axis and for a flow velocity of 1 µL/h.

To show the effect of the flow rate on the required channel length and final position, we repeat the simulations for the middle case, i.e., for particles starting 10 µm from the channel center (along the width axis), for three different flow velocities (shown in Figure 8). This middle case shows the worst-case scenario for particles situated in the central 50% of the channel width.

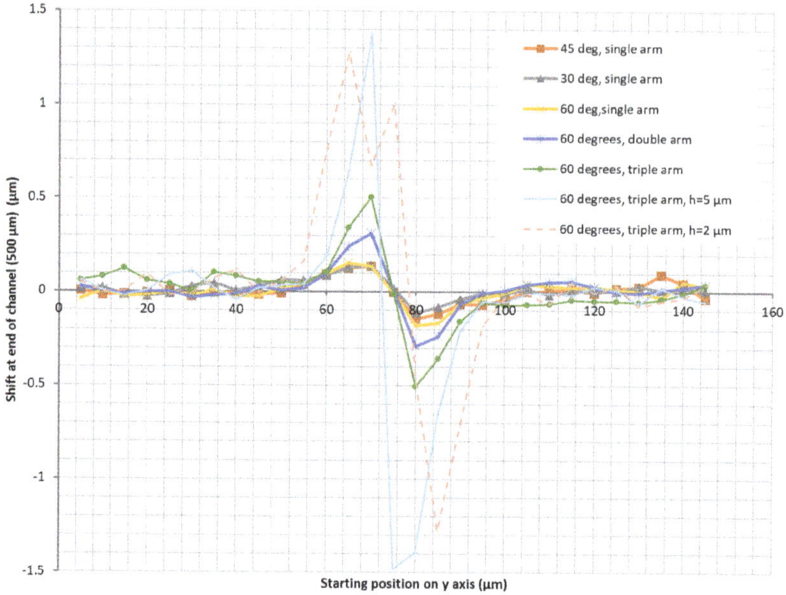

Figure 5. Particle displacement along the channel width after 500 µm of channel length under the influence of dielectrophoretic forces. Unless otherwise indicated, the channel height is 20 µm. The figure shows that the displacement is larger for smaller but unrealistic channel heights. Double refers to electrodes having only two "teeth" instead of the three shown in Figure 1b, which in this graph is indicated as "triple".

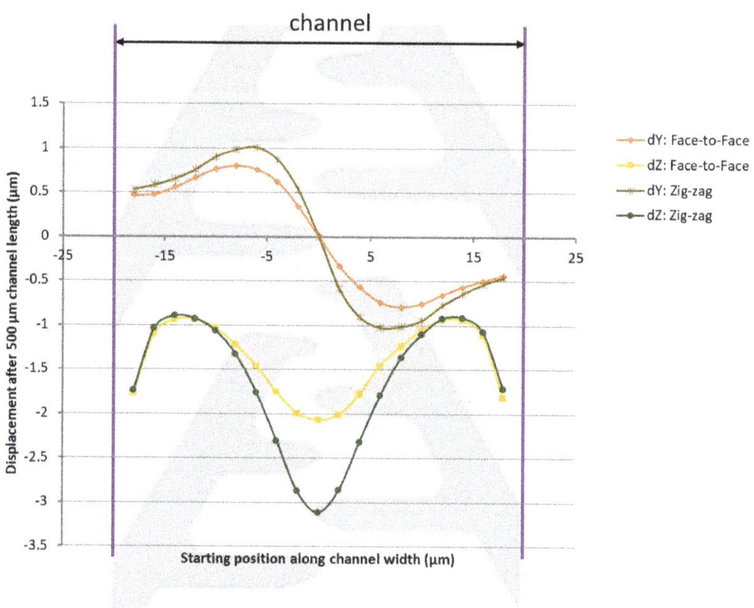

Figure 6. Particle displacement along the channel width (dY) and the channel height (dZ) after 500 µm of channel length under the influence of DEP forces.

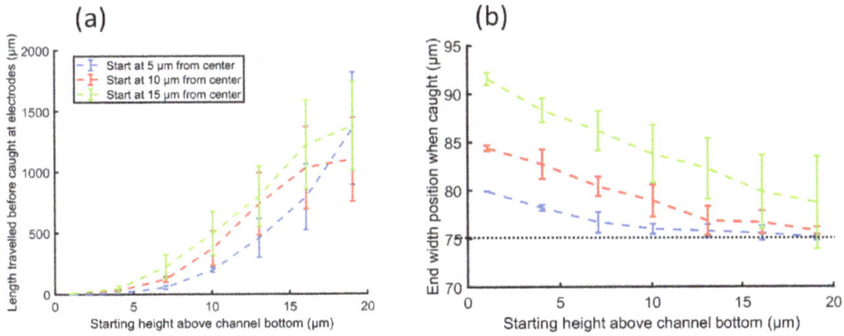

Figure 7. (**a**) The required channel length to catch particles on the electrode plane as a function of starting position along the height of the channel for three different starting positions along the channel width. (**b**) The end position along the width of the channel as a function of starting position along the channel height.

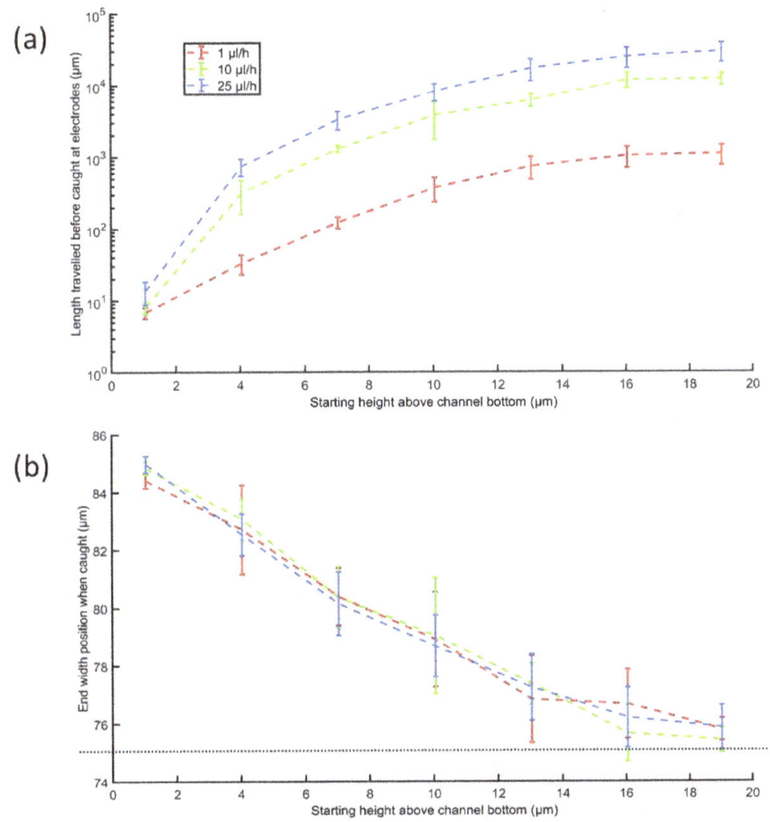

Figure 8. (**a**) Required channel length for catching the particles on the electrodes. (**b**) End position on the width axis. The channel middle is at y = 75 µm (dotted line).

3.3. Particle Focusing

As shown in Figure 9, 47 nm particles were successfully aligned using an AC voltage of 20 V peak-to-peak at a frequency of 200 kHz. When the voltage was turned off, the fluorescence was seen in the entire width of the channel, as shown in the signal intensity curve. However, when the voltage was turned on, there was a clear peak in the middle of the channel, indicating that the particles are focused in the channel center as expected.

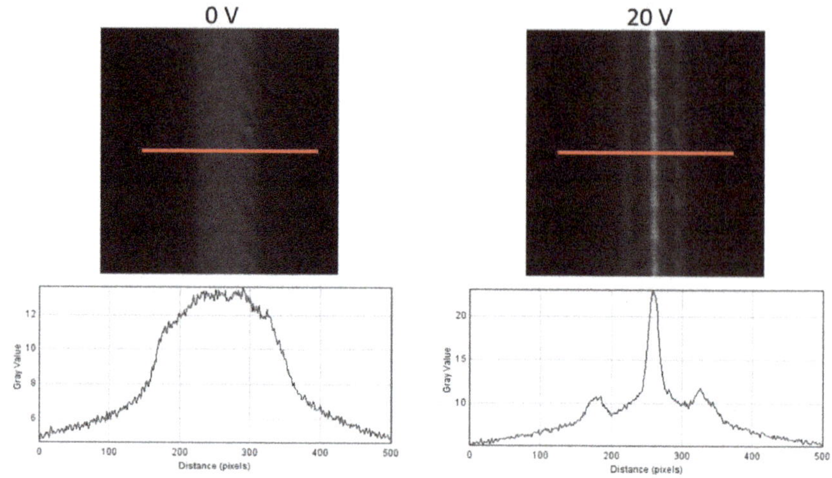

Figure 9. Particle focusing at 20 V peak-to-peak and a frequency of 200 kHz.

We observed that the particles tend to stick to the electrodes when they come into contact with them. In order to avoid this effect, we tested periodically switching the frequency on and off with a frequency between 1 Hz and 5 Hz. The results show that frequencies above 5 Hz reduce the focusing efficiency; therefore, a frequency of 3 Hz was selected as the optimal.

The DC chip showed equally good alignment results. Figure 10a shows the fluorescence image superimposed on the optical image for 84 nm polymer particles that are clearly focused in the middle of the channel with a voltage of 2 V applied at the middle electrode and a flow rate of 50 µL/h. After prolonged exposure (hours) to the DC voltage, the center gold electrode showed a significant amount of degradation and/or dirt accumulation that could not readily be removed (Figure 10b). In addition, 47 nm particles were aligned under the same conditions.

Figure 10. (**a**) Eighty-four-nanometer PS particles focused towards the center of the channel and directed towards the center outlet. Flow speed 50 µL/h at 2 V. (**b**) After prolonged (hours) exposure to 2 V potential, the center gold electrodes showed a significant amount of degradation/dirt accumulation.

3.4. *Upconcentration Evaluation*

Using ImageJ, we analyzed the velocity of up to 20 particles per outlet in order to determine the flow distribution in the three outlets. We expect that each of the side outlets should carry 45% of the flow, while the central outlet should only carry 10% of the flow. Figure 11 shows the measured velocities, calculated as described in Section 2.4. The error bars in Figure 11b are calculated based on the calculated standard deviations from the data in Figure 11a.

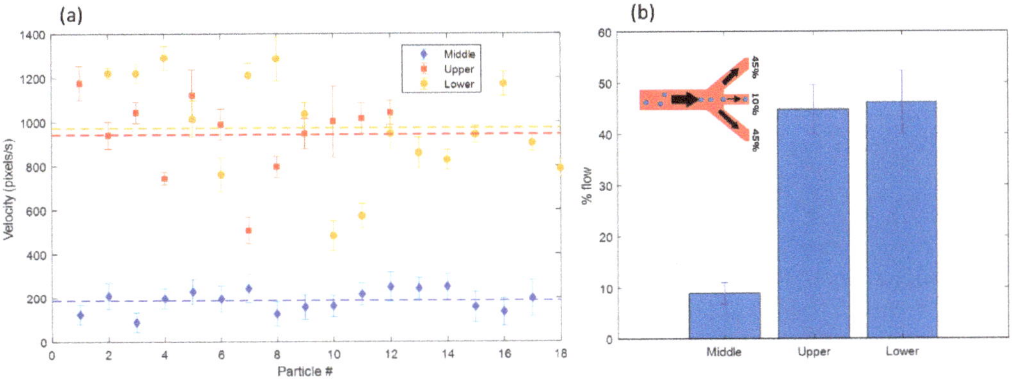

Figure 11. (**a**) Velocity of particles in the three outlets as calculated by ImageJ. The dotted lines indicate the average values for each outlet. (**b**) Percentage of the total flow entering each of the three outlets. The inlet shows the theoretically expected distribution.

The results show that the middle outlet carries 8.96% of the flow, while the two side channels carry 44.84% and 46.2%, respectively.

4. Discussion

Based on the optical measurements of the fabricated chips, the fabrication process can accurately reproduce the desired dimensions. The minimum distance between face-to-face electrodes was on average 30 nm larger than designed, while the minimum distance between zig-zag electrodes was on average 210 nm smaller than designed, something that only improves the chip performance, as the electric field is larger than expected for a given voltage. The DC electrode width is 240 nm larger than designed, meaning that the spacing between electrodes is smaller than expected, again resulting in higher electric fields. A certain deviation from the CAD designs is expected due to the lithography process, as process parameters such as contact mode, exposure time, resist height, and development time have an effect on the result, usually around 100–300 nm. The lithography mask itself is produced with a pixel size of 0.2 μm [18], which means that it is reasonable to expect deviations of this magnitude after fabrication.

Due to limitations in computational power, we have simulated the particle trajectories for a channel length of 500 μm. The simulations show that the zig-zag pattern is the most effective, which is also expected, as there are more regions of a high electric field in this design compared to the face-to-face electrode design. The simulations do not take Brownian motion into consideration, something that is hard to include in COMSOL in 3D designs for a meaningful amount of particles. Instead, an algorithm was developed that performs particle tracing, with certain limitations, based on COMSOL data in a unit cell of the geometry.

The graphs (Figures 7 and 8) show that even at a velocity of 25 μL/h, which is significantly higher than the originally calculated velocity of 1 μL/h, a channel length of no more than 30 mm is needed to catch the particles. Our prototype design has a channel length of 4–6 cm. The figure also shows that the end position along the channel width is independent of the flow velocity, which is expected considering the symmetric design of the electrodes across the channel length.

Our experiments were performed with 47 nm particles and show that focusing is indeed achieved well within the channel length of 4–6 mm with a flow rate of 2 μL/h. The DC chip is equally good at focusing the 47 nm particles even at higher flow rates and requires a significantly lower voltage. However, DC voltages come with a variety of issues. Particles need to be charged in order for the method to work, which is only the case at specific sample conditions (pH). Moreover, Joule heating is quite significant and can result in unwanted secondary flows that may disturb the focusing. Visual inspection of the electrodes after some hours of use shows that the electrodes are significantly degraded, possibly due to hydrolysis effects. Indeed, using voltages over 2 V immediately results in the development of bubbles, after which the electrodes show the same signs of degradation. This is not observed for the AC voltage electrodes, which is why these are preferred, even though they require a larger applied voltage. Although it is unlikely, we cannot completely exclude the possibility of electrolytic corrosion of the Ti/Au microelectrode. This happens at around the same voltages but usually at acidic or alkaline pH [19].

Apart from the DEP focusing forces, other phenomena can occur that will disturb, or maybe help, the focusing of the sub-nanometer particles. Electroosmotic fluid flow has been shown to be the dominant force in the 10 Hz–100 kHz frequency range [20], which means that it may not play a role in the presented system. Although it is not easy to predict the movement of a particle in our complicated electrode geometry, in general, electroosmosis tends to pull the particles away from the electrode edges and towards the surface of the electrodes, at least in a parallel plate electrode geometry. We would therefore expect the band of particles to broaden around the middle of the channel, more than the distance between the electrodes. As we collect particles in the middle 4 μm of the channel in the particle outlet, this effect, if at play, would not be an issue. Similarly, temperature gradients, and consequently also electrothermal flow, can occur due to the high voltage dissipated in the system, particularly if the conductivity of the fluid is high (~0.1 S/m) [20]. Again, order of magnitude calculations indicate that these forces are not significant, except when the particles are less than 1 μm from the electrodes in a highly conducting liquid. Considering the high degree of focusing that we observe in our experiments, it is likely that the used frequency of 200 kHz does not give rise to significant secondary effects.

The chip design also works as expected in terms of upconcentration. The center outlet (particle outlet) carries approximately 9% of the flow, while the two outer outlets on average carry 45.5% each. Considering the difficulties in ensuring equal pressure difference for the two outer outlet channels, the small discrepancy observed is insignificant, and in any case well within the statistical error. We note, however, that the next device iterations have a single outlet connecting the two outer channels, which ensures that they present the same hydraulic resistance to the fluid.

The flow rate used for the upconcentration experiments was 2 μL/h up to 50 μL/h, which is very slow. However, the electrode design can easily be adapted for massive parallelization. Figure 12 shows a schematic of the parallelization concept, along with a prototype of the multiple electrodes and 128 channels. Each layer can therefore provide a flow rate of 256 μL/h up to 6.4 mL/h, and by vertically stacking 10 such chips, we achieve a total flow rate of 2.5 mL/h up to 64 mL/h on a 35 cm^2 footprint. Depending on the application, this can be further enhanced. However, we acknowledge that for applications requiring the processing of several liters per hour, other methods of upconcentration need to be applied beforehand.

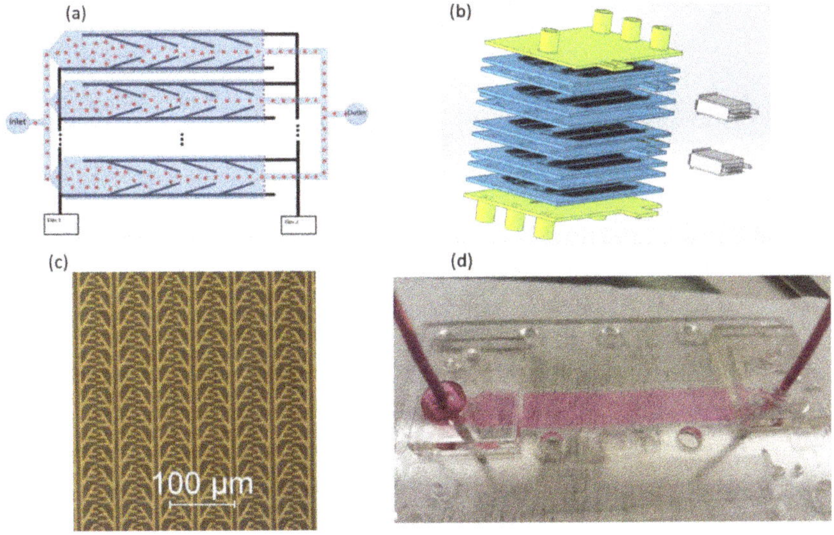

Figure 12. (**a**) A schematic of the parallelization concept. Only two electrodes provide voltage on all electrode pairs for an arbitrarily large number of parallel channels. All particle outlets are connected together in the chip plane, while all side outlets exit the chip from a common outlet in the bottom (not drawn). (**b**) A 3D drawing of the full chip, containing vertically stacked units, each containing 128 parallel channels. The dimensions of each chip are 65 mm × 48 mm. (**c**) Multiple electrode pairs fabricated by photolithography as described above. In this image, 6 parallel channels are shown. (**d**) A flow test of the microfluidic part of the device. For testing purposes, every 32 parallel channels were grouped together with one common inlet and outlet.

5. Conclusions

In conclusion, we have presented a continuous flow upconcentration system for nanoparticles using dielectrophoresis. The prototype was able to upconcentrate particles with sizes as low as 47 nm at a flow rate of 2 µL/h. We showed how the electrode design was optimized for achieving a maximum force on the particles and how the design is scalable and can be massively parallelized in order to achieve the much higher flow rates required for most applications.

Author Contributions: Conceptualization, N.R. and W.E.S.; methodology, M.D. and M.H.O.; software, M.D.; validation, M.D. and M.H.O.; formal analysis, M.D., M.H.O., N.R. and W.E.S.; investigation, M.D. and M.H.O.; resources, N.R. and W.E.S.; data curation, M.D.; writing—original draft preparation, M.D.; writing—review and editing, M.D., W.E.S., N.R. and M.H.O.; visualization, M.D. and M.H.O.; supervision, N.R. and W.E.S.; project administration, N.R.; funding acquisition, N.R. and W.E.S. All authors have read and agreed to the published version of the manuscript.

Funding: This research was funded by the EU FP7 project "AquaVir", grant agreement No. 604069.

Data Availability Statement: The data presented in this study are available on request from the corresponding author.

Acknowledgments: The authors thank Michael Døssing from Eir Diagnostics for the technical design of the stacked chip.

Conflicts of Interest: The authors declare no conflict of interest. The funders had no role in the design of the study; in the collection, analyses, or interpretation of data; in the writing of the manuscript; or in the decision to publish the results.

References

1. Sauter, E.R. Exosomes in blood and cancer. *Transl. Cancer Res.* **2017**, *6* (Suppl. S8), S1316–S1320. [CrossRef]
2. Akpe, V.; Kim, T.H.; Brown, C.L.; Cock, I.E. Circulating tumour cells: A broad perspective. *J. R. Soc. Interface* **2020**, *17*, 20200065. [CrossRef]
3. Zhou, Y.; Ma, Z.; Tayebi, M.; Ai, Y. Submicron Particle Focusing and Exosome Sorting by Wavy Microchannel Structures within Viscoelastic Fluids. *Anal. Chem.* **2019**, *91*, 4577–4584. [CrossRef] [PubMed]
4. Mashhadian, A.; Shamloo, A. Inertial microfluidics: A method for fast prediction of focusing pattern of particles in the cross section of the channel. *Anal. Chim. Acta* **2019**, *1083*, 137–149. [CrossRef] [PubMed]
5. Salafi, T.; Zhang, Y.; Zhang, Y. A Review on Deterministic Lateral Displacement for Particle Separation and Detection. *Nano-Micro Lett.* **2019**, *11*, 77. [CrossRef] [PubMed]
6. Salafi, T.; Zeming, K.K.; Zhang, Y. Advancements in microfluidics for nanoparticle separation. *Lab Chip* **2017**, *17*, 11–33. [CrossRef] [PubMed]
7. Benhal, P.; Quashie, D.; Kim, Y.; Ali, J. Insulator Based Dielectrophoresis: Micro, Nano, and Molecular Scale Biological Applications. *Sensors* **2020**, *20*, 5095. [CrossRef] [PubMed]
8. Pesch, G.R.; Du, F. A review of dielectrophoretic separation and classification of non-biological particles. *Electrophoresis* **2021**, *42*, 134–152. [CrossRef] [PubMed]
9. Shim, S.; Stemke-Hale, K.; Noshari, J.; Becker, F.F.; Gascoyne, P.R.C. Dielectrophoresis has broad applicability to marker-free isolation of tumor cells from blood by microfluidic systems. *Biomicrofluidics* **2013**, *7*, 011808. [CrossRef] [PubMed]
10. Waheed, W.; Sharaf, O.Z.; Alazzam, A.; Abu-Nada, E. Dielectrophoresis-field flow fractionation for separation of particles: A critical review. *J. Chromatogr. A* **2021**, *1637*, 461799. [CrossRef]
11. Demierre, N.; Braschler, T.; Linderholm, P.; Seger, U.; van Lintel, H.; Renaud, P. Characterization and optimization of liquid electrodes for lateral dielectrophoresis. *Lab Chip* **2007**, *7*, 355–365. [CrossRef] [PubMed]
12. Zhao, K.; Peng, R.; Li, D. Separation of nanoparticles by a nano-orifice based DC-dielectrophoresis method in a pressure-driven flow. *Nanoscale* **2016**, *8*, 18945–18955. [CrossRef] [PubMed]
13. Miura, D.; Taguchi, Y.; Nagasaka, Y. Nanoparticle Sorting Technique using Laser Induced Dielectrophoresis with Phase Modulated Interference. In Proceedings of the 2019 International Conference on Optical MEMS and Nanophotonics (OMN), Daejeon, Korea, 28 July–1 August 2019; pp. 100–101.
14. Shi, L.; Rana, A.; Esfandiari, L. A low voltage nanopipette dielectrophoretic device for rapid entrapment of nanoparticles and exosomes extracted from plasma of healthy donors. *Sci. Rep.* **2018**, *8*, 6751. [CrossRef] [PubMed]
15. Morgan, H.; Green, N.G. *AC Electrokinetics: Colloids and Nanoparticles*; Research Studies Press Ltd.: Baldock, UK, 2003.
16. Ermolina, I.; Morgan, H. The electrokinetic properties of latex particles: Comparison of electrophoresis and dielectrophoresis. *J. Colloid Interface Sci.* **2005**, *285*, 419–428. [CrossRef]
17. Dimaki, M.; Bøggild, P. Dielectrophoresis of carbon nanotubes using microelectrodes: A numerical study. *Nanotechnology* **2004**, *15*, 1095–1102. [CrossRef]
18. Delta Mask. Available online: https://www.deltamask.nl/frames.html (accessed on 10 May 2022).
19. Cherevko, S.; Zeradjanin, A.R.; Keeley, G.P.; Mayrhofer, K.J.J. A Comparative Study on Gold and Platinum Dissolution in Acidic and Alkaline Media. *J. Electrochem. Soc.* **2014**, *161*, H822. [CrossRef]
20. Salari, A.; Navi, M.; Lijnse, T.; Dalton, C. AC Electrothermal Effect in Microfluidics: A Review. *Micromachines* **2019**, *10*, 762. [CrossRef] [PubMed]

micromachines

MDPI

Article

Dielectrophoresis-Based Positioning of Carbon Nanotubes for Wafer-Scale Fabrication of Carbon Nanotube Devices

Joevonte Kimbrough [1], Lauren Williams [1], Qunying Yuan [2] and Zhigang Xiao [1,*]

[1] Department of Electrical Engineering and Computer Science, Alabama A&M University, Normal, AL 35762, USA; jkimbro3@bulldogs.aamu.edu (J.K.); lwilli81@bulldogs.aamu.edu (L.W.)

[2] Department of Biological and Environmental Science, Alabama A&M University, Normal, AL 35762, USA; qunying.yuan@aamu.edu

* Correspondence: zhigang.xiao@aamu.edu; Tel.: +1-256-372-5679; Fax: +1-256-372-5855

Abstract: In this paper, we report the wafer-scale fabrication of carbon nanotube field-effect transistors (CNTFETs) with the dielectrophoresis (DEP) method. Semiconducting carbon nanotubes (CNTs) were positioned as the active channel material in the fabrication of carbon nanotube field-effect transistors (CNTFETs) with dielectrophoresis (DEP). The drain-source current (I_{DS}) was measured as a function of the drain-source voltage (V_{DS}) and gate-source voltage (V_{GS}) from each CNTFET on the fabricated wafer. The I_{DS} on/off ratio was derived for each CNTFET. It was found that 87% of the fabricated CNTFETs was functional, and that among the functional CNTFETs, 30% of the CNTFETs had an I_{DS} on/off ratio larger than 20 while 70% of the CNTFETs had an I_{DS} on/off ratio lower than 20. The highest I_{DS} on/off ratio was about 490. The DEP-based positioning of carbon nanotubes is simple and effective, and the DEP-based device fabrication steps are compatible with Si technology processes and could lead to the wafer-scale fabrication of CNT electronic devices.

Keywords: dielectrophoresis (DEP); semiconducting carbon nanotubes; carbon nanotube field-effect transistors (CNTFETs)

Citation: Kimbrough, J.; Williams, L.; Yuan, Q.; Xiao, Z. Dielectrophoresis-Based Positioning of Carbon Nanotubes for Wafer-Scale Fabrication of Carbon Nanotube Devices. *Micromachines* **2021**, *12*, 12. https://dx.doi.org/10.3390/mi12010012

Received: 4 December 2020
Accepted: 23 December 2020
Published: 25 December 2020

1. Introduction

Carbon nanotube-based electronic devices such as carbon nanotube field-effect transistors (CNTFETs) and electronic circuits have been investigated extensively in the past years for the application of future nanoelectronic devices [1–15]. Since carbon nanotubes (CNTs) have a high aspect (ratio of the length to diameter) with a diameter of several nanometers and a length of several micrometers, the electrons or holes in the tubes are subject to a strong quantum-confinement effect in all directions perpendicular to the tube, making their transport in the tube a ballistic transport with less scattering and collisions [16,17]. These properties are not available in bulk electronic materials and can make carbon nanotube field-effect transistors (CNTFETs) operate at a high speed and low power without energy dissipating in the tube [18].

The wafer-scale fabrication of carbon nanotube electronic circuits can lead to the production and application of carbon nanotube electronic devices and has recently been reported by researchers [19–21]. In this paper, we report the wafer-scale fabrication of carbon nanotube field-effect transistors (CNTFETs) using dielectrophoresis (DEP), which is simple and unique compared to other methods [22,23]. Carbon nanotubes were positioned and assembled as the active channel material in the wafer-scale fabrication of carbon nanotube field-effect transistors (CNTFETs) using electric field-directed dielectrophoresis (DEP). Dielectrophoresis (DEP) is a translational motion of neutral matter caused by the polarization effects in a nonuniform electric field and has recently been investigated theoretically and experimentally for the deposition and alignment of CNTs [24–29]. It is also effective in assembling other materials such as liquid metal [30]. Aligned and dense carbon nanotubes (CNTs) can be obtained in the DEP process by optimizing the ac frequency, the trapping time, and the CNT solution concentration [31–33]. Impurities such as the metal impurity in CNTs, which come from the growth of CNTs due to the metal catalyst, could electrically short CNTFETs and cannot be tolerated in the fabrication of CNTFETs [34,35]. However, the impurities can be filtered in the pre-DEP process with solvents such as N-methyl pyrrolidone (NMP), making the DEP-based fabrication of CNTFETs appealing for obtaining high-performance carbon nanotube field-effect transistors and electronic devices. Since the DEP-based device fabrication steps are compatible with Si technology processes, they are capable of being further optimized by following process development protocols practiced by the semiconductor industry and can lead to the wafer-scale fabrication of CNT electronic devices and sensors.

2. Experimental Details

2.1. CNT-NMP Solution Preparation

Carbon nanotube (CNT) powder with 98% semiconducting CNTs (from the NanoIntergris Company) was used in this research. The CNT powder was dispersed ultrasonically in the Nmethyl-2-pyrrolidone (NMP) solution (from the Sigma-Aldrich Company) for the dielectrophoresis (DEP) process. The CNTs were subjected to a cleaning process for filtering any possible impurity in the CNTs before being used for the DEP process. 1 mg CNT powder was added to 20 ml NMP and sonicated for 10 min. The CNT solution was then centrifuged at 14,000 rpm for 30 min. The resultant supernatant was decanted for the sedimented carbon nanotubes, which were again subjected to the cleaning cycle with a fresh NMP. The cleaning process was repeated three times. Finally, the sedimented carbon nanotubes were added into a 30 ml NMP solution; the solution was sonicated for 5 min. After the sonication, the carbon nanotubes were separated and uniformly dispersed in the NMP solution, and the CNT-NMP solution was ready for the dielectrophoresis process in the following device fabrication.

2.2. FET Device Fabrication

A 3-inch diameter 350-μm thick n-type <100> silicon (Si) wafer (from Virginia Semiconductor, Inc.) was used as the substrate in this research. The Si wafer was about 350 μm thick and had a resistivity of about 3 Ω cm. The wafer was initially oxidized at 1100 °C for 120 min to grow a silicon dioxide (SiO_2) layer of about 1 μm thickness using wet oxidation. A PVD 75 (Kurt J. Lesker Company, Pittsburgh, PA, USA) e-beam/thermal evaporation system was used to grow chromium (Cr)/gold (Au) thin films for the fabrication of Cr/Au electrodes and device metallization. The process chamber had a background pressure of 2×10^{-7} Torr, and the film thickness was controlled by an INFICON deposition monitor. An ALD-150LX system (Kurt J. Lesker Company, Pittsburgh, USA) was used to grow the hafnium dioxide (HfO_2) thin film as the gate oxide in the fabrication of CNTFETs [36,37]. A set of four-piece photo masks was designed and used for the fabrication of the carbon nanotube field-effect transistors (CNTFETs). The photo masks were fabricated by Photoscience, Inc. Figure 1a–c show the fabrication steps of CNTFETs. The first mask was used to define and pattern 5-nm thick Cr/100-nm thick Au (99.99% pure from Alfa Aesar) source and drain electrodes with gaps of about 3 μm using standard ultraviolet (UV) photolithography and metal lift-off processes (Figure 1a). The thin Cr layer was used to enhance the adhesion of Au to the underlayer. The device wafer was designed to have 30 CNTFETs on it. After the first mask, 30 pairs of electrodes were fabricated on the wafer (Figure 2), and all the electrodes were connected in parallel to two major pads, which were applied with the ac voltages during the subsequent DEP process.

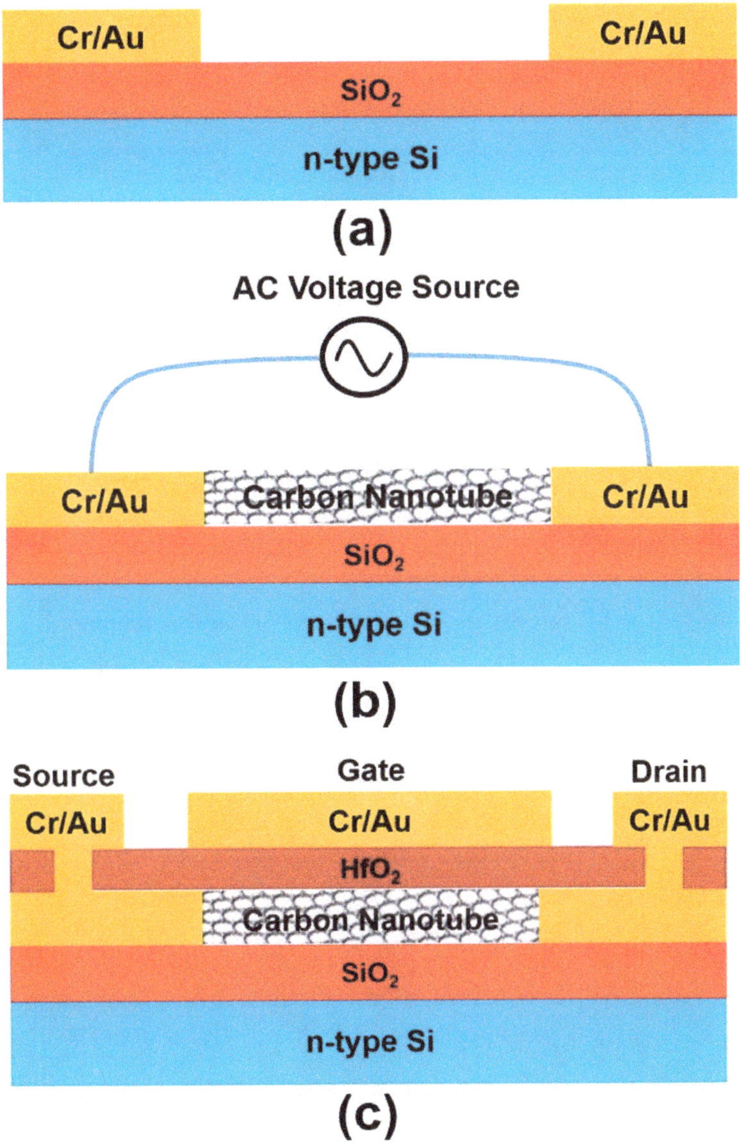

Figure 1. Carbon nanotube field-effect transistor (CNTFET) device fabrication steps. (**a**) Cr/Au electrode definition. (**b**) Alignment and assembling of CNTs with the DEP process under an applied sinusoidal voltage source between source and drain electrodes; the subsequent steps of the etching of Cr/Au films and the definition of HfO_2 gate oxide are not depicted in this figure. (**c**) Source, drain, and gate metal electrode definitions.

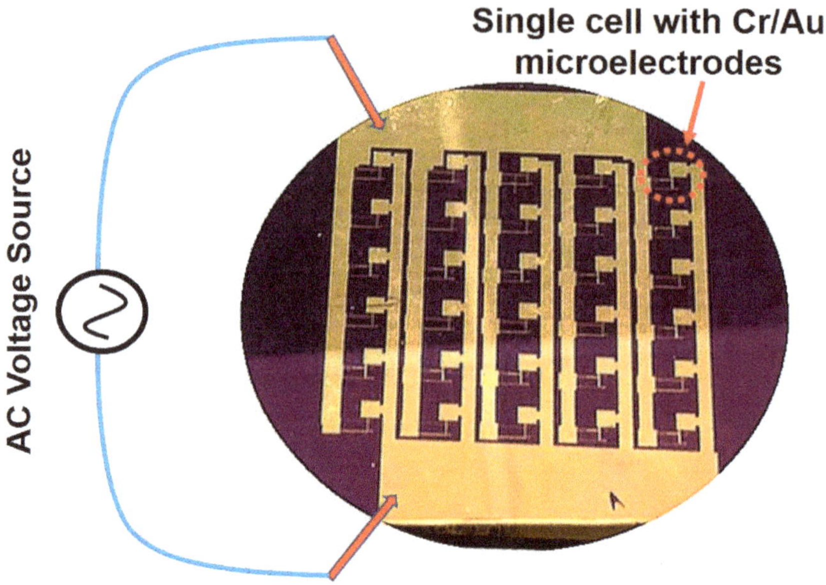

Figure 2. The wafer with 30 pairs of Cr/Au electrodes fabricated on it, where all the electrodes were connected in parallel to two major pads to which the ac voltages were applied during the subsequent DEP process.

After the fabrication of the Cr/Au electrodes on the wafer, the dielectrophoresis (DEP) process was then performed to deposit and align CNTs across the gap of electrodes (Figure 1b). A sinusoidal voltage source was connected to the two major pads of the wafer (Figure 2), and the wafer was then dipped into a glass beaker filled with the CNT–NMP solution prepared as described above. A sinusoidal voltage with a peak-to-peak voltage of 10 V and frequency of 1 MHz was applied to the electrodes for 10 min. After that, the wafer was taken out from the beaker. The solvent was blown off the wafer surface with nitrogen gas. The wafer surface was then cleaned with 2-Propanol and rinsed in deionized (DI) water for 1 min. After that, the wafer was dried with nitrogen gas. The CNTs were imaged in a JEOL JSM-6610LV scanning electron microscope (JEOL Ltd, Akishima, Japan). Figure 3a shows the scanning electron micrograph (SEM) image of CNTs positioned between a pair of electrodes in a CNT-NMP solution with a higher concentration of CNTs using the DEP process, while Figure 3b is the enlarged view of the CNTs across the gap of electrodes, showing a network-like profile of CNTs. Figure 4a,b show the SEM images of CNTs positioned between the electrodes in a CNT-NMP solution with a lower concentration of CNTs using the DEP process for two pairs of electrodes. As shown in Figure 4a,b, although they were positioned with the same DEP process conditions, the CNTs in Figure 4a,b had varying profiles because of the random and uncontrollable property of CNTs dispersed in the CNT-NMP solution. Figure 4 also shows that the positioned CNTs are not single-tube and are bundling tubes (a single tube should have a diameter of several nanometers). The CNT-NMP solution with a lower concentration of CNTs was used for the final device fabrication.

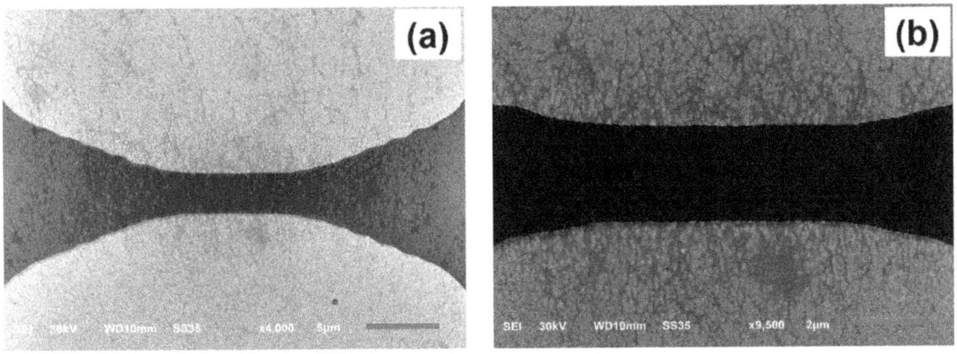

Figure 3. (**a**) SEM image of CNTs aligned and assembled between a pair of electrodes in a CNT-NMP solution with a higher concentration of CNTs using the DEP process; (**b**) enlarged view of the CNTs, showing a network-like profile of CNTs. The scale bar is 5 μm.

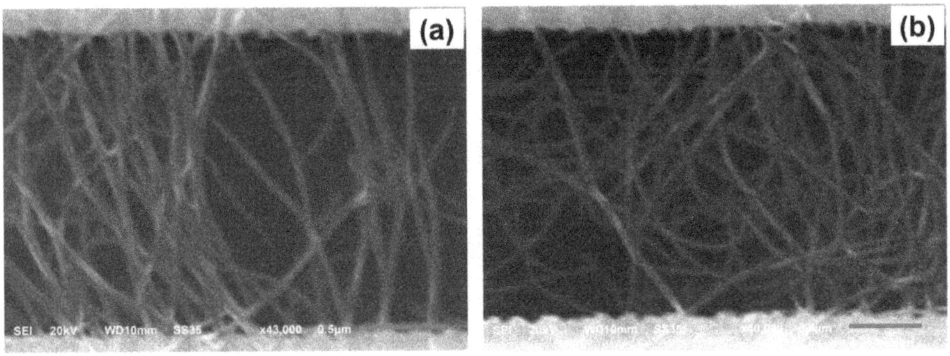

Figure 4. SEM images of CNTs aligned and assembled between the electrodes in two pairs of electrodes in a CNT-NMP solution with a lower concentration of CNTs using the DEP process, where both (**a**) and (**b**) show profiles of CNTs with some alignment of CNTs across the electrodes. The scale bar is 0.5 μm.

The second mask was applied to create patterns for etching the Cr/Au metal film that connected with the pairs of electrodes using the UV lithography and to separate the electrodes into individual pairs of electrodes. After the UV lithography process, each pair of electrodes, together with the CNTs between the gap, was covered by Shipley 1818 positive photoresist, but all other areas on the wafer were open without being covered by the photoresist. The wafer was dipped in a gold etchant (Transene, Danvers, MA, USA) for 30 s, then in DI water for 1 min, and in a chromium etchant (Transene, Danvers, MA, USA) for 30 s to remove all the Cr/Au except that protected by the photoresist. The wafer was finally rinsed in DI water for 1 min and was dried with nitrogen gas. After that, a 10-nm thick HfO_2 film was deposited over the whole wafer as the gate oxide using plasma-enhanced atomic layer deposition.

The third mask was applied to open a window through the HfO_2 film on the source/drain for the final source/drain metal contact using the UV lithography. After the UV lithography process, the wafer was covered by Shipley 1818 photoresist, except for a 30 μm × 30 μm small area on the source/drain. The wafer was then dipped in buffered oxide etchant (BOE) (Transene, Danvers, MA, USA) for 3 min to etch the HfO_2 through the patterned windows on the source/drain for the contact.

Finally, the fourth photoresist mask aligned to the opened windows was applied to define the source, drain, and gate metal contacts (100-nm thick Au/5-nm thick Cr) via e-beam evaporation and metal lift-off. Again, Cr was used to enhance the adhesion of Au to the underneath layer (HfO_2). Figure 1c shows the cross-sectional diagram of the CNTFET device, and Figure 5 shows the scanning electron micrograph of the top view of the fabricated CNTFET.

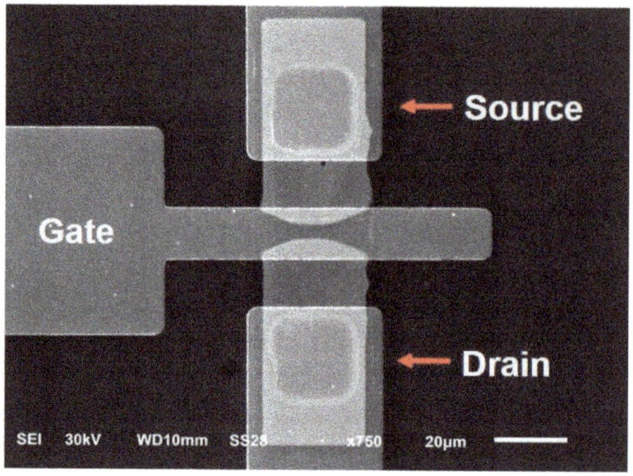

Figure 5. SEM image of a fabricated CNTFET.

2.3. *Electrical Measurements*

Electrical contacts to the substrate, source, drain and gate electrodes were performed at room temperature using a probe station (CM-170 from Signatone, Gilroy, USA) equipped with four triax micromanipulated probes. The wafer substrate remained electrically unconnected (floating) during the measurements. Electrical signals for the device characterization were sourced and measured using a precision semiconductor parameter analyzer (4156C from Agilent, Santa Clara, CA, USA).

2.4. *Electrical Breakdown Process*

To increase the I_{DS} on/off ratios with the electrical breakdown method [38,39], CNTFETs were fabricated using the same processes. The IV curves and transfer characteristics of the CNTFETs were first measured as fabricated; after that, the devices were subjected to an electrical breakdown process, which consisted of sweeping V_{DS} from -8 V to 8 V at different V_{GS} values that varied from -8 V to 8 V in steps of 0.5 V until the I_{DS} values decreased to values that were about five to 10 times smaller. Then, the IV curves and transfer characteristics of the CNTFETs were measured again.

3. Results and Discussion

After completion of the device fabrication, each carbon nanotube field-effect transistor (CNTFET) fabricated on the wafer was probed and measured one by one as described above. It was found that 26 (87%) of the 30 CNTFETs on the wafer were functional, while the other four were broken. Among the 26 functional CNTFETs, 30% of the CNTFETs had a drain-source current (I_{DS}) on/off ratio larger than 20 while 70% of the CNTFETs had a drain-source current (I_{DS}) on/off ratio under 20. The highest drain-source current (I_{DS}) on/off ratio was about 490. The reason why the electrical properties such as the I_{DS} on/off ratio varied in the CNTFETs was because the profile and number of CNTs positioned between the electrodes varied in the CNTFETs, as shown in Figure 4a,b. Both the profile and number of aligned CNTs can have effects on the electrical property of CNTFETs. The CNTFET with more CNTs had a lower channel resistance and therefore a higher I_{DS} value. The carbon nanotubes used in this research had about 98% semiconducting carbon tubes and about 2% metallic carbon tubes. The metallic tubes in the CNTs could be the major factor responsible for the variation of the I_{DS} on/off ratio in the CNTFETs. The metallic carbon nanotube has a zero band gap and functions like metal, and the resistance cannot be modulated with the applied gate voltage to CNTFETs. The semiconducting carbon nanotube contributes to the increase of the I_{DS} on/off ratio in a CNTFET, while the metallic carbon nanotube functions in an opposite way [40,41]. A higher number of metallic CNTs in the channel of CNTFET makes the modulation of channel resistance more difficult, resulting in a lower I_{DS} on/off ratio. Since the metallic carbon nanotubes could not be filtered in the DEP process, they were randomly assembled into the channel of CNTFET, resulting in the variation of the I_{DS} on/off ratio in CNTFETs.

Figure 6a–f show the drain-source current (I_{DS}) versus the drain-source voltage (V_{DS}) at various V_{GS} values (Figure 6a,c,e) and the I_{DS}-V_{GS} transfer curves of the CNTFET at various V_{DS} values (Figure 6b,d,f) for three functional devices, respectively. The I_{DS} on/off ratio, defined as the maximum I_{DS} value divided by the lowest one in the I_{DS}-V_{GS} transfer curve, is 8 for the device in Figure 6b, 27 for the device in Figure 6d, and 10 for the device in Figure 6f. All the devices present electrical properties of a p-channel field-effect transistor (FET), which are like those of the p-type carbon nanotube transistors reported by other researchers [42,43]. The I_{DS} is sensitive to the V_{GS} and can be modulated by V_{GS}, as shown in the figure. When increasing V_{GS} negatively, the I_{DS} values increase correspondingly. P-type transistors conduct holes when a negative voltage is applied to the gate. They do not conduct electrons, even at high positive gate voltages. The physical explanation for this is that the Fermi level at the contact metal-nanotube junction is closer to the valence band of the nanotube, leading to hole conduction and p-type behavior. The holes see a small barrier and can thus tunnel through, whereas the electrons see a much larger barrier and are not able to tunnel [44].

Figure 7a shows the drain-source current (I_{DS}) versus the drain-source voltage (V_{DS}) at various V_{GS} values for the functional CNTFET that had the highest I_{DS} on/off ratio among all the functional devices on the fabricated wafer, and it presents better IV curves, with the saturation like that of a silicon-based p-channel field-effect transistor [45]. Figure 7b shows the I_{DS}-V_{GS} transfer curves of the CNTFET at V_{DS} = 4 V, 3 V, and 2 V. The I_{DS} on/off ratio is 61 at V_{DS} = 4 V, 190 at V_{DS} = 3 V, and 490 at V_{DS} = 2 V. As described above, the reason why the CNTFET had a higher I_{DS} on/off ratio is possibly because less or no metallic tubes were assembled in the channel during the DEP process. The I_{DS}-V_{DS} curves in Figure 7a are comparable with those reported by other researchers [46,47].

Figure 8a–d shows the comparison of the electrical characteristics of a fabricated CNTFET, which were measured as fabricated or after being subjected to the electrical breakdown process. Figure 8a,b shows the electrical characteristics of the CNTFET, which were measured as fabricated. Figure 8a shows the drain-source current (I_{DS}) versus drain-source voltage (V_{DS}) and gate voltage (V_{GS}), while Figure 8b shows the transfer characteristics at V_{DS} = 5 V and 0.5 V, where the I_{DS} on/off ratio is 4 at V_{DS} = 5 V and 6 at V_{DS} = 0.5 V. Figure 8c,d shows the electrical characteristics of the CNTFET, which were measured after the CNTFET was subjected to the electrical breakdown process described in Section 2.4, where the I_{DS} on/off ratio is about 339 at V_{DS} = 5 V and 488 at V_{DS} = 0.5 V. Although it decreased the I_D values, the electrical breakdown process significantly increased the I_{DS} on/off ratios. This post-device fabrication process has been used to increase the I_{DS} on/off ratio of CNTFETs and decrease the leakage current [38,39]. In the CNTFET channel with the intermixed presence of metallic and semiconducting CNTs, the electrical breakdown process can selectively destroy, by Joule heating, the metallic CNTs responsible for the unwanted electrical characteristics while preserving semiconducting CNTs.

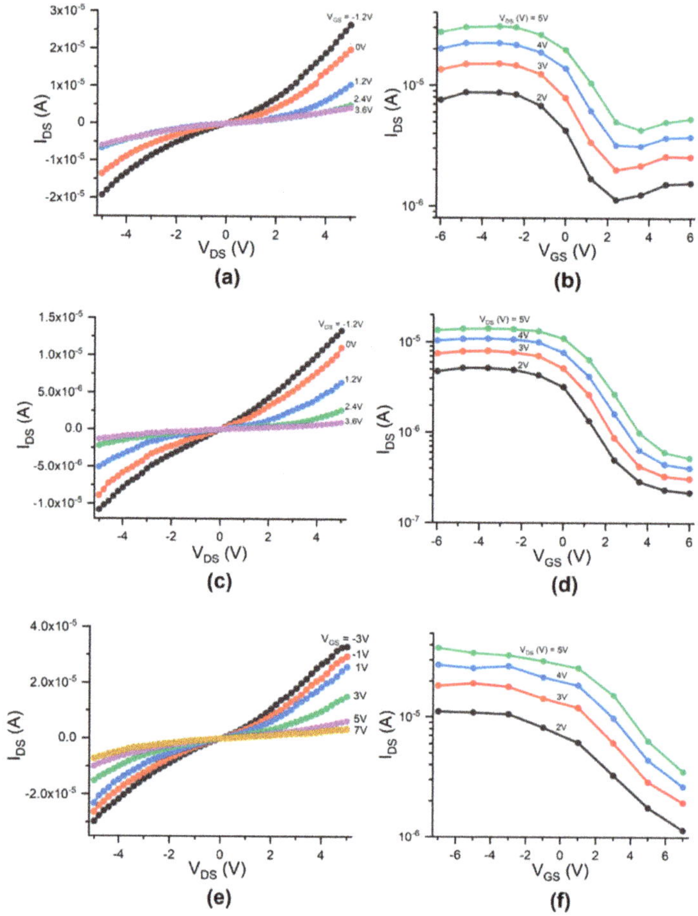

Figure 6. Electrical measurement results for three functional CNTFET devices: (**a**) Drain-source current (I_{DS}) versus drain-source voltage (V_{DS}) and gate voltage (V_{GS}) and (**b**) transfer characteristics at V_{DS} = +5 V, +4 V, +3 V, and +2 V for device one, where the I_{DS} on/off ratio is 8; (**c**) Drain-source current (I_{DS}) versus drain-source voltage (V_{DS}) and gate voltage (V_{GS}) and (**d**) transfer characteristics for device two, where the I_{DS} on/off ratio is 27; (**e**) Drain-source current (I_{DS}) versus drain-source voltage (V_{DS}) and gate voltage (V_{GS}) and (**f**) transfer characteristics for device three, where the I_{DS} on/off ratio is 10.

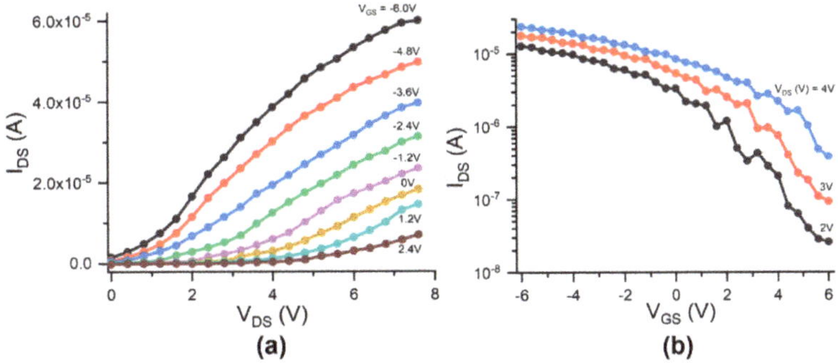

Figure 7. (**a**) Drain-source current (I_{DS}) versus drain-source voltage (V_{DS}) and gate voltage (V_{GS}) and (**b**) transfer characteristics at V_{DS} = 5 V, 4 V, 3 V, and 2 V for a functional CNTFET, respectively; the I_{DS} on/off ratio is 490.

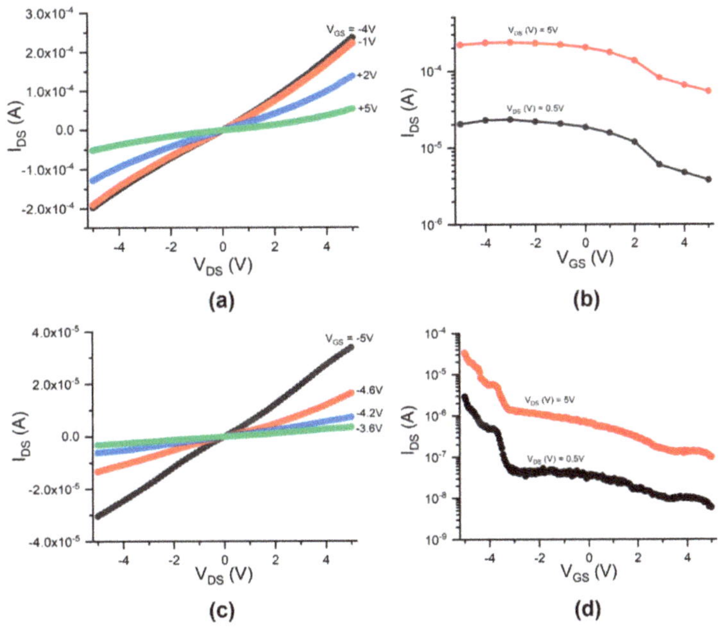

Figure 8. (**a**) Drain-source current (I_{DS}) versus drain-source voltage (V_{DS}) and gate voltage (V_{GS}) and (**b**) transfer characteristics at V_{DS} = 5 V and 0.5 V for a functional CNTFET, respectively, where the I_{DS} on/off ratio is 4 at V_{DS} = 5 V and 6 at V_{DS} = 0.5 V; (**c**,**d**) Similar curves as in (**a**,**b**) but measured after the CNTFET was subjected to the electrical breakdown process, where the I_{DS} on/off ratio is 339 at V_{DS} = 5 V and 488 at V_{DS} = 0.5 V.

4. Summary

We present the wafer-scale fabrication of carbon nanotube field-effect transistors (CNTFETs) with carbon nanotubes as the active channel material by using the dielectrophoresis (DEP) process. A device wafer with 30 CNTFETs was fabricated and characterized successfully, and it was found that 87% of the fabricated CNTFETs was functional and that, among the functional CNTFETs, 30% of the CNTFETs had an I_{DS} on/off ratio larger than 20 while 70% of the CNTFETs had an I_{DS} on/off ratio under 20. The highest I_{DS} on/off ratio was 490. The variation of I_{DS} on/off ratios in the CNTFETs is mainly due to the effect of the metallic carbon nanotubes in the channel. If the source carbon nanotubes are pure semiconducting tubes or have fewer metallic tubes, the I_{DS} on/off ratios should be much higher and the variation of the ratios should be much smaller. The electrical breakdown was used to improve the I_D on/off ratios of the CNTFET, and it was found that it significantly increased the I_{DS} on/off ratios of the fabricated CNTFET. The DEP-based positioning of carbon nanotubes (CNTs) in the fabrication of CNTFETs is simple and effective. The DEP-based method can be easily extended for the assembly of other nanomaterials in the development of the wafer-scale fabrication of devices. The DEP-based device fabrication steps are compatible with Si technology processes, can be further optimized following process development protocols practiced by the semiconductor industry, and can lead to the wafer-scale fabrication of CNT electronic devices and sensors.

Author Contributions: Z.X. designed the experiments, fabricated the materials and devices, measured the electrical properties of the devices, analyzed the data, and wrote the manuscript. J.K. performed the fabrication and characterization of the devices. L.W. performed the fabrication and characterization of the devices. Q.Y. performed the experiment in the preparation of the CNT solution and the dielectrophoresis process and contributed to writing the manuscript. All authors have read and agreed to the published version of the manuscript.

Funding: This research was supported by U.S. Department of Energy, Office of Basic Energy Sciences, under Contract No. DE-SC00112704; National Science Foundation under Grant No. CBET-1740687; and Department of Defense under Grant No. W911NF-16-1-0554 and W911NF-17-1-0474.

Institutional Review Board Statement: Not applicable.

Informed Consent Statement: Not applicable.

Data Availability Statement: The data presented in this study are available on request from the corresponding author.

Acknowledgments: Research carried out in part at the Center for Functional Nanomaterials, Brookhaven National Laboratory, which is supported by the U.S. Department of Energy, Office of Basic Energy Sciences, under Contract No. DE-SC00112704; the research is supported by National Science Foundation under Grant No. CBET-1740687 and Department of Defense under Grant No. W911NF-16-1-0554 and W911NF-17-1-0474.

Conflicts of Interest: The authors declared no conflicts of interest.

References

1. Lemieux, M.C.; Roberts, M.; Barman, S.; Jin, Y.W.; Kim, J.M.; Bao, Z. Self-Sorted, Aligned Nanotube Networks for Thin-Film Transistors. *Science* **2008**, *321*, 101–104. [CrossRef]
2. Javey, A.; Guo, J.; Farmer, D.B.; Wang, Q.; Wang, D.; Gordon, R.G.; Lundstrom, A.M.; Dai, H. Carbon Nanotube Field-Effect Transistors with Integrated Ohmic Contacts and High-κ Gate Dielectrics. *Nano Lett.* **2004**, *4*, 447–450. [CrossRef]
3. Seidel, R.V.; Graham, A.P.; Kretz, J.; Rajasekharan, B.; Duesberg, G.S.; Liebau, M.; Unger, E.; Kreupl, F.; Hoenlein, W. Sub-20 nm Short Channel Carbon Nanotube Transistors. *Nano Lett.* **2005**, *5*, 147–150. [CrossRef]
4. Wang, H.; Cobb, B.; Van Breemen, A.; Gelinck, G.G.; Bao, Z. Highly Stable Carbon Nanotube Top-Gate Transistors with Tunable Threshold Voltage. *Adv. Mater.* **2014**, *26*, 4588–4593. [CrossRef]
5. Qiu, C.; Zhang, Z.; Zhong, D.; Si, J.; Yang, Y.; Peng, L.-M. Carbon Nanotube Feedback-Gate Field-Effect Transistor: Suppressing Current Leakage and Increasing On/Off Ratio. *ACS Nano* **2015**, *9*, 969–977. [CrossRef] [PubMed]
6. Wanga, H.; Weib, P.; Lic, Y.; Hanb, J.; Leec, H.; Naabd, B.; Liub, N.; Wange, C.; Adijantob, E.; Teec, B.; et al. Tuning the threshold voltage of carbon nanotube transistors by n-type molecular doping for robust and flexible complementary circuits. *Proc. Natl. Acad. Sci. USA* **2014**, *111*, 4776. [CrossRef] [PubMed]
7. Sun, D.-M.; Timmermans, M.Y.; Tian, Y.; Nasibulin, A.G.; Kauppinen, E.I.; Kishimoto, S.; Mizutani, T.; Ohno, Y. Flexible high-performance carbon nanotube integrated circuits. *Nat. Nanotechnol.* **2011**, *6*, 156–161. [CrossRef] [PubMed]
8. Ding, L.; Zhang, Z.; Liang, S.; Pei, T.; Wang, S.; Li, Y.; Zhou, W.; Liu, J.; Peng, L.-M. CMOS-based carbon nanotube pass-transistor logic integrated circuits. *Nat. Commun.* **2012**, *3*, 677. [CrossRef]
9. Kim, B.; Jang, S.; Geier, M.L.; Prabhumirashi, P.L.; Hersam, M.C.; Dodabalapur, A. High-Speed, Inkjet-Printed Carbon Nanotube/Zinc Tin Oxide Hybrid Complementary Ring Oscillators. *Nano Lett.* **2014**, *14*, 3683–3687. [CrossRef]
10. Zaumseil, J. Single-walled carbon nanotube networks for flexible and printed electronics. *Semicond. Sci. Technol.* **2015**, *30*. [CrossRef]

11. Chen, H.; Cao, Y.; Zhang, J.; Zhou, C. Large-scale complementary macroelectronics using hybrid integration of carbon nanotubes and IGZO thin-film transistors. *Nat. Commun.* **2014**, *5*, 4097. [CrossRef] [PubMed]
12. Park, S.; Vosguerichian, M.; Bao, Z. A review of fabrication and applications of carbon nanotube film-based flexible electronics. *Nanoscale* **2013**, *5*, 1727. [CrossRef] [PubMed]
13. Zhang, Z.; Liang, X.; Wang, S.; Yao, K.; Hu, Y.; Zhu, Y.; Chen, Q.; Zhou, W.; Li, Y.; Yao, Y.; et al. Doping-Free Fabrication of Carbon Nanotube Based Ballistic CMOS Devices and Circuits. *Nano Lett.* **2007**, *7*, 3603–3607. [CrossRef]
14. Geier, M.L.; Prabhumirashi, P.L.; McMorrow, J.J.; Xu, W.; Seo, J.-W.T.; Everaerts, K.; Kim, C.H.; Marks, T.J.; Hersam, M.C. Subnanowatt Carbon Nanotube Complementary Logic Enabled by Threshold Voltage Control. *Nano Lett.* **2013**, *13*, 4810–4814. [CrossRef]
15. Chen, Z.; Appenzeller, J.; Lin, Y.-M.; Sippel-Oakley, J.; Rinzler, A.G.; Tang, J.; Wind, S.J.; Solomon, P.M.; Avouris, P. An Integrated Logic Circuit Assembled on a Single Carbon Nanotube. *Science* **2006**, *311*, 1735. [CrossRef]
16. Wei, B.Q.; Vajtai, R.; Ajayan, P.M. Reliability and current carrying capacity of carbon nanotubes. *Appl. Phys. Lett.* **2001**, *79*, 1172–1174. [CrossRef]
17. White, C.T.; Todorov, T.N. Quantum electronics: Nanotubes go ballistic. *Nature* **2001**, *411*, 649. [CrossRef]
18. Datta, S. *Electronic Transport Properties in Mesoscopic Systems*; Cambridge University Press: Cambridge, UK, 1995.
19. Shulaker, M.M.; Hills, G.; Patil, N.; Wei, H.; Chen, H.-Y.; Wong, H.-S.P.; Mitra, S. Carbon nanotube computer. *Nat. Cell Biol.* **2013**, *501*, 526–530. [CrossRef]
20. Xiao, Z.; Elike, J.; Reynolds, A.; Moten, R.; Zhao, X. The fabrication of carbon nanotube electronic circuits with dielectrophoresis. *Microelectron. Eng.* **2016**, *164*, 123–127. [CrossRef]
21. Jariwala, D.; Sangwan, V.K.; Lauhon, L.J.; Marks, T.J.; Hersam, M.C. Carbon nanomaterials for electronics, optoelectronics, photovoltaics, and sensing. *Chem. Soc. Rev.* **2013**, *42*, 2824–2860. [CrossRef]
22. Peng, L.-M.; Zhang, Z.; Wang, S. Carbon nanotube electronics: Recent advances. *Mater. Today* **2014**, *17*, 433–442. [CrossRef]
23. Shulaker, M.M.; Van Rethy, J.; Wu, T.F.; Liyanage, L.S.; Wei, H.; Li, Z.; Pop, E.; Gielen, G.; Wong, H.-S.P.; Mitra, S. Carbon Nanotube Circuit Integration up to Sub-20 nm Channel Lengths. *ACS Nano* **2014**, *8*, 3434–3443. [CrossRef] [PubMed]
24. Benedict, L.X.; Louie, S.G.; Cohen, M.L. Static polarizabilities of single-wall carbon nanotubes. *Phys. Rev. B* **1995**, *52*, 8541–8549. [CrossRef] [PubMed]
25. Zhang, Z.-B.; Liu, X.; Campbell, E.E.B.; Zhang, S.-L. Alternating current dielectrophoresis of carbon nanotubes. *J. Appl. Phys.* **2005**, *98*, 56103. [CrossRef]
26. Li, P.; Xue, W. Selective Deposition and Alignment of Single-Walled Carbon Nanotubes Assisted by Dielectrophoresis: From Thin Films to Individual Nanotubes. *Nanoscale Res. Lett.* **2010**, *5*, 1072–1078. [CrossRef] [PubMed]
27. Yang, B.; Yang, Z.; Zhao, Z.; Hu, Y.; Li, J. The assembly of carbon nanotubes by dielectrophoresis: Insights into the dielectrophoretic nanotube–nanotube interactions. *Phys. E Low-Dimens. Syst. Nanostruct.* **2014**, *56*, 117–122. [CrossRef]
28. Naieni, A.K.; Nojeh, A. Effect of solution conductivity and electrode shape on the deposition of carbon nanotubes from solution using dielectrophoresis. *Nanotechnology* **2012**, *23*, 495606. [CrossRef]
29. Xiao, Z.; Sharma, H.; Zhu, M.Y.; Pearson, T. Dielectrophoresis-assisted deposition and alignment of single-walled carbon nanotubes for electronic-device fabrication. *J. Vac. Sci. Technol. A* **2010**, *28*, 750–754. [CrossRef]
30. Krisnadi, F.; Nguyen, L.L.; Ankit; Ma, J.; Kulkarni, M.R.; Mathews, N.; Dickey, M.D. Directed Assembly of Liquid Metal–Elastomer Conductors for Stretchable and Self-Healing Electronics. *Adv. Mater.* **2020**, *32*, 2001642. [CrossRef]
31. Naieni, A.K.; Nojeh, A. Dielectrophoretic deposition of carbon nanotubes: The role of field frequency and its dependence on solution conductivity. *Microelectron. Eng.* **2014**, *114*, 26–30. [CrossRef]
32. Shekhar, S.; Stokes, P.; Khondaker, S.I. Ultrahigh Density Alignment of Carbon Nanotube Arrays by Dielectrophoresis. *ACS Nano* **2011**, *5*, 1739–1746. [CrossRef] [PubMed]
33. Sarker, B.K.; Shekhar, S.; Khondaker, S.I. Semiconducting Enriched Carbon Nanotube Aligned Arrays of Tunable Density and Their Electrical Transport Properties. *ACS Nano* **2011**, *5*, 6297–6305. [CrossRef] [PubMed]
34. Tittmann-Otto, J.; Hermann, S.; Kalbacova, J.; Hartmann, M.; Toader, M.; Rodriguez, R.D.; Schulz, S.; Zahn, D.R.T.; Gessner, T. Effect of cleaning procedures on the electrical properties of carbon nanotube transistors—A statistical study. *J. Appl. Phys.* **2016**, *119*, 124509. [CrossRef]
35. Liu, B.; Liu, J.; Li, H.-B.; Bhola, R.; Jackson, E.A.; Scott, L.T.; Page, A.J.; Irle, S.; Morokuma, K.; Zhou, C. Nearly Exclusive Growth of Small Diameter Semiconducting Single-Wall Carbon Nanotubes from Organic Chemistry Synthetic End-Cap Molecules. *Nano Lett.* **2015**, *15*, 586. [CrossRef]
36. Xiao, Z.; Williams, L.; Kisslinger, K.; Sadowski, J.T.; Camino, F. Fabrication of field-effect transistors with transfer-free nanostructured carbon as semiconducting channel material. *Nanotechnology* **2020**, *31*, 485203. [CrossRef]
37. Xiao, Z.; Kisslinger, K.; Chance, S.; Banks, S. Comparison of Hafnium Dioxide and Zirconium Dioxide Grown by Plasma-Enhanced Atomic Layer Deposition for the Application of Electronic Materials. *Crystals* **2020**, *10*, 136. [CrossRef]
38. Collins, P.G.; Arnold, M.S.; Avouris, P. Engineering Carbon Nanotubes and Nanotube Circuits Using Electrical Breakdown. *Science* **2001**, *292*, 706–709. [CrossRef]
39. Otsuka, K.; Inoue, T.; Chiashi, S.; Maruyama, S. Selective removal of metallic single-walled carbon nanotubes in full length by organic film-assisted electrical breakdown. *Nanoscale* **2014**, *6*, 8831–8835. [CrossRef]

40. Xiao, Z.; Camino, F.E. The fabrication of carbon nanotube field-effect transistors with semiconductors as the source and drain contact materials. *Nanotechnol.* **2009**, *20*, 135205. [CrossRef]
41. Xiao, Z.; Camino, F.E. An Effective Switching-Off Mechanism for High-Performance Carbon Nanotube Field-Effect Transistors. *J. Nanoelectron. Optoelectron.* **2010**, *5*, 26–29. [CrossRef]
42. Avouris, P. Carbon nanotube electronics. *Chem. Phys.* **2002**, *281*, 429. [CrossRef]
43. Avouris, P.; Martel, R.; DeRycke, V.; Appenzeller, J. Carbon nanotube transistors and logic circuits. *Phys. B Condens. Matter* **2002**, *323*, 6–14. [CrossRef]
44. Martel, R.; DeRycke, V.; Appenzeller, J.; Wind, S.; Avouris, P. Carbon nanotube field-effect transistors and logic circuits. In Proceedings of the ACM/IEEE Design Automation Conference, New Orleans, LA, USA, 10–14 June 2002; pp. 94–98.
45. Xiao, Z.; Kisslinger, K. Electron-beam-evaporated thin films of hafnium dioxide for fabricating electronic devices. *J. Vac. Sci. Technol. B* **2015**, *33*, 042001. [CrossRef]
46. Schroter, M.; Claus, M.; Sakalas, P.; Haferlach, M.; Wang, D. Carbon Nanotube FET Technology for Radio-Frequency Electronics: State-of-the-Art Overview. *IEEE J. Electron Devices Soc.* **2013**, *1*, 9–20. [CrossRef]
47. Franklin, A.D.; Koswatta, S.O.; Farmer, D.B.; Smith, J.T.; Gignac, L.; Breslin, C.M.; Han, S.-J.; Tulevski, G.S.; Miyazoe, H.; Haensch, W.; et al. Carbon Nanotube Complementary Wrap-Gate Transistors. *Nano Lett.* **2013**, *13*, 2490–2495. [CrossRef]

micromachines

MDPI

Article

Design of Driving Waveform Based on a Damping Oscillation for Optimizing Red Saturation in Three-Color Electrophoretic Displays

Zichuan Yi [1], Weibo Zeng [1], Simin Ma [1], Haoqiang Feng [1,2,*], Wenjun Zeng [1,2], Shitao Shen [2], Lingling Shui [1,2], Guofu Zhou [2] and Chongfu Zhang [1]

1 College of Electron and Information, University of Electronic Science and Technology of China Zhongshan Institute, Zhongshan 528402, China; yizichuan@zsc.edu.cn (Z.Y.); zengweibo133@126.com (W.Z.); masimin0402@163.com (S.M.); zwjcareer@163.com (W.Z.); Shuill@m.scnu.edu.cn (L.S.); cfzhang@uestc.edu.cn (C.Z.)
2 South China Academy of Advanced Optoelectronics, South China Normal University, Guangzhou 510006, China; shenshitao@m.scnu.edu.cn (S.S.); guofu.zhou@m.scnu.edu.cn (G.Z.)
* Correspondence: haoqiang.feng@m.scnu.edu.cn; Tel.: +86-0760-8831-4612

Abstract: At present, three-color electrophoretic displays (EPDs) have problems of dim brightness and insufficient color saturation. In this paper, a driving waveform based on a damping oscillation was proposed to optimize the red saturation in three-color EPDs. The optimized driving waveform was composed of an erasing stage, a particles activation stage, a red electrophoretic particles purification stage, and a red display stage. The driving duration was set to 360 ms, 880 ms, 400 ms, and 2400 ms, respectively. The erasing stage was used to erase the current pixel state and refresh to a black state. The particles' activation stage was set as two cycles, and then refreshed to the black state. The red electrophoretic particles' purification stage was a damping oscillation driving waveform. The red and black electrophoretic particles were separated by changing the magnitude and polarity of applied electric filed, so that the red electrophoretic particles were purified. The red display stage was a low positive voltage, and red electrophoretic particles were driven to the common electrode to display a red state. The experimental results showed that the maximum red saturation could reach 0.583, which was increased by 27.57% compared with the traditional driving waveform.

Keywords: electrophoretic displays; driving waveform; damping oscillation; particles separation; red saturation

Citation: Yi, Z.; Zeng, W.; Ma, S.; Feng, H.; Zeng, W.; Shen, S.; Shui, L.; Zhou, G.; Zhang, C. Design of Driving Waveform Based on a Damping Oscillation for Optimizing Red Saturation in Three-Color Electrophoretic Displays. *Micromachines* **2021**, *12*, 162. https://doi.org/10.3390/mi12020162

Academic Editor: Rodrigo Martinez-Duarte
Received: 23 January 2021
Accepted: 5 February 2021
Published: 7 February 2021

1. Introduction

As the carrier of human–computer information interaction, a flat panel display is very important in modern life. In recent years, electronic paper displays, which are a new type of display, have occupied a certain share in the display market due to its advantages such as large viewing angle, light in mass, low power consumption, repetitive erasing, and readability under sunlight [1–5]. As a kind of electronic paper display, electrophoretic displays (EPDs) have an excellent performance in the field of device manufacturing, which is expected to become one of the mainstream technologies for next-generation displays [6–8]. The charged particles in traditional EPDs are black electrophoretic particles with a positive charge and white electrophoretic particles with a negative charge [9]. They can be driven to the top or bottom of a pixel to display different states by applying an electric field [10–12]. However, the black and white particles cannot perfectly express the content of pictures because of the signal color. Hence, a multi-color EPD is urgently needed.

The three-color EPD technology could cover people's requirements for multi-color electronic papers. In recent years, a three-color EPD has been reported [13]. Electrophoretic particles of three colors were successfully driven in this work, and the EPD can be driven for displaying corresponding colors by different voltage sequences. However, the red and black electrophoretic particles in electrophoretic fluid have the same polarity of charge, and they are driven in the same direction when a driving electric field is applied. The disadvantage of this design is an insufficient color saturation, especially red. Worse still, low display quality can be caused such as ghost image and flicker, which are also problems in traditional EPD displays [14–17]. In order to improve the display quality of EPDs, a lot of work has been done. The display quality of EPDs has obviously improved by synthetizing new material, manufacturing new devices, and optimizing the driving waveform, etc. [18–20]. As optimizing the driving waveform has an obvious improvement in the display quality, researchers have done many works in this aspect. For example, a driving waveform based on a delay response was proposed, and the hysteresis characteristic curve of an EPD was measured [21]. Moreover, Johnson et al. [22] proposed a driving waveform in which a reference state was designed, where the EPD could display the next gray scale more regularly and accurately, and this method could greatly improve the display quality of EPDs. At present, the red saturation of three-color EPDs is insufficient and it could be optimized by improving driving waveforms.

In this paper, an insufficient red saturation of the three-color EPD was improved by designing a new driving waveform. This driving waveform included an addition stage of a damping oscillation; this stage could separate the red and black electrophoretic particles more completely. The saturation is greatly improved when the EPD displayed a red state. At the same time, ghost image and fringe phenomena can be weakened effectively by optimizing other stages of the driving waveform.

2. Principle of Electrophoretic Displays (EPDs)

The structure of a three-color EPD is shown in Figure 1. White electrophoretic particles, black electrophoretic particles, and red electrophoretic particles are wrapped into a microcapsule, and three-color particles have different polarities. The white electrophoretic particles are negatively charged, and the red and black electrophoretic particles are positively charged [13]. Among these particles, the red and black electrophoretic particles are different in charged, mass, and volume. Therefore, they can be successfully separated to display different colors by controlling the magnitude and duration of the driving voltage [23]. As shown in Figure 1a, black electrophoretic particles are driven to the top in a pixel with a high positive polarity voltage. As shown in Figure 1b, white electrophoretic particles are driven to the top in a pixel with a negative polarity voltage. As shown in Figure 1c, red electrophoretic particles are driven to the top in a pixel with a low positive polarity voltage.

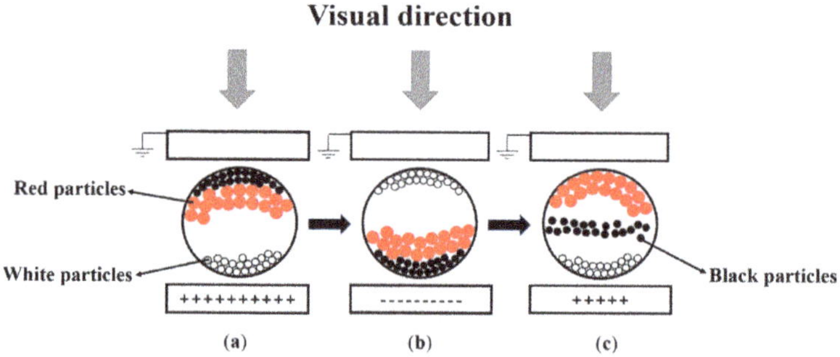

Figure 1. A schematic diagram of a three-color electrophoretic display (EPD). The top is a common electrode, the middle is the microcapsule that contains three-color particles, and the bottom is a pixel electrode. (**a**) When the applied external voltage is a high positive voltage, the pixel is black. (**b**) When the applied external voltage is a negative voltage, the pixel is white. (**c**) When the applied voltage is a low positive voltage, the pixel is red. The particles are in a static state if there is no driving voltage, which is called the bistable state of EPDs.

The driving waveform refers to a voltage sequence applied to pixels, which seriously affects the display quality of EPDs. A traditional driving waveform is composed of an erasing stage, a particle activation stage, and a target state display stage [24,25]. Changes in each stage can affect the display quality of EPDs. There are many details that need to be improved in the driving waveform for a good display quality such as ghost image, fringe phenomena, flicker, insufficient color saturation, etc. [15,26–28]. In the original stage, particles are distributed irregularly in microcapsules. Therefore, it is necessary to rearrange particles to prepare for the next gray scale of EPDs [29]. In this stage, the original state can be driven to a state where electrophoretic particles are regularly arranged to the same state. The migration rate of electrophoretic particles can be decreased when electrophoretic particles remain in the same state for a long time. Therefore, the driving waveform of the particle activation stage is designed. In this stage, pixels are refreshed to black state first, and refreshed to white state immediately. This process can make sure that particles must be activated during the driving process and reach steady state. Next, the target gray scale can be driven by an applied electric field.

Charged particles are subjected to an electric field force in an electric field. As shown in Equation (1):

$$F_q = q \times E \tag{1}$$

where F_q is the electric field force; q is the electric charge of electrophoretic particles; and E is the applied electric field.

Since electrophoretic particles are dispersed in colloidal solution, the movement of electrophoretic particles is also hindered by Stokes force [30]. Its expression is shown in Equation (2):

$$F_d = 6\pi\mu v r \tag{2}$$

where F_d is the Stokes force; μ is the liquid viscosity coefficient; v is the motion relative rate between particles and fluids; and r is the sphere radius. The combined force of the electric field force and the Stokes force is used as the driving force for the movement of electrophoretic particles. As shown in Equation (3):

$$F = F_q - F_d = m\frac{dv}{dt} \tag{3}$$

where F is the driving force that can drive electrophoretic particles; m is the mass of an electrophoretic particle; and $\frac{dv}{dt}$ is the acceleration of electrophoretic particles.

The moving distance of electrophoretic particles in a microcapsule can be calculated according to the integral of speed and time. As shown in Equation (4):

$$S = \int c dt = \int \sigma E dt \tag{4}$$

where S is the moving distance of electrophoretic particles in a microcapsule; c is the moving speed of electrophoretic particles; σ is the electrophoretic mobility of electrophoretic particles; and E is the applied electric field.

Particles are subjected to random pulse signals in the electric field, as shown in Equation (5):

$$f(t) = \sum_{n=1}^{N} a_n[u(t - nT) - u(t - nT - \tau)] \tag{5}$$

where *f(t)* is a random amplitude rectangular pulse signal; a_n is the rectangular pulse amplitude; t is the time; T is the pulse signal period; and τ is the pulse width.

The key step for increasing the red saturation is that red and black electrophoretic particles in microcapsules can be separated completely, which can be completed by designing a damping oscillation driving waveform according to the nature of particles. A damping oscillation refers to the process in which the amplitude of vibration can be gradually decreased over time. As shown in Equations (6) and (7), the characteristic of a damped oscillation is that the amplitude gradually is decreased. Therefore, the intensity of the electric field applied to the pixel gets lower and lower in the driving waveform. Driving black electrophoretic particles requires a high electric field, and driving red electrophoretic particles requires a low electric field. Therefore, the red and black electrophoretic particles can be successfully separated.

$$F_r = -Bc \tag{6}$$

$$X(t) = Ae^{-\delta t}\cos(\omega t + \varphi) \tag{7}$$

where F_r is the damped force; B is the drag coefficient; $Ae^{-\delta t}$ is the amplitude; and ω is the angular frequency. In this process, the mean value of time is defined as Equation (8):

$$\overline{X(t)} = \lim_{T\to\infty}\frac{1}{2T}\int_{-T}^{T} X(t)dt \tag{8}$$

3. Design of Driving Waveforms

To improve the red saturation in three-color EPDs, an optimized driving waveform was designed and divided into four stages: (1) An erasing stage; (2) a particle activation stage; (3) a red electrophoretic particles purification stage, and (4) a red display stage. As shown in Figure 2, the first stage was the erasing stage, which could erase the original pixel state and refresh the white or black state. A positive 15 V could be applied for 200 ms to reset the original state, and the EPD could display a black state. Similarly, a negative 15 V was applied to reset the original state, and the EPD could display a white state. Then, 0 V was applied after 15 V voltage for 80 ms. This duration was set to buffer the electrophoretic particles, and red, black, and white particles were separated at this stage. The second stage was a particle activation stage. The EPDs were driven multiple times to reach the optical limit, which could increase the activity of particles and further eliminate ghost images. At this stage, electrophoretic particles could be driven to move in the colloidal solution with several cycles by an applied voltage. A cycle included a positive 15 V and a negative 15 V, and the duration was 340 ms. Then, the red and black electrophoretic particles were driven to the top of a pixel by a positive 15 V voltage for 200 ms at the end of this stage. The third stage was a red electrophoretic particle purification stage. The purpose of this stage was to separate red and black electrophoretic particles for displaying a red state. Two kinds of particles can be driven at different speeds in microcapsules with the same applied voltage. Hence, the damping oscillation driving waveform had a good effect on separating red and black particles, and the red particles were purified greatly. The fourth stage was a red state display stage, and a low positive voltage was applied during the whole stage.

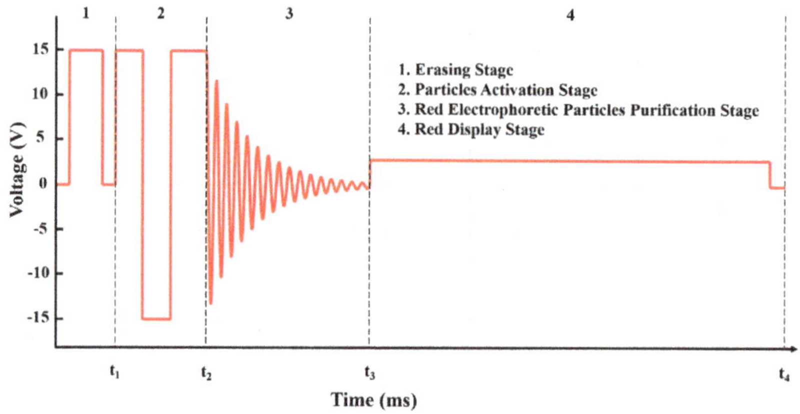

Figure 2. Optimized driving waveform. This is composed of four stages: (**1**) Erasing stage; (**2**) particle activation stage; (**3**) red electrophoretic particles purification stage; and (**4**) red display stage.

4. Results and Discussion

4.1. Construction of Test Platform

In this work, the experimental system is shown in Figure 3 and includes a function generator, a signal amplifier, a colorimeter, a microcapsule EPD, and a computer. The microcapsule EPD device was designed by us, and made by foundry (Dalian Longning Technology Co. Ltd., Dalian, China). The driving waveform used in the experiment was first designed by MATLAB (2017, MathWorks, Natick, MA, USA). Subsequently, it was converted to generate a txt format file. Then, the tfw format file was output by Arbexpress (Version 3.4, Tektronix. Inc, Beaverton, OR, USA). Next, the file was burnt into the function generator by a Universal Serial BUS (USB). Finally, the driving waveform was output from the signal amplifier to drive an EPD to display gray scales.

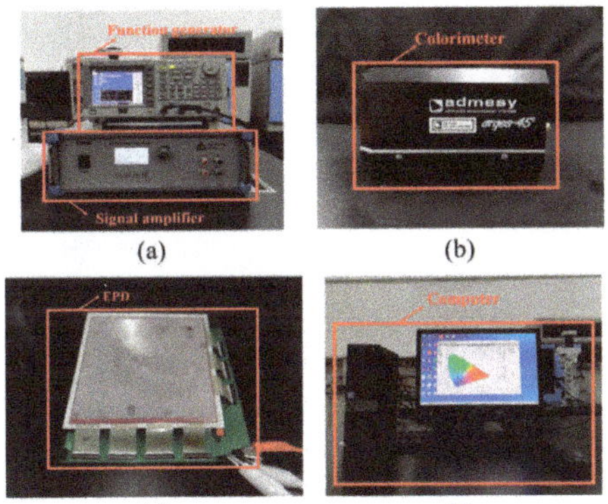

Figure 3. An electrophoretic display (EPD) testing system. (**a**) A function generator and a signal amplifier. (**b**) A colorimeter. (**c**) A microcapsule EPDs. (**d**) A computer.

The workflow of the entire system is shown in Figure 4. All equipment and instruments were first connected. Then, the saturation data acquisition system was activated, and the relevant parameters were set, which included averaging and trial. Averaging was used to set the time interval for obtaining red saturation data, and the trial was used to set the number of measurements that could appear on the color saturation diagram. Then, the system was effectively calibrated. Finally, the EPD was driven to display relevant colors.

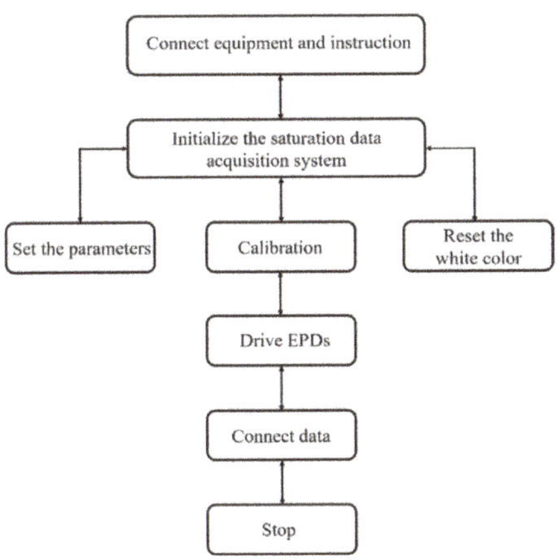

Figure 4. Flow chart of the saturation data acquisition.

4.2. Erasing Stage Optimization

Generally, the first stage of the driving waveform is an erasing stage, which is used to erase previous states. In this work, the first stage of the driving waveform was designed to erase the original gray scale and then refresh the EPD to a black state. As shown in Figure 5a, the duration of this stage was set to 360 ms. Compared with the traditional driving waveform, the erasing stage could erase to a white state and the polarity of the applied voltage was the opposite, as shown in Figure 5b. The experimental results showed that when the erasing stage was a black state, the red saturation was 0.583, and the chromaticity diagram is shown in Figure 5c. However, when the erasing stage was a white state, the red saturation was 0.448, and the chromaticity diagram is shown in Figure 5d. This is because when the driving waveform is designed to erase to a white state, white particles were near the common electrode, the red and black electrophoretic particles were at the bottom of a microcapsule. Therefore, it took a lot of time to drive red particles from the bottom to the common electrode in the particle activation and display red color stages.

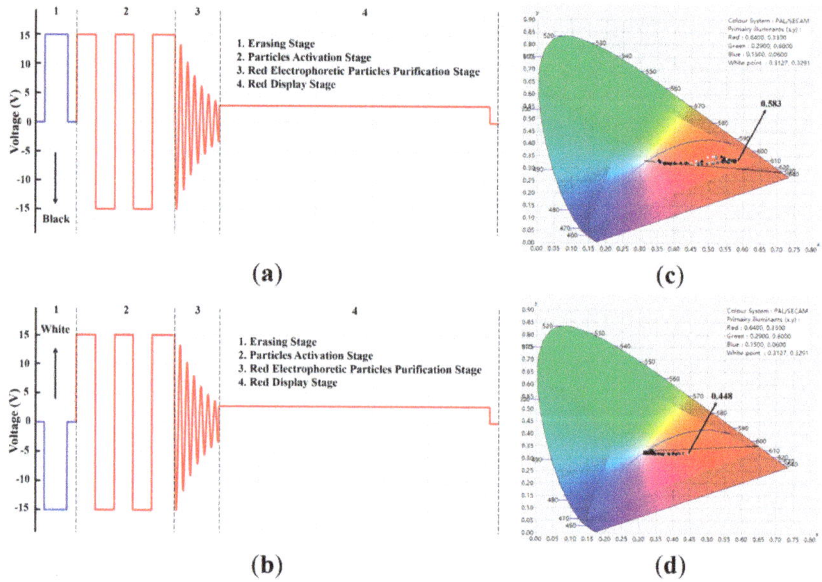

Figure 5. The maximum red saturation when the optimized driving waveform was used in the erasing stage. (**a**) Driving waveform design when it was erased to a black state. (**b**) Driving waveform design when it was erased to a white state. (**c**) The maximum red saturation when the EPD was erased to a black state. (**d**) The maximum red saturation when the EPD was erased to a white state.

4.3. Particle Activation Stage Optimization

According to the traditional driving waveform, the particle activation stage has an important contribution to reducing ghost images. At this stage, different cycles were tested. The duration of a cycle was designed to be 340 ms, and the duration for resetting to the black state was 200 ms. The maximum red saturation with different cycles are shown in Figure 6. The experimental results showed that the maximum red saturation could reach 0.583 when the stage was two cycles. As the cycle was increased, the maximum red saturation was gradually decreased. This is because electrophoretic particles could not return to their original position due to the resistance of the colloidal solution when the EPD was driven to the black state or the white state. After multiple cycles, the distance was increased so that white electrophoretic particles could not be driven to the bottom in a microcapsule.

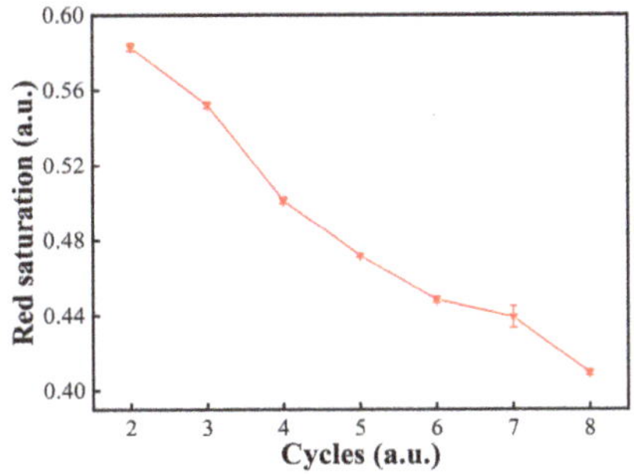

Figure 6. The relationship between cycles and the maximum red saturation.

4.4. Waveform Design of the Damping Oscillation

Since the red and black electrophoretic particles have the same polarity charge, it was unrealistic to separate them by simply changing the driving voltage polarity. However, the nature of red and black electrophoretic particles is different. A damping oscillation driving waveform can not only change the polarity, but also change the magnitude of the electric field. Hence, the red and black electrophoretic particles can be separated by the damping oscillation driving waveform. As shown in Figure 7, we tested the effect of different damping oscillation durations on the red saturation. It could be seen that the maximum red saturation could be obtained when a short duration was designed for the damping oscillation, and the maximum red saturation could be gradually improved when the duration of the damping oscillation was increased to 400 ms. When the duration exceeded 400 ms, the maximum red saturation gradually became worse. This was because the damping oscillation duration was too short, so the red and black electrophoretic particles were not completely separated. Then, the black particles were under red particles when the target state was red so that the red saturation was low. In addition, when the damping oscillation duration was too long, the red and black electrophoretic particles were mixed again. Hence, the red saturation was also low.

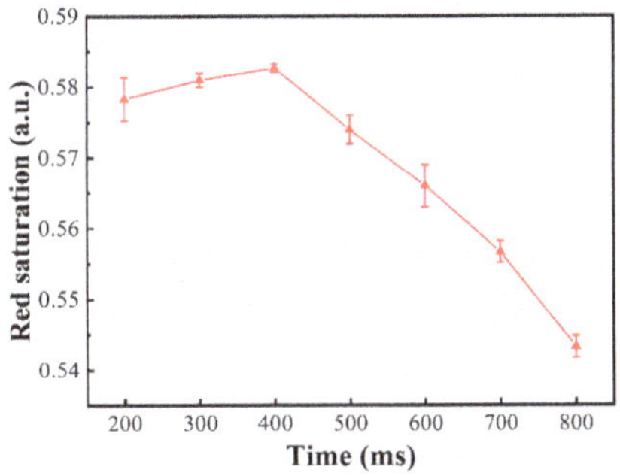

Figure 7. The relationship between the damping oscillation duration and the maximum red saturation of the target state.

4.5. Red Display Stage Optimization

In order to obtain a high red saturation, the driving waveform was optimized in two aspects: driving voltage and duration. As shown in Figure 8, the influence of different driving voltages on the red saturation was verified by designing different durations. The experimental results showed that the maximum red saturation values were different with different durations. The maximum red saturation was gradually improved when the duration was increased from 2000 ms to 2400 ms and the maximum red saturation showed a downward trend when it exceeded 2400 ms. This is because a long duration would drive black electrophoretic particles toward the common electrode. In addition, the red saturation showed an upward trend when the driving voltage was increased from 2.5 V to 2.9 V, and the red saturation was gradually decreased when it exceeded 2.9 V. Comprehensively, the optimal parameters were the driving voltage of 2.9 V and the duration of 2400 ms.

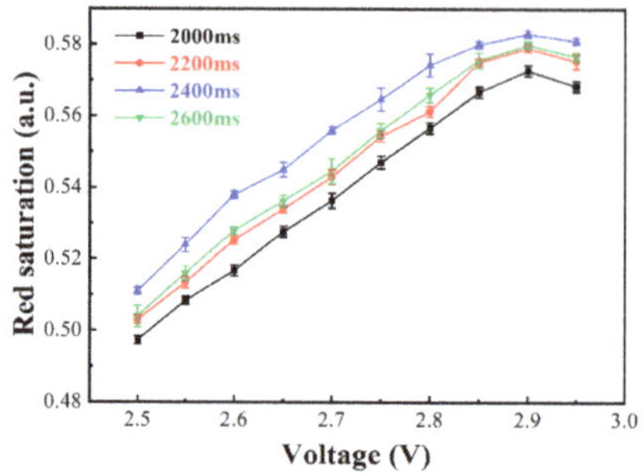

Figure 8. The relationship between the maximum red saturation of the target state and the driving voltage with different durations.

We tested the maximum red saturation based on the traditional driving waveform and the optimized driving waveform, respectively. As shown in Figure 9, the experimental results showed that the maximum red saturation was increased from 0.457 to 0.583, therefore the red saturation increased by 27.57%.

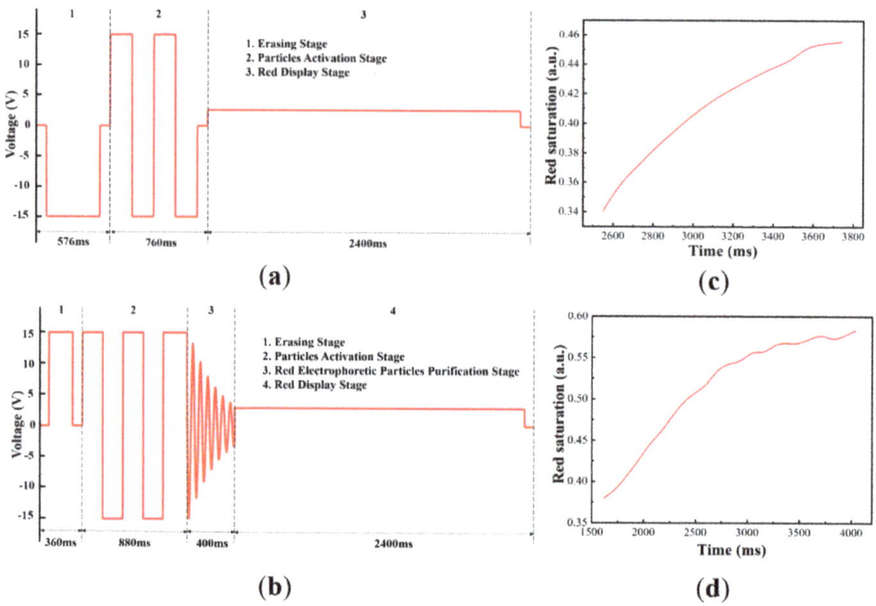

Figure 9. (**a**) The traditional waveform which was composed of three stages: (1) The erasing stage, (2) particle activation stage, (3) target state display stage. The driving waveform duration of three stages was set to 576 ms, 760 ms, and 2400 ms, respectively [13]. (**b**) The optimized driving waveform was composed of four stages: (1) The erasing stage, (2) particle activation stage, (3) red electrophoretic particles purification stage, and (4) red display stage. The driving waveform duration of the four stages were set to 360 ms, 880 ms, 400 ms, and 2400 ms, respectively. (**c**) The relationship between the red saturation and red display stage based on the traditional driving waveform. (**d**) The relationship between the red saturation and red display stage based on the optimized driving waveform.

5. Conclusions

In this paper, a driving waveform that could optimize red saturation was proposed for three-color EPDs. The damping oscillation driving waveform could separate red and black electrophoretic particles very well. Compared with traditional driving waveforms, the red saturation of the optimized driving waveform could be effectively improved by 0.126. At the same time, the ghost image could be reduced and the steady state of particles could be improved. The concept of the damping oscillation can provide effective design ideas for the design of driving waveforms for color EPDs, which can provide a better and more comfortable visual experience for users.

Author Contributions: Z.Y., W.Z. (Weibo Zeng), and H.F. designed this project. H.F., W.Z. (Weibo Zeng), and S.M. carried out most of the experiments and data analysis. W.Z. (Wenjun Zeng) and S.S. performed part of the experiments and helped with discussions during manuscript preparation. L.S., G.Z., and C.Z. gave suggestions on project management and provided helpful discussions on the experimental results. All authors have read and agreed to the published version of the manuscript.

Funding: This research was funded by the Guangdong Basic and Applied Basic Research Foundation (no. 2020A1515010420), the Key Research Platforms and Research Projects in Universities and Colleges of Guangdong Provincial Department of Education (no. 2018KQNCX334), the Zhongshan Innovative Research Team Program (no. 180809162197886), the Guangdong Provincial Key Laboratory of Optical Information Materials and Technology (no. 2017B030301007), and the Project for Innovation Team of Guangdong University (no. 2018KCXTD033).

Institutional Review Board Statement: Not application.

Informed Consent Statement: Not application.

Data Availability Statement: Data is contained within the article.

Conflicts of Interest: The authors declare no conflict of interest.

References

1. Heikenfeld, J.; Drzaic, P.; Yeo, J.S.; Koch, T. Review paper: A Critical review of the present and future prospects for electronic paper. *J. Soc. Inf. Disp.* **2011**, *19*, 2. [CrossRef]
2. Yi, Z.C.; Liu, L.W.; Wang, L.; Li, W.; Shui, L.L.; Zhou, G.F. A driving system for fast and precise gray-scale response based on amplitude–frequency mixed modulation in tft electrowetting displays. *Micromachines* **2019**, *10*, 732. [CrossRef]
3. Dou, Y.Y.; Chen, L.; Li, H.; Tang, B.; Henzen, A.; Zhou, G.F. Photolithography fabricated spacer arrays offering mechanical strengthening and oil motion control in electrowetting displays. *Sensors* **2020**, *20*, 494. [CrossRef] [PubMed]
4. Lee, J.K.; Kim, S.S.; Park, Y.I.; Kim, C.D.; Hwang, Y.K. In-cell adaptive touch technology for a flexible e-paper display. *Solid State Electron.* **2011**, *56*, 159–162. [CrossRef]
5. Inoue, S.; Kawai, H.; Kanbe, S.; Saeki, T.; Shimoda, T. High-resolution microencapsulated electrophoretic display (EPD) driven by poly-Si TFTs with four-level grayscale. *IEEE Trans. Electron. Devices* **2002**, *49*, 1532–1539. [CrossRef]
6. Yi, Z.C.; Huang, Z.Y.; Lai, S.F.; He, W.Y.; Wang, L.; Chi, F.; Zhang, C.F.; Shui, L.L.; Zhou, G.F. Driving waveform design of electrowetting displays based on an exponential function for a stable grayscale and short driving time. *Micromachines* **2020**, *11*, 313. [CrossRef] [PubMed]
7. Chrstophersen, M.; Phlips, B.F. Recent patents on electrophoretic displays and materials. *Recent Pat. Nanotechnol.* **2019**, *4*, 137–149. [CrossRef]
8. Qin, Z.; Wang, H.I.; Chen, Z.Y.; Chen, C.H.; Tien, P.L.; Liu, M.H.; Liu, S.C.; Hung, C.M.; Tsai, C.C.; Huang, Y.P. Digital halftoning method with simultaneously optimized perceptual image quality and drive current for multi-tonal electrophoretic displays. *Appl. Optics* **2020**, *59*, 201–209. [CrossRef] [PubMed]
9. Comiskey, B.; Albert, J.D.; Yoshizawa, H.; Jacobson, J. An electrophoretic ink for all-printed reflective electronic displays. *Nature* **1998**, *394*, 253–255. [CrossRef]
10. Shen, S.T.; Gong, Y.X.; Jin, M.L.; Yan, Z.B.; Xu, C.; Yi, Z.C.; Zhou, G.F.; Shui, L.L. Improving electrophoretic particle motion control in electrophoretic displays by eliminating the fringing effect via driving waveform design. *Micromachines* **2018**, *9*, 143. [CrossRef]
11. Kao, W.C.; Chen, H.Y.; Liu, Y.H.; Liou, S.C. Hardware engine for supporting gray-tone paintbrush function on electrophoretic papers. *J. Disp. Technol.* **2014**, *10*, 138–145. [CrossRef]
12. Yang, S.H.; Lin, F.C.; Huang, Y.P.; Shieh, H.P.D. Ghosting reduction driving method in electrophoretic displays. *SID Symp. Dig. Tech. Pap.* **2012**, *43*, 1361–1364. [CrossRef]
13. Kao, W.C.; Tsai, J.C. Driving method of three particle electrophoretic displays. *IEEE Trans. Electron. Device* **2018**, *65*, 1023–1028. [CrossRef]

14. He, W.Y.; Yi, Z.C.; Shen, S.T.; Huang, Z.Y.; Liu, L.W.; Zhang, T.Y.; Li, W. Driving waveform design of electrophoretic display based on optimized particle activation for a rapid response speed. *Micromachines* **2020**, *11*, 498. [CrossRef] [PubMed]
15. Yang, B.R.; Wang, Y.C.; Wang, L. The design considerations for full-color e-paper. *Adv. Disp. Technol. VII.* **2017**, *10126*, UNSP1012602. [CrossRef]
16. Wang, L.; Yi, Z.C.; Jin, M.L.; Shui, L.L.; Zhou, G.F. Improvement of video playback performance of electrophoretic displays by optimized waveforms with shortened refresh time. *Displays* **2017**, *49*, 95–100. [CrossRef]
17. Yi, Z.C.; Bai, P.F.; Wang, L.; Zhang, X.; Zhou, G.F. An electrophoretic display driving waveform based on improvement of activation pattern. *J. Cent. South Univ.* **2014**, *21*, 3133–3137. [CrossRef]
18. Wang, W.; Zhang, A.; Jiang, Y.F.; Lan, D.S.; Lu, F.H.; Zheng, L.L.; Zhuang, L.; Hong, R.J. Large-scale preparation of size-controlled Fe_3O_4@SiO_2 particles for electrophoretic display with non-irdescent structural colors. *RSC Adv.* **2019**, *9*, 498–506. [CrossRef]
19. Cao, J.X.; Qin, Z.; Zeng, Z.; Hu, W.J.; Song, L.Y.; Hu, D.L.; Wang, X.D.; Zeng, X.; Chen, Y.; Yang, B.R. A convolutional neural network for ghost image recognition and waveform design of electrophoretic displays. *IEEE Trans. Consum. Electron.* **2020**, *66*, 356–365. [CrossRef]
20. Kao, W.C.; Liu, J.J.; Chu, M.I.; Wang, Y.K.; Yang, T.H. Photometric calibration for image enhancement of electrophoretic displays. In Proceedings of the IEEE International Symposium on Consumer Electronics (ISCE 2010), Braunschweig, Germany, 7–10 June 2010.
21. Kao, W.C.; Chang, W.T.; Ye, J.A. Driving waveform design based on response latency analysis of electrophoretic displays. *J. Disp. Technol.* **2012**, *8*, 596–601. [CrossRef]
22. Johnson, M.T.; Zhou, G.F.; Zehner, R.; Amundson, K.; Henzen, A.; VandeKamer, J. High-quality images on electrophoretic displays. *J. Soc. Inf. Disp.* **2006**, *14*, 175–180. [CrossRef]
23. Kim, J.-M.; Kwangjoon, K.; Lee, S. Multilevel driving waveform for electrophoretic displays to improve grey levels and response characteristics. *Electron. Lett.* **2014**, *50*, 1925–1927. [CrossRef]
24. Cho, K.M. An analysis of reflectivity and response time by charge -to-mass of charged particles in an electrophoretic displays. *Trans. Electr. Electron. Mater.* **2016**, *17*, 212–216. [CrossRef]
25. Lu, C.M.; Wey, C.L. A controller design for high-quality images on microcapsule active-matrix electrophoretic displays. *J. Inf. Disp.* **2012**, *13*, 21–30. [CrossRef]
26. Xu, K.P.; Huang, J.Y.; Liang, H.J.; He, Z.W.; Gao, M.Y. An FPGA based frame rate enhancer for LCD display in embedded systems. *Adv. Des. Res. Manuf.* **2012**, *605*, 2095–2099. [CrossRef]
27. Wang, L.; Yi, Z.C.; Peng, B.; Zhou, G.F. An improved driving waveform reference grayscale of electrophoretic display. In Proceedings of the Volume 9672, AOPC 2015: Advanced Display Technology; and Micro/Nano Optical Imaging Technologies and Applications, Beijing, China, 5 May 2015; Volume 9672, pp. 967204–967209. [CrossRef]
28. Duan, F.B.; Bai, P.F.; Henzen, A.; Shui, L.L.; Tang, B.A.; Zhou, G.F. An adaptive generation method for electrophoretic display driving waveform design. *J. Soc. Inf. Disp.* **2016**, *24*, 676–685. [CrossRef]
29. Kao, W.C.; Ye, J.A.; Lin, F.S.; Cheng, P.Y.; Sprague, R. Configurable timing controller design for active matrix electrophoretic display. *IEEE Trans. Consum. Electron.* **2009**, *55*, 1–5. [CrossRef]
30. Yu, D.; An, J.H.; Bae, J.Y.; Jung, D.; Kim, S.; Ahn, S.D.; Kang, S.; Suh, K.S. Preparation and characterization of acrylic-based electronic inks by in situ emulsifier-free emulsion polymerization for electrophoretic displays. *Chem. Mater.* **2004**, *16*, 4693–4698. [CrossRef]

MDPI
St. Alban-Anlage 66
4052 Basel
Switzerland
Tel. +41 61 683 77 34
Fax +41 61 302 89 18
www.mdpi.com

Micromachines Editorial Office
E-mail: micromachines@mdpi.com
www.mdpi.com/journal/micromachines

www.ingramcontent.com/pod-product-compliance
Lightning Source LLC
Chambersburg PA
CBHW040743250926
51850CB00007BB/1396
9783036573014